PROGRESS IN DNA METHYLATION RESEARCH

PROGRESS IN DNA METHYLATION RESEARCH

HANS P. NEUMANN
EDITOR

Nova Biomedical Books
New York

For permission to use material from this book please contact us:
Telephone 631-231-7269; Fax 631-231-8175
Web Site: http://www.novapublishers.com

Library of Congress Cataloging-in-Publication Data

Progress in DNA methylation research / Hans P. Neumann (editor).
p. ; cm.
Includes bibliographical references and index.
ISBN-13: 978-1-60021-722-7 (hardcover)
ISBN-10: 1-60021-722-2 (hardcover)
1. DNA--Methylation. I. Neumann, Hans P.
[DNLM: 1. DNA Methylation. 2. Epigenesis, Genetic. QU 58.5 P9635 2007]
QP624.5.M46P76 2007
572.8'6--dc22

2007015417

Published by Nova Science Publishers, Inc. ✢ New York

CONTENTS

PREFACE

DNA methylation is a type of chemical modification of DNA that can be inherited without changing the DNA sequence. As such, it is part of the epigenetic code and is the most characterized epigenetic mechanism. DNA methylation involves the addition of a methyl group to DNA — for example, to the number 5 carbon of the cytosine pyrimidine ring. DNA methylation is probably universal in eukaryotes. In humans, approximately 1% of DNA bases undergo DNA methylation. In adult somatic tissues, DNA methylation typically occurs in a CpG dinucleotide context; non-CpG methylation is prevalent in embryonic stem cells.In plants, cytosines are methylated both symmetrically (CpG or CpNpG) and asymmetrically (CpNpNp), where N can be any nucleotide. The methylation status of specific cytosines can be determined using methods based on bisulfite sequencing. This new book presents leading-edge research advances in the field.

Chapter I - High degree of nuclear DNA (nDNA) methylation is a specific feature of plant genomes, they do contain 5-methylcytosine (m^5C) and N^6-methyladenine (m^6A). More than 30% m^5C is located in CNG sequences. Specific changes in DNA methylation accompany the entire life of a plant starting from seed germination up to the death programed or induced by various agents and factors of biological or abiotic nature. Modulation of DNA methylation is one of the possible modes of the hormonal action in plant. DNA methylation in plants is species-, tissue-, organelle- and age-specific; it is involved in the control of all genetic functions including transcription, replication, DNA repair, gene transposition and cell differentiation. DNA methylation is engaged in gene silencing and parental imprinting, it controls transgenes and foreign DNA.

Plants have much more complicated and sophisticated system of the multicomponent and sometimes even conjugated genome (nuclear DNA) methylations compared with animals; besides, unlike animals, they have the plastids with their own unique DNA modification system that may control plastid differentiation and functioning; DNA methylation in plant mitochondria is performed in other fashion compared with it in nuclei. The nuclear DNA methylation system is controlled by three major families of cytosine DNA-methyltransferase genes, at least. In contrast to animals the inactivation of major maintenance methyltransferase MET1 (similar to animal Dnmt1) has no significant consequences for plant survival. Other plant cytosine DNA-methyltransferases have no analogs in animals. Some of them (DRM) are responsible for *de novo* DNA methylation including asymmetric sequences.

Plant gene may be methylated at both adenine and cytosine residues; specific adenine DNA-methyltransferase was described. Adenine DNA methylation may influence cytosine modification and *vice versa.* Anyway, two different systems of the genome modification based on methylation of adenines and cytosines coexist in higher plants.

The specific endonucleases discriminating between methylated and unmethylated DNA are present in plants. Thus, plants may have restriction-modification system.

There are peculiar complicated controls for growth and development by DNA methylations in plants; they are well coordinated with other epigenetic signals modulating chromatin organization.

Chapter II - Treating gDNA with bisulfite selectively deaminates non-methylated cytosine (C) moieties thus converting C into uracil (U). Subsequent PCR results in amplicons wherein thymidine (T) replaces previously unmethylated C (as well as replacing pre-existing Ts), and C replaces 5-methylcytosine (5mC). Since methylation of C almost always occurs in the context of CpG motifs (5mCpG), it is possible to determine the positions and relative extent of methylation of gDNA by analyzing DNA following bisulfite conversion. Such methods include methylation specific priming, sequencing, cloning and sequencing, single-base extension, oligonucleotide ligation assays (OLA), fluorogenic (e.g., TaqMan®) assays, and hybridization. Bisulfite conversion and efficient, selective PCR amplification are fundamental to each of these sequence detection methods in order to obtain accurate methylation analysis. During our investigation of improving sequencing methodology, and development of a novel fragment analysis method by capillary electrophoresis (CE), we have compiled data that illustrate unique challenges of analyzing the 3-base bisulfite-converted genome using techniques that had been originally optimized for 4-base genomes. We will summarize our findings in a comprehensive and practical manner aimed at providing guidelines to readers for achieving successful methylation analysis by CE. We will provide protocols, as well as real-life examples, for readers to rationalize and troubleshoot on their own.

Chapter III - DNA methylation can alter the genetic code without changing the base sequence of DNA. Changes in DNA methylation accompany the development and progression of diseases including cancer, aging, imprinting, instability of DNA repeat elements, neurodevelopment and other disorders. Environmental factors, including diet, are known to play a role in DNA methylation and other associated epigenetic changes in the chromatin. This chapter focuses on the contribution of the environment toward DNA methylation, with an emphasis on diet and nutrition. The human diet contains several nutrients that have the potential to influence DNA methylation. Vitamins such as folic acid, vitamin B6 and vitamin B12, and the amino acids choline and methionine are capable of providing methyl groups for DNA methylation. These and other nutrients can impact DNA methylation at several key points. Dietary methyl donors may regulate the methyl supply for formation of S-adenosylmethionine (SAM), a key intermediate required for DNA methylation. Dietary components also modify the activity of DNA methyltransferases that methylate DNA and methyl CpG binding proteins that bind to CpG groups of methylated DNA. One area of DNA methylation that is receiving increase attention is DNA demethylation. There is little information available about how environmental factors influence the removal of methyl groups from DNA. We will discuss the effect of environmental factors on these aspects of DNA methylation. Other topics to be discussed include the relationship between diet, global DNA hypomethylation and gene-specific DNA hypermethylation, which is not clear, especially in disease states such as cancer. In addition, the permissive effect of the environment on the expression of polymorphic genes present in pathways leading to DNA methylation will be covered in this chapter. Finally, we will review how early environmental exposures might leave a DNA methylation imprint that affects disease risk in later life. Changes in DNA methylation do not occur in the absence of environmental influences, and in some cases the environment can play a key role in the process of DNA methylation.

Chapter IV - In contrast to a large number of publications as to zebrafish mutants with defects in early morphogenesis, no zebrafish mutants affecting DNA methylation have been reported, hence little is known about DNA methylation in the fish. But I believe that zebrafish will be recognized as an excellent model organism for studying DNA methylation machinery in vertebrates, considering shared characteristics of cytosine methylation between mammals and the fish. Here, I first introduce approaches and methodologies established for zebrafish developmental biology, which can be applicable to the analyses of roles and mechanisms of DNA methylation in zebrafish. Next, I show a few examples of *de novo* methylation and demethylation that are observed in zebrafish embryos. I finally describe two methyltransferase genes in the zebrafish genome, *dnmt1* and *dnmt7*, whose expression patterns and functions are already demonstrated.

Chapter V - The current nuclear transfer and microinsemination techniques enable us to produce animals even from adult somatic cells and immature sperm such as a round spermatid. On the other hand, the offspring production rates, when using those unnatural cells, are significantly lower than that of normally fertilized embryos. Even though somatic cell clones have been produced in many mammals including cow, mouse and pig after Dolly, the first cloned sheep, the success rates are extremely low regardless of the species. The mechanism underlying the inefficiency is not completely understood; nonetheless studies have revealed complicated anomalies and epigenetic aberrancy associated with somatic cell clones. First, this chapter will overview the epigenetic aberrancy focusing on abnormal DNA methylation in somatic cell clones and the resulting phenotypes. Second, a new nuclear transfer technique will be introduced which has been recently developed to increase the success rates 2-5 fold in mice. This new technique is simply to treat somatic cloned embryos with trichostatin A, a histone deacetylase inhibitor. This success indicates that the ability of oocytes to reprogram the somatic cell nuclei can be artificially improved leading to higher success rates. Taken together, I would finally propose the concept of 'assisted reprogramming technology' for future reproductive biotechnologies.

Chapter VI - The extensive genetic analysis of human diseases began more than ten years ago; however, some diseases are explained not by genetic changes but by epigenetic changes. One of the best examples is cancer. It is well recognized that cancer is caused by the transcriptional silencing of tumor suppressor genes via genetic changes; however, one important aspect of cancer genomics is the transcriptional silencing of tumor suppressor genes caused by epigenetic changes such as DNA methylation. Furthermore, epigenetic silencing is a more frequent mechanism of gene inactivation than genetic change in some cancers. Emerging evidence suggests that epigenetic changes are involved in common diseases such as schizophrenia, autism, and diabetes.

Genome-wide DNA methylation profiling has become one of the most exciting frontiers both in epigenetics and in genome sciences. It has the potential to find epigenetic mutations in common diseases. We developed the first genome-wide DNA methylation profiling technology called restriction landmark genomic scanning (RLGS) fifteen years ago. Several thousand loci in a genome can be detected simultaneously by using two-dimensional gel electrophoresis combined with methylation-sensitive restriction enzyme digestion. RLGS has successfully identified imprinted genes and several putative tumor suppressor genes. However, this method cannot be applied to all genes in the genome and requires some skill. Therefore, we developed the microarray-based integrated analysis of methylation by isoschizomers (MIAMI) to solve these problems. This method uses the methylation-sensitive restriction enzyme *Hpa* II for methylation detection and the methylation-insensitive isoschizomer *Msp* I for control to exclude effects other than methylation. Methylation differences between the samples are detected as differences in fluorescence intensities of Cy3 and Cy5 of DNA probes on a microarray. Genome tiling-arrays can be used for this method; therefore, DNA methylation can be scanned throughout the genome.

Applications of RLGS and MIAMI will be described in this chapter. A comparison with other genome-wide profiling will also be discussed.

Chapter VII - Neural differentiation is a complex process involving the differential expression of several thousand genes. Most of these genes have been identified in recent years by genome-wide microarray analysis. The challenge now is to understand the mechanisms by which the genome orchestrates this process. Although transcription factor networks play a key role, it is now evident that other levels of regulation are also involved. Of these, epigenetic modifications of DNA and histones is emerging as a major component, not just in terms of gene silencing, but also in the differential expression of developmental genes. To better understand the mechanisms involved in the differentiation of neurons and astrocytes, we initially identified key gene clusters through microarray analysis, such as genes playing a role in RA binding, embryogenesis, regulation of neural fate, differentiation of neurons, etc. Using QPCR to supplement the microarray data we have categorized these genes into stage-specific groups, such as precursor-specific, progenitor cell state-specific, astrocyte-specific and neuron-specific. A common aspect of these genes is that many of them are subject to epigenetic control. In this chapter we will review DNA methylation, particularly in the context of neurodevelopment and also indicate to what extent these marks in the genome become disrupted in disease states.

Chapter VIII - The analysis performed in *HeLa* cells demonstrates that post-synthetic methylation of almost the half population of mitochondrial DNA (mtDNA) is cell-cycle (CC) dependent, in harmony with the timing of post-synthetic methylation of macromolecules functioning in gene expression. At variance with nuclear DNA (nDNA), which is synthesized and methylated in the S phase, ribosomal, transfer and messenger RNAs as well as ribosomal and nuclear SDS-soluble proteins are synthesized in G_1 and G_2 but methylated in S. The replicating mtDNA exhibits two fractions: the first is detectable during S in synchrony with nDNA replication; the second is detectable during G_2 when replication of nDNA is switched-off. The methylation process involves the mtDNA revealed in S, while the mtDNA revealed in G_2 remains unmethylated. This diversification unequivocally accounts for heterogeneity of mitochondria, in a cell world with which they interact maintaining a large margin of autonomy.

Chapter IX - The presence of symmetric methylation in the promoter sequences and its effect on transcriptional inactivation in cancer has been well documented. Asymmetric cytosine methylation, very frequent in plants, is associated with transient or permanent gene inactivation. Evidence on this type of methylation in mammalian organisms is scarce and mostly concerns epigenetic modifications in embryonic stem cells or inactivation of viral sequences by the mammalian host. In two reports however, asymmetric methylation has been also shown in cancerous cells, in *p53,* either in promoter (CCG) or oligoC sequences in exon 5.

The IARC's *p53* somatic mutation database is an excellent source of genetic information reflecting the positive association between cytosine methylation and genetic instability (C>T and G>A mutations) in carcinogenesis. Analysis of the distribution and the frequency of silent and nonsense mutations could be of particular importance in the elucidation of the possible role of asymmetric non-CpG cytosine methylation in genetic instability. Recently, we have shown that in certain cancers, the existence of high rates of silent transitions observed in specific exonic non-CpG cytosines and the frequently observed G:C>A:T transitions in codons of low functional impact could be indicative for the presence of asymmetric cytosine methylation.

The data presented here reveal that using the same tools of mutation analysis, i.e. distribution and frequency of silent and nonsense mutations and comparison with the rates of transitions at CpG sequences (which are always methylated), it can be concluded that the presence of extensive asymmetric methylation in several other cancers, oral, skin, esophagus stomach and in sarcomas besides lung and breast cancer is also highly probably involved in carcinogenesis. The frequency of G:C>A:T mutations at CpG and non-CpG G:C base pairs increases later in life compared to all other silent and nonsense mutations. Asymmetric cytosine methylation in some sporadic cancers (i.e. adrenal, ovarian) is probably focused in specific sites while in colon/colorectal and hemopoietic cancers this modification is probably absent. Extensive asymmetric methylation is also experimentally shown by direct sequencing of $NaHSO_3$-modified DNA in oral and lung cancer specimens. Furthermore, we provide experimental evidence for the expression, in the same lung cancer specimens, of aberrant transcripts.

In conclusion, detailed analysis of *p53* somatic mutation database and experimental evidence are in support of the frequent presence of spread, or focused exonic asymmetric cytosine methylation in certain cancers which might be related to the frequent presence of aberrant transcripts, experimentally detected in lung cancer.

In: Progress in DNA Methylation Research
Editor: Hans P. Neumann, pp. 1-90
ISBN: 978-1-60021-722-7

Chapter I

PECULIARITIES OF DNA METHYLATION IN PLANTS

Boris F. Vanyushin and Vasili V. Ashapkin

Belozersky Institute of Physical and Chemical Biology, Lomonosov Moscow State University, Moscow 119992, Russia; E-mail: vanyush@belozersky.msu.ru.

ABSTRACT

High degree of nuclear DNA (nDNA) methylation is a specific feature of plant genomes, they do contain 5-methylcytosine (m^5C) and N^6-methyladenine (m^6A). More than 30% m^5C is located in CNG sequences. Specific changes in DNA methylation accompany the entire life of a plant starting from seed germination up to the death programed or induced by various agents and factors of biological or abiotic nature. Modulation of DNA methylation is one of the possible modes of the hormonal action in plant. DNA methylation in plants is species-, tissue-, organelle- and age-specific; it is involved in the control of all genetic functions including transcription, replication, DNA repair, gene transposition and cell differentiation. DNA methylation is engaged in gene silencing and parental imprinting, it controls transgenes and foreign DNA.

Plants have much more complicated and sophisticated system of the multicomponent and sometimes even conjugated genome (nuclear DNA) methylations compared with animals; besides, unlike animals, they have the plastids with their own unique DNA modification system that may control plastid differentiation and functioning; DNA methylation in plant mitochondria is performed in other fashion compared with it in nuclei. The nuclear DNA methylation system is controlled by three major families of cytosine DNA-methyltransferase genes, at least. In contrast to animals the inactivation of major maintenance methyltransferase MET1 (similar to animal Dnmt1) has no significant consequences for plant survival. Other plant cytosine DNA-methyltransferases have no analogs in animals. Some of them (DRM) are responsible for *de novo* DNA methylation including asymmetric sequences.

Plant gene may be methylated at both adenine and cytosine residues; specific adenine DNA-methyltransferase was described. Adenine DNA methylation may influence cytosine modification and *vice versa.* Anyway, two different systems of the genome modification based on methylation of adenines and cytosines coexist in higher plants.

The specific endonucleases discriminating between methylated and unmethylated DNA are present in plants. Thus, plants may have restriction-modification system.

There are peculiar complicated controls for growth and development by DNA methylations in plants; they are well coordinated with other epigenetic signals modulating chromatin organization.

Life is measured by the rapidity of change, the succession of influences that modify the being
George Eliot *(1819–80), English novelist*

INTRODUCTION

A specific feature of plant genomes is high degree of the nuclear DNA (nDNA) methylation, they do contain 5-methylcytosine (m^5C) and N^6-methyladenine (m^6A).These additional bases appear in plant DNA as a result of methylation with specific enzymes DNA-methyltransferases that transfer methyl groups from the universal methyl donor, S-adenosyl-L-methionine (SAM, or AdoMet) onto cytosine and adenine residues located in specific DNA sequences. Main target sequence to be methylated is CG but more than 30% m^5C in plant genome is located in CNG sequences. m^5C was found in DNA of all archegoniate (mosses, ferns, gymnosperms and others) and flowering plants (dicots, monocots) investigated. As a rule, DNA of gymnosperms contains less m^5C than DNA of flowering plants (Vanyushin and Belozersky, 1959; Vanyushin et al. 1971). The species differences of phylogenetic significance in frequencies of methylated CNG sequences in genomes of plants are clearly pronounced (Kovarik et al. 1997; Fulnecek et al. 2002).

DNA methylation in plants is tissue-, organelle- and age-specific. The tissue specificity of DNA methylation established first in animals (Vanyushin et al. 1970) and than in plants (Vanyushin et al. 1979) demonstrated that DNA methylation is associated with cellular differentiation.There are many data available now indicating that methylation patterns of total DNA or distinct genes in various tissues of one and the same plant are different (Bianchi and Viotti, 1988; Lo Schiavo et al. 1989; Riggs and Chrispeels, 1999; Palmgren et al. 1991; Kutueva et al. 1996; Rossi et al. 1997; Ashapkin et al. 2002; Chopra et al. 2003). The m^5C content in DNA from different plant tissues is associated with a flowering gradient (Chvojka et al. 1978). Gene silencing associated with DNA methylation is tissue specific also; methylation of a β-glucuronidase reporter gene in the transgenic rice plant accompanied by loss of expression was initially restricted to the promoter region and observed in the vascular bundle tissue only, the expression character was similar to that of a promoter with deleted vascular bundle expression element (Klotti et al. 2002). The first comprehensive DNA methylation mapping of an entire genome of *Arabidopsis thaliana* showed that pericentromeric heterochromatin, repetitive sequences, and regions producing small interfering RNAs are heavily methylated. Interestingly, over one-third of expressed genes are methylated within transcribed regions and only about 5% genes within promoter regions. Genes methylated in transcribed regions are highly expressed and constitutively active, whereas promoter-methylated genes show a greater degree of tissue-specific expression (Zhang et al. 2006).

Specific changes in DNA methylation accompany the entire life of a plant, starting from seed germination up to the death programmed or induced by various agents and factors of biological or abiotic nature. In fact, the ontogenesis and the life itself are impossible without DNA methylation, because this genome modification in plants, like in other eukaryotes, is involved in a control of all genetic functions including transcription, replication, DNA repair, gene transposition and cell differentiation. DNA methylation controls plant growth and development. On the other hand, plant growth and development are regulated by specific phytohormones, and modulation of DNA methylation is one of the modes of the hormonal action in plant.

Plant DNA methylation has many things in common with it in animals but it has some distinguished specific features and even surprises. First, the share of methylated CNG and methylated asymmetric DNA sequences in plant genomes is much higher than that in animals. In general, plants have a more complicated and sophisticated system of genome methylations (including interactive one) compared with animals. Few plant cytosine DNA-methyltransferases have no analogs in animals. Some plant DNA-methyltrasferases are unique, unlike respective animal enzymes they contain the conservative ubiquitin association (UBA) domain and seem to be controlled in a cell cycle by the ubiquitin-mediated protein degradation pathway or (and) the ubiquitinization may alter the cellular localization of these enzymes due to respective external signals, the cell cycle, or transposon or retroviral activity. Interestingly, the plant DNA-methyltransferase activity seems to be directy influenced by plant growth regulators. Besides, unlike animals, the plant kingdom representatives have specific organelles plastids (chloroplasts, chromoplasts, leucoplasts, amyloplasts and others) with their own DNA modification systems that may control plastid differentiation and functioning. DNA methylation in plant mitochondria is performed in a different fashion compared with nuclei. Contrary to animals, N^6-methyladenine is present in plant mtDNA, whereas m^5C, known for animal mtDNA, in plant mtDNA is not found. Thus, in general, the systems of DNA modifications in cytoplasmic organelles in plants and animals are different. Unlike animals, plants seem to have a restriction-modification (R-M) system. Anyhow, plants supply us with unique systems or models of living organisms that help us to understand and decipher the intimate mechanisms and the functional role of enzymatic genome modifications and functioning in eukaryotes.

Some features and regularities of DNA methylation in plants are described in this chapter, which cannot be a comprehensive elucidation of many complicated problems associated with this genome modification in the plant kingdom. An interested reader may find the intriguing details of plant DNA methylation and its biological consequences also in available reviews (Fedoroff 1995; Meyer 1995; Richards 1997; Dennis et al. 1998; Finnegan et al. 1998b; Colot and Rossignol 1999; Kooter et al. 1999; Finnegan et al. 2000; Finnegan and Kovac 2000; Matzke et al. 2000; Sheldon et al. 2000; Wassenegger 2000; Bender 2001; Chaudhury et al. 2001; Martienssen and Colot 2001; Paszkowski and Whitham 2001; Vaucheret and Fagard 2001; Bourc'his and Bestor 2002; Kakutani 2002; Li et al. 2002;

Wassenegger 2002; Liu and Wendel 2003; Stokes 2003; Vinkenoog et al. 2003; Bender, 2004; Matzke et al. 2004; Montgomery 2004; Yi et al. 2004; Scott and Spielman 2004; Steimer et al. 2004; Tariq and Paszkowski 2004; Gendrel and Colot, 2005; Vanyushin, 2005, 2006).

IS THE CYTOSINE DNA METHYLATION AT ALL IMPORTANT?

Cytosine methylation of plant DNA is implicated in epigenetic silencing of repeated transgenes (Matzke and Matzke, 1995), repeated endogenous genes (Bender and Fink, 1995; Ronchi et al. 1995) and transposable elements (Brutnell and Dellaporta, 1994; Martienssen and Baron, 1994; Schlappi et al. 1994). The better part of existing knowledge in this field was obtained by genetic analyses.

Kakutani and coauthors were first to obtain a number of the DNA hypomethylation mutants in *Arabidopsis thaliana* (Vongs et al. 1993; Kakutani et al. 1995). The DNA methylation levels at both CpG and CpNpG sites seemed to be equally affected, the general methylation level being somewhat 30% of the wild-type value in homozygous mutant plants. The respective locus was logically named as *DDM1* (for decrease in DNA methylation). Initial phenotypic analysis of *ddm1* homozygous mutants did not reveal any evident morphological abnormalities, which seemed to be in a striking contrast to known effects of hypomethylation mutations in mice, where similar ~70% reduction of genomic DNA methylation leads to early embryonic lethality (Li et al. 1992). More careful phenotypic and biochemical characterization of *ddm1* mutants disclosed two important points. The first one was that the methylation activity for both CpG and CpNpG substrates is not affected, that proved the *DDM1* locus not to encode a DNA-methyltransferase. The second one was that indeed there are some phenotypic changes, namely *ddm1* homozygotes exhibited altered leaf shape, increased cauline leaf number and a delay in the onset of flowering when compared to non-mutant siblings in a segregating population. A high incidence of morphological abnormalities was noted in the *ddm1* homozygous lines propagated by repeated self-pollination (Kakutani et al. 1996). The onset of the abnormalities was strictly associated with the *ddm1* mutations. Similar morphological defects were caused by *ddm1* mutations in, at least, two genetic backgrounds of *Arabidopsis thaliana*, Columbia and Landsberg erecta. Moreover, these severe developmental defects were seen in selfed lines carrying independently isolated *ddm* alleles arguing against any contribution from additional mutations closely linked to *ddm1*. After six generations of selfpollination the plants exhibiting aberrant morphology including reduction or increase in apical dominance, short internode length, late flowering, small leaf size, increased cauline leaf number and reduced fertility were found in all *ddm1/ddm1* lines. In addition, some lines displayed plants with abnormal flowers. Namely, plants with reduced sepal number (3 out of 14 *ddm1/ddm1* selfed lines) and hooked and partially unfused carpels were noted. After 7 generations of selfpollination a high degree of sterility or seedling lethality was observed (5 of 14 *ddm1/ddm1* lines). While there were differences in the spectrum of the phenotypes among *ddm1/ddm1* lines, some abnormal characters frequently occurred together. One such combination was an increase in apical dominance and in cauline leaf number, and a delay in time to flowering. Another commonly seen combination (''ball''syndrome) was the reduced apical dominance, twisted leaves, and small plant size. The severity of the ball syndrome was progressive with more pronounced phenotypes exhibited by plants in families resulting from higher numbers of self-pollinations. The ball syndrome was shown to be inherited as a simple Mendelian monogenic trait. Crosses between *ddm1/ddm1* phenotypic ball plants (strain Columbia) and wild-type Columbia plants yielded plants with normal phenotypes and intermediate ball phenotypes. F2 generations derived by selfing the phenotypically intermediate plants contained plants with normal, intermediate, and severe ball phenotypes with a 1:2:1 ratio, respectively, suggesting the segregation of a semidominant lesion. Inheritance of the ball phenotype in the F2 generation was independent of the segregation of the *ddm1* mutation itself. Starting with a *DDM1/ddm1* severe ball F2 plant, several severe ball *DDM1/DDM1* lines were obtained, in which no normal plants were seen through three generations of self-pollination. The *ddm1* mutation was mapped to distal portion of the lower arm of chromosome 5, whereas the locus responsible for the ball phenotype (*BAL*) - to the lower arm of chromosome 4. Similar results were obtained for the inheritance of another complex trait, designated ''clam,'' which appeared frequently in *ddm1/ddm1* selfed lines. This trait is characterized by a small, compressed rosette, reduced internode length and reduced fertility. The inheritance of the clam phenotype indicated that the trait is caused by a monogenic recessive lesion. The locus responsible for the clam phenotype (*CLM*) is also unlinked to the *DDM1* locus and maps to the center of chromosome 3. The precision of the global DNA methylation measurements in *ddm1/ddm1* lines during consequitive self-pollination generations was too low to reliably detect small changes in the methylation levels. But examination of specific genomic regions by Southern blot analysis did revealed a progressive reduction in cytosine methylation.

Accumulated loss of multiple methylation sites at a single locus could explain the delayed onset and progressive severity of the morphological defects. The variation in phenotypic severity seen among siblings in selfed populations could be due, in part, to continued creation of new epimutations (loss of methylated sites) in somatic tissue followed by transmission to and segregation in the next generation. Several considerations suggest that the loss of cytosine methylation is indeed responsible for the delayed onset of morphological phenotypes. Phenotypes resembling the *ddm1* induced delayed-onset defects were seen in transgenic *A. thaliana* expressing the cytosine methyltransferase antisense constructs (Ronemus et al. 1996; Finnegan et al. 1996). The sets of mutant phenotypes observed in self-pollinated *ddm1* lines and in those DNA-methyltransferase (MET1) antisense plants were basically the same: leaves with margins curled upward, increase in stamen, leaf and shoot numbers, and, last but not least, delay in flowering initiation. The delayed onset of flowering is a most frequently observed phenotype in both groups of DNA hypomethylation mutants. Similarly to other phenotypes, it became quite evident after several generations of self-pollinations. Upon outcrossing of such selfed late-flowering lines to normally flowering ones the late-flowering phenotype segregates in the ratio 3:1 consistent with Mendelian monogenic dominant trait. Analyses of a number of the independent *ddm1* lines showed the locus responsible for this late-flowering phenotype to reside between DNA markers *RPS2* and *AG* (closer to the first one) in the bottom arm of chromosome 4, which is quite different to location of *DDM1* locus itself (chromosome 5). This position of late-flowering locus coincides with that of *FWA*, a gene known to be involved in flowering timing. The dominant property of the late-flowering trait is consistent with a straightforward suggestion that a hypomethylation-induced activation of previously suppressed gene occurs.

The DNA-methyltransferase inhibitors, 5-azacytidine and 5-aza-2'-deoxycytidine, inhibited adventitious shoot induction in *Petunia* leaf cultures; cytosine methylation at CCGG and CGCG sites within a MADS-box gene and a CDC48 homolog, among others, shows strong positive correlation with adventitious shoot bud induction (Prakash et al. 2003). Application of the hypomethylation drugs 5-azacytidine or dihydroxypropyladenine to transgenic tobacco lines resulted in about 30% reduced methylation of cytosines located in a non-symmetrical sequences in the 3'-untranslated region of the neomycin phosphotransferase II (nptII) reporter gene, this hypomethylation was accompanied by up to 12-fold increase in NPTII protein level (Kovarik et al. 2000b). 5-azacytidine sharply accelerated apoptotic DNA fragmentation in the coleoptiles of wheat seedlings exposed to this compound, it can be caused by DNA demethylation and, correspondingly, by derepression and induction of various apoptogenic factors, including, for example, caspases, endonucleases and regulatory proteins (Vanyushin et al. 2002). The treatment of plants with 5-azaCyt is responsible for dwarfism in rice (Sano, 2002) and an increased storage protein content in wheat seeds (Vanyushin et al. 1990), both are inherited in few generations. In the transgenic rice seedlings the *bar* gene expression induced by 5-azaCyt treatment disappears in about 20-50 days (Kumpatla and Hall, 1998); this means that plants have a tendency and ability to reestablish an initial genome methylation pattern that was distorted by the drug. Treatment with 5-aza-2'-deoxycytidine resulted in the development of altered morphologies in the synthetic allotetraploids of *Arabidopsis* and *Cardaminopsis arenosa* (Madlung et al. 2002).

DNA methylation controls flowering in plants that need vernalization (exposure to cold) to initiate flowering. Vernalization accompanied by DNA demethylation may be substituted for 5-azacytidine treatment or MET1 inactivation (antisense) that promote flowering in vernalization-responsive *Arabidopsis* plants (Burn et al. 1993; Finnegan et al. 1998a); DNA methylation regulates transcription of *FLC*, a repressor of flowering (Finnegan et al. 1998a). *FLC* is a key gene in the vernalization response; plants with high FLC expression respond to vernalization by downregulating *FLC* and, thereby, flowering at an earlier time. The downregulation of *FLC* by low temperatures is maintained throughout vegetative development but is reset at each generation. A small gene cluster including *FLC* and its two flanking genes is coordinately regulated in response to vernalization (Finnegan et al. 2004). It is remarkable that foreign genes inserted into the cluster also acquire the low-temperature response. At other chromosomal locations, *FLC* maintains its response to vernalization and imposes a parallel response on a flanking gene, thus, *FLC* contains sequences that confer changes in gene expression extending beyond *FLC* itself, perhaps, through chromatin modification (Finnegan et al. 2004).

Cold stress induces DNA demethylation in various plants, it may, in particular, be associated with cold-dependent expression of specific proteins. When maize seedlings were exposed to cold stress, a genome-wide demethylation occurred in root tissues (Steward et al. 2002). One particular 1.8-kb fragment (ZmMI1) containing a part of the coding region of a putative protein and part of a retrotransposon-like sequence was demethylated and transcribed only under cold stress. Interestingly, cold stress induced severe DNA demethylation in the nucleosome core but not in the linkers; methylation and demethylation were periodic in nucleosomes (Steward et al. 2002).

It is known that the transposition frequency of Tam3 in *Antirrhinum majus*, unlike that of most other cut-and-paste-type transposons, is tightly controlled by temperature: Tam3 transposes rarely at 25°C, but much more frequently at 15°C, the temperature shift induced a remarkable change of the methylation state of unique to Tam3 sequences in the genome: higher temperature resulted in hypermethylation, whereas lower temperature resulted in reduced methylation. The methylation state was reversible within a single generation in response to a temperature shift (Hashida et al. 2003). Differences in the methylation pattern were observed in DNA of spring and winter wheat (*Triticum aestivum*), as well as in unvernalized and vernalized wheat plants. Winter wheat DNA was more highly methylated than spring wheat DNA; changes in the methylation pattern were observed at the end and after vernalization. Thus, there is not only a vernalization-induced demethylation related to flower induction, but there is also a more general and non-specific demethylation of sequences unrelated to flowering (Sherman and Talbert, 2002).

DNA methylation in plants is involved in parental imprinting and regulation of the developmental programme (Finnegan *et al.* 2000). In sexual species, endosperm typically requires a ratio of two maternal genomes to one paternal genome for normal development but this ratio is often altered in apomicts, suggesting that the imprinting system is altered as well; DNA methylation is one mechanism by which the imprinting system could be altered to allow endosperm development in apomicts (Spielman et al. 2003). Analysis of inbred lines and their reciprocal crosses in maize identified a large number of conserved, differentially methylated DNA regions (DMRs) that were specific to the endosperm. DMRs were hypomethylated upon maternal transmission, whereas upon paternal transmission the methylation levels were similar to those observed in embryo and leaf. Maternal hypomethylation was extensive and it offers a likely explanation for 13% reduction in m^5C content in DNA of the endosperm compared with leaf tissue (Lauria et al. 2004). In the maize endosperm the genes for α-zeins and α-tubulins methylated in sporophytic diploid tissues become undermethylated in the triploid endosperm, and the demethylation correlating with gene expression is often restricted to two chromosomes of maternal origin (Lund et al. 1995a, b). In *Arabidopsis* the paternally inherited *MEA* alleles are transcriptionally silent in both young embryo and endosperm; *MEA* gene imprinted in the *Arabidopsis* endosperm encodes a SET-domain protein of the Polycomb group that regulates cell proliferation by exerting a gametophytic maternal control during seed development; *ddm1* mutations are able to rescue *mea* seeds by functionally reactivating paternally inherited MEA alleles during seed development; thus, the maintenance of the genomic imprint at the *mea* locus requires zygotic DDM1 activity (Vielle-Calzada et al. 1999). Imprinting of the *MEA* Polycomb gene is controlled in the female gametophyte by antagonism between two DNA-modifying enzymes, MET1 methyltransferase and DME glycosylase (Xiao et al. 2003). DME DNA glycosylase activates maternal *MEA* allele expression in the central cell of the female gametophyte, the progenitor of the endosperm. Maternal mutant *dme* or *mea* alleles result in seed abortion; mutations that suppress *dme* seed abortion are found to be resided in the MET1 methyltransferase gene. *MET1* functions upstream of, or at, *MEA* and is required for DNA methylation of three regions in the *MEA* promoter in seeds (Xiao et al. 2003). Parental imprinting in *A. thaliana* involves the activity of the DNA *MET1* gene; plants transformed with an antisense *MET1* construct have hypomethylated genomes and show alterations in the behavior of their gametes in crosses with wild-type plants; a hybridization barrier between diploid *A. thaliana* (when used as a seed parent) and tetraploid *A. arenosa* (when used as a pollen parent) can be overcome by increasing maternal ploidy but restored by hypomethylation; thus, hypomethylation restores the hybridization barrier through paternalization of endosperm; manipulation with DNA methylation can be sufficient to erect hybridization barriers, offering a potential mechanism for speciation and a mean of controlling gene flow between species (Bushell et al. 2003). The *Arabidopsis FWA* gene displays imprinted (maternal origin-specific) expression associated with heritable hypomethylation of repeats around transcription starting sites in endosperm. The *FWA* imprint depends on the maintenance DNA-methyltransferase MET1 and is not established by allele-specific *de novo* methylation but by maternal gametophyte-specific gene activation, which depends on a DNA glycosylase gene, *DEMETER* (Kinoshita et al. 2004).

Due to known reaction of the oxidative m^5C deamination conjugated with cytosine methylation (Mazin et al. 1985), DNA methylation is an essential mutagenic factor that is responsible for a well known phenomenon of CG and CNG suppressions that are common for many plant genes (Lund et al. 2003). Thus, DNA methylation is an important factor of plant evolution.

DNA methylation may be essentially modulated by various biological (viral, bacterial fungal, parasitic plant infections) or abiotic factors that may influence a plant growth and development. Interestingly, the Chernobyl radiation accident resulted in a global DNA hypermethylation in some plants investigated (Kovalchuk et al. 2003). Fungal infections most strongly distort methylation in repetitive but not unique sequences in plant genome (Guseinov and Vanyushin, 1975). By such way fungi, viruses and other infective agents may switch over the gene transcription program in the host plant mostly in a favor of respective infective agent. On the other hand, plants are able to modify viral DNAs that are not integrated into the plant genome. In few days after inoculation into turnip leaves the unencapsidated cauliflower mosaic virus DNA was found to be in methylated state at almost all HpaII/MspI sites (Tang and Leisner, 1998). In fact, proper DNA methylation may stabilize foreign DNA in host plant (Rogers and Rogers, 1992). The foreign DNA introduced into barley cells was able to persist through at least two plant generations. Transformation of barley cells was defined by showing initiation of transcription at the proper site on the barley promoter for the chimeric gene in aleurone tissue from both a primary transformant and its progeny, and by tissue-specific expression (aleurone greater than leaf) in the progeny; this persistence through many multiples of cell division is considered as formally equivalent to transformation, regardless of whether the DNA was chromosomally integrated or carried as an episome, but did not necessarily represent stable integration into the genome since the foreign DNA was frequently rearranged or lost (Rogers and Rogers, 1992). The foreign DNA was most stable when plasmid DNA used in transformation lacked adenine methylation but had complete methylation of cytosine residues in the CG at Hpa II sites; adenine methylation alone was associated with marked foreign DNA instability. Thus, barley cells have a system that identifies DNA lacking the proper methylation pattern and causes its loss from actively dividing cells (Rogers and Rogers, 1992). These intriguing data on foreign DNA methylation in plant cells may resemble host modification phenomenon that is common in prokaryotes.

Thus, in fact, cytosine DNA methylation controls plant growth and development. Similarly to animals (Holliday and Pugh, 1975; Razin and Riggs, 1980; Bird, 1992; Razin, 1998), specific cytosine DNA methylation in plants controls practically all genetic processes including transcription, replication, DNA repair, cell differentiation and, in particular, is involved in specific gene silencing and transposition.

ARE TRANSPOSABLE SEQUENCES SILENCED BY CYTOSINE METHYLATION?

When *ddm1* mutation was introduced into an *Arabidopsis* cell line carrying inactivated tobacco retrotransposon Tto1, this element became hypomethylated and transcriptionally and transpositionally active; therefore, the inactivation of retrotransposons and the silencing of genes have mechanisms in common (Hirochika et al. 2000). Plant S1 SINE (short interspersed elements) retroposons mainly integrate in hypomethylated DNA regions and are targeted by methylases; methylation can then spread from the SINE into flanking genomic sequences, creating distal epigenetic modifications. This methylation spreading is vectorially directed upstream or downstream of the S1 element, suggesting that it could be facilitated, when a potentially good methylatable sequence is single stranded during DNA replication, particularly, when located on the lagging strand. Replication of a short methylated DNA region could thus lead to the *de novo* methylation of upstream or downstream adjacent sequences (Arnaud et al. 2000).

DNA methylation influences the mobility of transposons. The influence seems to be associated, particularly, with different affinity for Ac transposase binding to holo-, hemi-, and unmethylated transposon ends. In petunia cells a holomethylated Ds is unable to excise from a nonreplicating vector and replication restores excision. A Ds element hemimethylated on one DNA strand transposes in the absence of replication, whereas hemimethylation of the complementary strand causes an inhibition of Ds excision; in the active hemimethylated state the Ds ends have a high binding affinity for the transposase, whereas binding to inactive ends is strongly reduced (Ros and Kunze, 2001). DNA methylation in the Tam3 end regions in *Antirrhinum* tended to suppress the excision activity, and the degree of methylation was dependent on the chromosomal position (Kitamura et al. 2001).

Mutator-like element with long terminal inverted repeats (TIR-MULEs) have been shown to be quiescent and not transposing in *Arabidopsis* strain Columbia (Singer et al. 2001). Quiescence is correlated with DNA methylation and a lack of transcription. In contrast, in Landsberg *erecta*, where they have lower levels of DNA methylation the TIR-MULEs are slightly transcribed and transpose occasionally. In the loss-of-function *ddm1* mutants the transposon methylation was eliminated in both strains and *AtMu1* was activated resulting in high levels (10%-20% per generation) of transposition. Given the predicted function of *DDM1* that encodes a SWI2/SNF2-like protein (Jeddeloh et al. 1999) the chromatin remodelling seems to be an important process for maintenance of DNA methylation and genome integrity. Chromatin remodeling and DNA methylation are, therefore, likely required for transcriptional, as well as transpositional repression of potentially active autonomous elements.

One of the *ddm1*-induced developmental abnormalities is a result of *CAC1* transposon insertion. In particular, it was found in a study of mutated gene responsible for *clam (clm)* phenotype, which is characterized by lack of shoots and petioles elongation (Miura et al. 2001). The phenotype unstable initially (phenotypically normal sectors were occasionally observed) was eventually stabilized in subsequent generations and inherited as a recessive Mendelian trait that could be mapped genetically.The respective locus (*clm*) was narrowed to a 64-kb region on chromosome 3 (BAC clone T3A5, GenBank AL132979). This region contains gene for 22-α-hydroxylase (*DWF4*), protein mediating the biosynthesis of brassinosteroid, a regulator of cell elongation. Complementation tests indicated that *clm* is indeed allelic to *dwf4* mutation. The sequencing of *DWF4* gene from stable *clm* plants revealed the presence of a 4-bp insertion in the second exon that converted TAG sequence to TAGCTAG at +527 position from the translation start. A stop codon appearance resulted in protein truncation after 149 amino acids. This is quite compelling cause of stable *clm* phenotype but it could not account for the instability of the *clm* phenotype in initial generations. An insertion of several-kilobase sequence in *DWF4* gene was found in the unstable *clm* plants by Southern blot analysis. Partial sequencing of insert showed the exact match to unique 8479-bp sequence in chromosome 2 (GenBank AC005897) bearing all features typical of the CACTA family of transposons, including conserved terminal inverted repeats CACTACAA and an internal ORF for putative transposase. Therefore, transposition of a full-length *CAC1* element from chromosome 2 to the *DWF4* gene on chromosome 3 appeared to be responsible for unstable *clm* phenotype. This was further confirmed by sequencing of *DWF4* gene in the sectors that have been reverted to normal phenotype: restoration of the insertion site to the normal structure was found. There are three additional sequences similar to *CAC1* in *Arabidopsis* genome designed as *CAC2* to *CAC4*. Southern blot analyses of a dozen of *ddm1* lines after 6-7 self-pollinated generations reveals a high incidence of the *CAC* element transpositions and increase in their copy-numbers by several-fold. Both *CAC1* and *CAC2* were found to transpose to unlinked loci throughout genome. On the contrary, such transpositions of *CAC* elements were never observed in the self-pollinated wild-type *DDM1* lines. The mobilization of *CAC* elements, therefore, seems to be a direct consequence of their demethylation and transcription activation in *ddm1* background.

Thus, cytosine methylation of transposable sequences seems to be a major mechanism inactivating transcription and transposition of these potentially dangerous elements in the plant genomes.

Are Multicopy Genes Silenced by Cytosine Methylation?

An ideal model system for studying the role of cytosine DNA methylation in gene expression in *Arabidopsis* is an endogenous methylated gene, *MePAI2,* whose silenced, fluorescent phenotype can be easily monitored by visual inspection throughout the development of the plant (Bender and Fink, 1995). Furthermore, the intensity of the fluorescent phenotype, which reflects the level of *MePAI2* silencing, can be evaluated. *PAI2* is one of four *PAI* sister genes in the Wassilewskija (WS) strain of *Arabidopsis* that encodes the third enzyme in the tryptophan biosynthetic pathway, phosphoribosylanthranilate isomerase (*PAI*). In WS the four *PAI* genes are located at three unlinked sites in the genome. All four genes are heavily cytosine-methylated over their regions of shared DNA sequence similarity. The combined expression of the four methylated *PAI* (*MePAI*) genes provides just enough PAI activity for a normal plant phenotype. However, in a mutant where two tandemly arrayed *PAI* genes (*MePAI1-MePAI4*) are deleted, the two remaining genes (*MePAI2* and *MePAI3*) provide insufficient *PAI* activity for normal development. A striking PAI-deficient phenotype is displayed by the **Δ***pai1-pai4* deletion mutants (blue fluorescent ones under UV light due to accumulation of early intermediates in the tryptophan pathway, anthranilate and anthranilate-derived compounds). Several lines of evidence suggest that the residual methylation of the *PAI2* gene in the fluorescent *pai* mutant is associated with PAI-deficient phenotypes. First, the fluorescent *pai* mutant gives rise to spontaneous nonfluorescent revertant progeny at 1%-5% per generation, and in these revertant lines there is a substantial hypomethylation of both *PAI2* and *PAI3* (Bender and Fink 1995). Spontaneous partial revertant lines with intermediate levels of fluorescence have also been isolated, and these lines display partial hypomethylation. Furthermore, growth of the fluorescent *pai* mutant in the presence of the cytosine methyltransferase-inhibiting compound 5-azacytidine relieves the silenced fluorescent phenotype (Bender and Fink 1995). As the *PAI3* gene has very low expression levels even when unmethylated, the *MePAI2* locus seems to be the critical determinant for the blue fluorescent PAI-deficient phenotype. Therefore, *MePAI2* serves as a facile reporter for the methylation-correlated gene silencing in *Arabidopsis.* To assess the effect of the DNA hypomethylation mutation *ddm1* on *PAI2* gene silencing, *ddm1* was introduced into the fluorescent *pai* mutant background by crossing a fluorescent *pai* mutant (**Δ***pai1-pai4*/**Δ***pai1-pai4; MePAI2*/*MePAI2* in the WS background) and a homozygous *ddm1* mutant strain (*ddm1*/*ddm1* in the Columbia strain). The F_2 fluorescent segregants were homozygous for the recessive **Δ***pai1-pai4* deletion and the recessive, methylated, and silenced *MePAI2* locus from the *pai* mutant parent.These were further screened with a polymorphic marker, m555, tightly linked to the *ddm1* mutation (within 1 cM) to determine the *ddm1* genotype of each line. One representative fluorescent segregant that was heterozygous for the m555 marker (and thus heterozygous *DDM1*/*ddm1*) was used for subsequent detailed analysis. The strongly fluorescent phenotype (71% F_3 plants) corresponded to plants that carried the wild-type *DDM1* WS allele (*DDM1*/*DDM1* and *DDM1*/*ddm1*). All plants that displayed the nonparental weakly fluorescent phenotype (26%) were homozygous for the *ddm1* Columbia allele. One of three nonfluorescent plants (1%) was homozygous for the *ddm1* allele, whereas the remaining two (2%) carried the WS *DDM1* allele and represent spontaneous nonfluorescent revertants of the *MePAI2* silent state, which were previously determined to segregate from the fluorescent *pai* mutant at 1%-5% per generation (Bender and Fink 1995). Therefore, plants homozygous for the recessive *ddm1* mutation display an immediate suppression of the fluorescent silenced *pai* phenotype. From this segregating F_3 family two *pai ddm1* mutant lines were started, as well as a sibling *pai DDM1* control line. Inbreeding *pai ddm1* mutants led to a progressive loss of residual *PAI2* gene silencing. This inbreeding effect is specific to *ddm1* mutants because no significant changes in fluorescence levels were seen upon inbreeding the *pai DDM1* control line. To investigate whether the *ddm1* mutation affects *PAI2* gene silencing through a reduction in DNA methylation, Southern blot analysis with cytosine methylation-sensitive restriction enzymes was used. The *PAI* genes in the fluorescent *pai DDM1* control DNA samples showed moderate to heavy methylation of all sites investigated. DNA from the spontaneous nonfluorescent revertant line *REV2* had hypomethylated restriction sites in *PAI2* and slight residual methylation of sites in *PAI3*. In contrast, the *ddm1* mutation caused a complex pattern of DNA hypomethylation of *PAI2* and *PAI3*. For example, an *Hpa*II-*Msp*I (CCGG) site within the transcribed region of the *PAI3* gene was progressively hypomethylated in one *ddm1* mutant line, whereas hypomethylation of *PAI2* but not of *PAI3* has taken place in another. Methylation of *Sau*3AI- *Dpn*II sites (GAT^{m}C) within the transcribed regions of *PAI2* and *PAI3* was also reduced in the *ddm1* mutant lines but the hypomethylation was incomplete, indicating further that the changes in methylation of different sites are independent. The detailed methylation analysis by bisulfite-mediated conversion of cytosines revealed that in weakly fluorescent *pai ddm1* double mutants there is a mixture of differentially methylated DNA alleles, whereas in nonfluorescent inbred progeny of the *pai ddm1* double mutant there is very little residual *PAI* gene methylation. In the fluorescent *pai DDM1* mutant the cytosine methylation occurs at symmetrical CpG and CpNpG sites and at asymmetrically disposed cytosines in the *PAI2* upstream region. The most heavily methylated allele from the fluorescent *pai DDM1* mutant had approximately half of the m^{5}C residues at asymmetric sites, whereas less methylated alleles contained predominantly symmetrical modification sites. In all sequenced alleles, methylation was heaviest from ~80-bp upstream of the transcription start site extending into the transcribed region of the *PAI2* gene. Also, in all sequenced alleles no methylation was observed in a region >210 bp upstream of the transcription start site. This is consistent with previous Southern blot analysis data showing that *PAI* methylation in the *pai* mutant and parental WS does not spread significantly beyond the boundaries of the shared sequence similarity among sister *PAI* genes (Bender and Fink 1995). Four of five sequenced alleles from the spontaneous nonfluorescent revertant strain *REV2* had essentially no methylation, whereas the fifth allele is hypermethylated. Again, this sequencing data are consistent with previous Southern blot analysis of methylation patterns in *REV2,* which indicate that slight residual methylation of the *PAI2* gene can occur in this line (Bender and Fink 1995). The *ddm1* mutation caused a reduction in methylated sites throughout the *PAI2* upstream region as compared with the *pai DDM1* fluorescent strain. In DNA prepared from weakly fluorescent *pai ddm1* double mutant plants, 7 of 10 *PAI2* alleles sequenced had no or very low levels of methylation, 2 of 10 alleles had moderate methylation, and 1 of 10 alleles remained heavily methylated. In the low and moderately methylated alleles, only 2 of 25 methylated sites were in asymmetric positions, whereas in the one heavily methylated allele 15 of 33 methylated sites were in asymmetric positions. Inbreeding the *pai ddm1* mutants led to an almost complete loss of DNA methylation in the *PAI2* upstream region. The pattern of progressive hypomethylation of the *PAI2* promoter in *ddm1* lines and their expression data suggest that the loss of *PAI2* gene silencing is connected to the methylation loss. It seems that mixture of differentially methylated alleles in the weakly fluorescent *pai ddm1* double mutant reflects the fluorescence sectoring phenotype with extensively methylated alleles corresponding to the weakly fluorescent sectors and the sparsely methylated alleles corresponding to nonfluorescent sectors. Dissected weakly fluorescent and nonfluorescent sectors from weakly fluorescent *pai ddm1* double mutants were used for DNA extracting and Southern blot analysis of methylation patterns. This analysis revealed that *PAI* genes from fluorescent sectors have higher methylation than *PAI* genes prepared from nonfluorescent sectors. This is consistent with a correlation between DNA methylation and gene silencing even within one and the same plant tissue. The *Hpa*II/*Msp*I assay revealed that introducing the *ddm1* mutation to WS plants posessing full *PAI* gene family has only a weak effect on methylation of the *PAI1-PAI4* inverted repeat locus but a strong hypomethylation effect on the singlet *PAI2* and *PAI3* genes (Bartee and Bender, 2001). This basic methylation pattern was established by the second generation although the *PAI2* and *PAI3* genes became progressively less methylated over four subsequent generations of inbreeding. The *met1* mutation had an intermediate hypomethylation effect on the inverted repeat *PAI1-PAI4* locus but only a weaker effect on the *PAI2* and *PAI3* genes. Both WS *ddm1* and WS *met1* inbred lines progressively accumulated a number of morphological defects and reduced fertility, as previously observed in the Col strain background. In particular, the most inbred WS *ddm1* line developed flowers with unfused carpels; it was late flowering and displayed a number of floral abnormalities. The antisense *MET1* transgene and the *met1* missense mutation have similar effects on the WS *PAI* gene methylation in second DNA generation. A *ddm1 met1* double mutant in the WS strain background was produced by crossing between WS *ddm1* and WS *met1* lines and by using polymorphisms associated with the methylation mutations to identify double mutant recombinants. A majority of plants in the segregating population from this cross were late flowering and/or sterile, presumably due to accumulation of methylation changes during long inbreeding regime of the parental strains. However, two independent double mutant individuals were recovered that were fertile when newly segregated.The double mutants had a number of morphological defects and became completely sterile by the second generation. The Southern blot analysis of the second generation progeny of each double mutant lineage showed that the WS *ddm1 met1* double mutants displayed strong hypomethylation of *PAI2* and *PAI3* but weak hypomethylation of the *PAI1-PAI4* locus, similarly to the *ddm1* single mutant. Thus, the combined methylation mutations were not sufficient to remove *PAI* methylation after two generations of inbreeding. To understand the effects of *ddm1* and *met1* mutations on *PAI* methylation at the nucleotide sequence level, the sodium bisulfite genomic sequencing was performed on the promoter regions of the *PAI1* inverted repeat gene and the *PAI2* singlet gene in genomic DNA prepared from four generation of inbred plants. In WS *ddm1* the methylation patterns for the *PAI1* promoter at the inverted repeat locus was found to be similar to those of the same region in parental WS: within the region of *PAI* sequence identity the symmetrical CG and CNG as well as asymmetrical cytosines are methylated and there is no significant spread of methylation into upstream heterologous sequences. The primary difference between WS and WS *ddm1 PAI1* methylation patterns is in methylation density that is moderately (by 27%) lower in WS *ddm1*. In contrast, for the singlet *PAI2* gene in WS the *ddm1* methylation is reduced to 32% of parental WS methylation levels. For both *PAI1* and *PAI2* sequences the *ddm1* mutation reduces both symmetrical (CG and CNG) and asymmetrical cytosine methylations although there is a stronger effect on non-CG methylation. In WS *met1* both the *PAI1* and *PAI2* promoters have <50% residual methylation relative to parental WS. As it was observed for *ddm1*, methylation of both symmetrical and asymmetrical cytosines is affected by the *met1* mutation. Thus, cytosine methylation does regulate expression of repeated sequences in the plant genome. All known types of target sequences (CpG, CpNpG and asymmetrical) appear to be involved.

Is Gene Silencing Always Associated with their Methylation?

A recently isolated suppressor (*mom1* mutation) of the transcriptional gene silencing in plants principally differs from *ddm1* and *met1* by its rather more specific action. Unlike these hypomethylation mutations, *mom*1 does not lead to any evident phenotypical abnormalities even after inbreeding for nine generations (Amedeo et al. 2000). Moreover, it causes reactivation of genes silenced by methylation without changes in their methylation. Last but not least, its action seems to be readily reversible: resilencing of genes activated by *mom1* occurs immediately upon introduction of wild-type *MOM* allele in F1 heterozygous plants. The mRNA populations from two lines of the 2-week-old *Arabidopsis* seedlings were compared by subtractive hybridization and subsequent direct Nothern blot analysis. The first one was the parental line carrying a silent hygromycin-resistance transgene locus; the second line was its mutant derivative *mom1* with transgene locus reactivated. The study revealed a striking genotype-dependent differential expression of two sequences. The respective cDNA clones were named TSI-A and TSI-B (for transcriptionally silent information). Both were abundant in *mom1* RNA but undetectable in parental line. TSI expression was also found in other mutant lines affecting gene silencing including those with general DNA hypomethylation (*ddm1*, *met1*, and antisense-*MET1*). Multiple copies of TSI-A- and TSI-B–homologous sequences were found to be present in the genome of *Arabidopsis* by Southern blot analyses. Their copy numbers were estimated to be ~200. TSI repeats appeared to be heavily methylated in wild-type *Arabidopsis* and parental line with a silent hygromycin-resistance locus and less methylated in mutants with genome-wide decreased DNA methylation. Importantly, TSI sequences remained methylated in the *mom1* mutant in spite of their expression. This is analogous to the maintenance of methylation of the reactivated transgene in the *mom1* background (Amedeo et al. 2000). Both TSI repeats are concentrated in the pericentromeric regions of *Arabidopsis* chromosomes. This was confirmed by cytological analysis of pachytene chromosomes with *in situ* fluorescent hybridization. Close examination revealed their presence in the heterochromatic domains flanking the centromeric 180-bp tandem repeat region. Both TSIs and all sequenced cDNA clones of their transcription products were mapped to the 3'- terminal half of the *Athila* element, a retrotransposon located at the pericentromeric region of chromosome 5. Two open reading frames (ORF) of *Athila* (ORF1 and ORF2) have no homology with proteins usually encoded by retroelements or with any other known protein, and neither transposition nor even transcriptional activity has ever been observed. However, only a subset of *Athila*-like sequences but not the *Athila* element itself was found to be reactivated in the mutant background. The homology of TSI to putative retrotransposon also raised the question of whether other endogenous retroelements are transcriptionally reactivated. Since reverse transcriptase is the most highly conserved protein encoded by retroviruses and retrotransposons, the degenerated primers in a conserved region of the reverse transcriptase gene were used to investigate whether members of the *Ta* retrotransposon family are transcribed in the mutant background. Although the expected 268-bp fragment was readily amplified from genomic DNA, no amplification was achieved in RT-PCR with *mom1* RNA as template. Thus, retrotransposons are not generally reactivated in the *mom1* mutant. TSI expression in the wild-type background was not detected either in young seedlings (2 weeks old) or in different tissues of mature wild-type *Arabidopsis* plants (roots, shoots, leaves, flowers, and siliques). The application of various stress treatments (increased salinity, UV-light irradiation, pathogen infection) also did not activate TSI in wild-type plants. Furthermore, TSI expression was not detected in freshly initiated callus cultures, and transcriptional suppression of TSI was stable even after several *in vitro* passages of the callus culture. The only exception known so far is a fast-growing long-term suspension cell culture derived from wild-type *Arabidopsis*. These cells expressed TSI to approximately the same extent as the mutants, it indicates on the release of epigenetic TSI silencing under these conditions. The TSI expression was found to be activated in the *cmt3-7* mutant line (Lindroth et al. 2001). Expression of the Ta3 retrotransposon that was previously found to be transcriptionally silent in both wild type plant and *ddm1* mutant (Hirochika et al. 2000) was easily detected in *cmt3-7* but no expression was observed in the wild-type line *clk-st*. Two additional retrotransposons, *Evelknievel* and *Tar17*, which were previously shown to be reactivated in the *ddm1* mutant (ibid), were found to remain silenced in *cmt3*. Together, these results demonstrate that *CMT3* is required for maintaining gene silencing at a subset of retrotransposon sequences. The differential reactivation of gene expression observed in the *cmt3* mutants suggests a model where different loci may depend preferentially on either CpNpG or CpG methylation as the main mechanism of gene silencing. For instance, *SUP* and the *Ta3* retrotransposon appear to depend more significantly on CpXpG methylation, whereas *FWA* and possibly *Tar17* rely more on CpG methylation. *Athila* sequences require both types of methylation because *Athila*-related transcripts are activated in both *cmt3* and *met1* mutants.

In general, DNA methylation represents only a part of a complicated multi-step process of gene silencing, though a very important one. Silenced state of genes is usually correlated with their hypermethylation, whereas hypomethylation leads to transcription reactivation. Nevertheless, the other steps of gene silencing also contribute to the maintanance of the silent state. Their breakage, for example, by *mom1* mutation, may lead to partial or full reactivation of transcription even when the affected gene remains to be heavily methylated.

Are the Epigenetic Changes Inheritable?

In both plants and mammals, epigenetic control of gene expression is often correlated with change in cytosine methylation of the affected locus. Mammalian epigenetic phenomena, in particular, parental imprinting and X-chromosome inactivation, are developmentally regulated, and the "resetting" of epigenetic status occurs in each generation. The methylation patterns in mammalian genome undergo reorganization (Monk et al. 1987) by extensive demethylation and "*de novo*" methylation during gametogenesis and early development (Yoder et al. 1997). In contrast, the epigenetic states of plant genes such as the *Arabidopsis*

SUPERMAN gene (Jacobsen and Meyerowitz, 1997), *PAI* genes (Bender and Fink, 1995), maize transposable elements (McClintok, 1967; Brutnell and Dellaporta, 1994; Martienssen and Baron, 1994; Schlappi et al. 1994) and repeated transgenes of tobacco (Park et al. 1996) are often stably inherited through generations.

The F_1 heterozygotes (*DDM1/ddm1*) produced by crossing a *ddm1* homozygote to a wild-type plant contain m^5C at levels halfway between those of two parents (Vongs et al. 1993). Southern blot analysis using the methylation-sensitive restriction enzyme, *Hpa*II, of the genomes of *ddm1/ddm1* mutant, wild-type and the F_1 plants showed that about half of the DNA in F_1 was hypomethylated as in *ddm1* mutant, and the rest was normally methylated as in the wild-type parent (Kakutani et al. 1999). The centromere repeats are known to be highly methylated in genome of the *Arabidopsis thaliana* wild-type plants; this can be readily seen by their complete resistance to cleavage with methylation-sensitive restriction endonucleases, namely, *Hpa*II. All examined progeny from a F_1 *DDM1/ddm1* x *DDM1/DDM1* cross and a reciprocal *DDM1/DDM1* x F_1 *DDM1/ddm1* cross contain some hypomethylated centromere repeats although the degree of hypomethylation differs from plant to plant. Similarly, all selfed F_2 progeny from the F_1 *DDM1/ddm1* plant showed a hypomethylated ladder of bands.This means that the methylation status is not determined by the *DDM1* genotype alone. Hypomethylated chromosome segments originating from a *ddm1* mutant plant seem to remain hypomethylated during meiosis and mitosis, resulting in hypomethylation of half chromosomes in F_1.The backcrossed heterozygote parent had fully methylated chromosomes due to dilution of the hypomethylated chromosomes by the normally methylated ones during repeated backcrossing. The presence of hypomethylated chromosomes in all F_2 progeny indicates that hypomethylated chromosome segments can be inherited independently of the *ddm1* mutation. The observation that the methylation level of F_1 is precisely intermediate between that of two parents suggests that rate of *de novo* methylation of unmethylated chromosome segments from a *ddm1* mutant parent is extremely low even in the wild-type *DDM1* backgrounds. An important conclusion from these results is that *ddm1*-induced hypomethylation in the majority of sequences in the *Arabidopsis* genome, both repeated and single-copy ones, can be stably inherited through both mitotic and meiotic cell divisions. This indicates that epigenetic information in the form of differential DNA methylation can be transmitted between plant generations (Kakutani et al. 1999).

Maintenance of CpG methylation is essential for epigenetic inheritance during plant gametogenesis: depletion of the *Arabidopsis* MET1 results in immense epigenetic diversification of gametes. This diversity seems to be a consequence of passive postmeiotic demethylation leading to gametes with random sets of fully demethylated and hemimethylated sites in DNA, followed by remethylation of hemimethylated sites once MET1 is again supplied in a zygote (Saze et al. 2003).

Cytosine DNA-Methyltransferases: How Many of Them Are Needed?

Traditionally, two different kinds of DNA-methyltransferase activities are recognized: (1) a "*de novo*" (Bestor et al. 1988), that transfers methyl groups to DNA irrespective of its previous methylation, and (2) a "maintenance" activity that methylates cytosines in proximity with 5-methylcytosines on the complementary strand (Holliday and Pugh 1975; Riggs 1975). The cytosine methylation in short symmetrical sequences, CpG and CpNpG, and the hemimethylated substrate preference of extractable methyltransferase activities were early indications of a maintenance methylation system that could preserve methylation patterns after DNA replication. Cytosine methyltransferases can also be categorized on the basis of enzyme structure and similarity of conserved amino acid motifs. Colot and Rossignol (1999) differentiated five various groups of DNA-methyltransferases on the basis of these criteria, named after prototypic genes/enzymes in each class: Dnmt1 (Bestor et al. 1988), pmt1/Dnmt2 (Wilkinson et al. 1995; Okano et al. 1998), Dnmt3 (Okano et al. 1999), chromomethyltransferases (CMT; Henikoff and Comai, 1998), and Masc1 (Colot and Rossignol 1999). The mammalian Dnmt3 (Okano et al. 1999; Dodge et al. 2002; Yokochi and Robertson, 2002) and fungal (*Ascobolus*) Masc1 (Malagnac et al. 1997) enzymes have been demonstrated to be *de novo* methyltransferases, while the Dnmt1 family members are thought to function primarily as maintenance methyltransferases (Li *et al.* 1992; Pradhan *et al.* 1999). Cao *et al.* (2000) identified maize (*Zmet3*) and *Arabidopsis* (*DRM*) genes encoding proteins closely related to Dnmt3 methyltransferases. Organisms can possess representatives of multiple methyltransferase classes. *Arabidopsis* has at least 10 genes that could encode DNA-methyltransferases, more than any other eukaryotic genome sequenced so far (Figure 1).

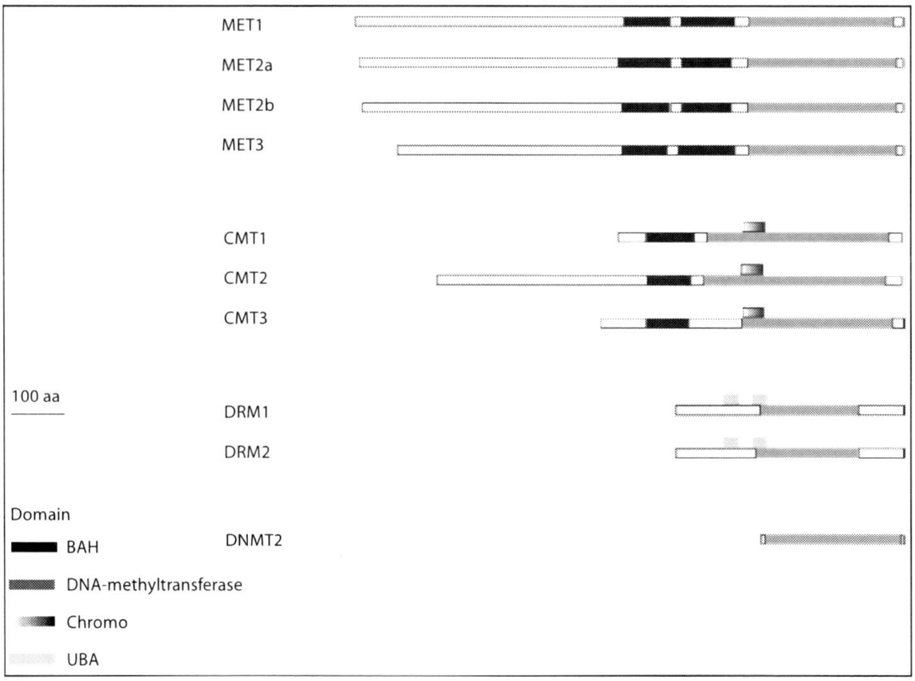

Figure 1. Four families of cytosine DNA-methyltransferases in *Arabidopsis thaliana*. Specific domains are indicated by boxes differently shaded: BAH - bromo-adjacent homology domain; Chromo - chromodomain; UBA - ubiquitin association domain. DNA-methyltransferase domain contains all canonical motifs characteristic of cytosine DNA-methyltransferase catalytic region (see text for more detailed description).

The MET class consists of genes related to the mammalian *Dnmt1* (Finnegan and Kovac, 2000). Three *MET1*-related genes, *MET2a*, *MET2b*, and *MET3*, remain to be functionally characterized. A second type of methyltransferases, the CMT "chromomethylase" class, is also related to Dnmt1 except that a novel chromodomain amino acid motif is inserted between two canonical methyltransferase motifs, I and IV (Henikoff and Comai, 1998). A third class of *Arabidopsis* methyltransferases is the "domain rearranged methyltransferases" or DRM class, which is most related to Dnmt3, except that the canonical methyltransferase motifs are organized in a novel order (Cao et al. 2000). It is attractive to think that, like the mammalian Dnmt3 enzymes, they might be involved in the establishment of methylation patterns. Finally, one gene (GenBank accession number At5g25480) resembles *Dnmt2*, a highly conserved but enigmatic putative mammalian methyltransferase gene, as well as its homologs in *Schizosaccharomyces pombe* and *Drosophila*, and it seems to be a tRNAAsp methyltransferase that specifically methylates cytosine 38 in the tRNA anticodon loop (Goll et al. 2006).

1. MET1 is a Major CpG-Specific Maintenance Plant DNA-Methyltransferase

MET1 is an *Arabidopsis thaliana* DNA-methyltransferase gene cloned by homology to mouse Dnmt1 gene (Finnegan and Dennis, 1993). It represents a member of small gene family and was mapped to a locus of chromosome 5 (position 68.9) nonallelic to *DDM1*. This gene is actively expressed in seedlings, vegetative and floral tissues, especially in their meristematic zones, and seems to code for a major maintenance DNA-methyltransferase in plants (Ronemus et al. 1996). Inhibition of *MET1* expression by the antisense construct of it's cDNA under the control of constitutive 35S promoter leads to variable degree of hypomethylation of both repetitive and single-copy sequences in genomes of transgenic plants as compared with wild-type seedlings. Namely, in most affected lines the general levels of m^5C and Cm^5CGG in a number of repetitive and single-copy sequences is reduced by about 70% that seems to be quite comparable to *ddm1* mutants. The methylation of external cytosine in CCGG sites is less affected. Normal patterns of development, especially those connected with flowering, are severely and pleiotropically perturbed in such strongly hypomethylated lines. In their outcrosses to wild-type plants the severe affected phenotypes cosegregate with transgene, whereas slightly diminished but still evident pleiotropic phenotypes are seen in progeny that has lost the transgene itself but retained significant levels of DNA hypomethylation. On the contrary, phenotypic revertants seen among outcross progeny have near to wild-type levels of genome methylation. In other words, the genome hypomethylation caused by antisense transgene seems to be sufficient to maintain the developmental perturbations in the absence of transgene itself, whereas genomic remethylation is required and sufficient for restoring the normal phenotype.

Finnegan et al. (1996) produced a number of transgenic plants containing antisense construct of *MET1* cDNA and showing various (from 10 to 90%) levels of reduction in cytosine methylation, predominantly in CpG dinucleotides. Family with the lowest level of cytosine methylation had three copies of transgene inherited as a single locus. F2 and F3 antisense plants from this family had 10–30% wild-type methylation. Methylation levels varied between progeny of the sibling F2 plants and within F3 lines. In general, plants homozygous for the transgene construct had the lowest levels of DNA methylation; it suggests that methyltransferase activity was inversely correlated with antisense expression. Methylation levels in a homozygous line remained the same over three generations. F3 plants that did not inherit the antisense construct also showed reduced levels of DNA methylation relative to wild type, although the methylation level, at 40–65% of normal, was higher than in homozygous siblings (20–30% of wild type). When these antisense-null plants were selfed, the methylation levels in progeny were still lower than normal ones. Morphological abnormalities in both vegetative and reproductive structures were observed in plants from the F2 and subsequent generations. The abnormal plants had decreased stature, smaller rounded leaves, leaves with margins curled toward the upper leaf surface, decreased fertility and reduced apical dominance resulting in a bushy appearance. Antisense plants had shorter roots with more branching at the root crown. Phenotypes were variable with individual plants displaying some or all these characteristics. The severity of the abnormal phenotype correlated with the extent of DNA demethylation. In families that had a smaller reduction in DNA methylation level the phenotypes were similar but less severe. *Arabidopsis* flowers have four different organs arranged in concentric rings or whorls; there are four sepals in the outer whorl, four petals in the second whorl, six stamens (the male reproductive organs) in the third whorl, and, in whorl four, single female reproductive organ consisting of two fused carpels. Floral homeotic genes specify organ identity and their function is restricted to defined domains on the floral bud that are coincident with the organ whorls. Some flowers in the antisense transgenic plants showed homeotic transformations of floral organs, with the flowers resembling those described in plants of floral homeotic mutants. The flowers with an increased number of stamens and reduced carpel tissue similar to the *superman* (*sup*) mutant were observed, however, when carpels developed, they contained ovules of normal morphology. When organs in the inner two whorls were transformed into petals or staminoid petals, organ number increased (like in *superman agamous* mutants). In other flowers, where sepals were replaced by carpelloid tissue, the number of organs in whorls two and three decreased (as in *apetala2* mutants). Sometimes extra flowers developed in place of a floral organ or in the internode between floral organs (similar to mutation in *apetala1*). Flowers on a single plant showed a spectrum of these phenotypic abnormalities and flowers formed later in development were more severely affected. Floral abnormalities were most common and diverse in families with the lowest level of DNA methylation. Abnormal flowers were observed in all antisense families with reduced methylation, and similar phenotypes were observed in plants with equivalent methylation levels. This, together with the observation that some antisense-null plants with 40–65% of normal methylation produced abnormal flowers, proves that the floral abnormalities are indeed a result of decreased DNA methylation. In addition to the morphological abnormalities observed in the antisense plants the timing of the transition from the vegetative to the reproductive phase of development was altered, this indicates that DNA methylation is involved in the timing of developmental processes. The floral homeotic transformations observed in the DNA-methyltransferase antisense plants suggested that the expression of floral genes may be abnormal. Indeed, it was found that the floral homeotic genes *AGAMOUS* and *APETALA3* are expressed in leaves of methyltransferase antisense plants, whereas in wild-type plants the expression of these genes is confined to restricted domains of the floral bud. Ectopic expression could result directly from demethylation of their promoter elements or of transcription factor genes that regulate them, or from an alteration in the chromatin structure surrounding these genes. The phenotype of the *Arabidopsis curly leaf* (*clf*) mutant with curled leaves, early flowering and *ap2*-like homeotic transformation in late flowers is similar to that of some DNA-methyltransferase antisense plants. In both *clf* mutants and methyltransferase antisense plants the floral homeotic genes, *AG* and *AP3*, are expressed ectopically. The *CLF* gene encodes a protein homologous to a member of the *Drosophila* polycomb-group proteins indicating that chromatin structure is important in regulating plant gene expression. The *Drosophila* polycomb (Pc-G) and trithorax (trx-G) group proteins affect higher order chromatin structure, and polycomb group proteins are involved in long-term gene repression by formation of stable chromatin complexes (Moehrle and Paro, 1994). The similarity of phenotypic abnormalities in *clf* mutants and methyltransferase antisense plants suggests that DNA methylation may act in a concert with a Pc-Gytrx-G-like system to stabilize determined states of gene expression in *Arabidopsis*. Thus, DNA methylation by MET1 seems to be an essential component in determining the processes of developmental phase transitions and meristem determinacy.

2. Chromomethylase is a Plant Specific DNA-Methyltransferase

The *CMT* class of cytosine DNA-methyltransferases seems to be unique to plants and consists of three related genes in *Arabidopsis* (Henikoff and Comai, 1998). Based on sequence alignment with known structures of DNA-associated cytosine methyltransferases, the chromodomain should lie along a face of the catalytic domain that is nearly perpendicular to DNA substrate. A phylogenetic tree based on block alignments places the *A. thaliana* chromomethylase along the higher eukaryotic branch separated weakly from bacterial enzymes but clearly separated from the *A. thaliana MET1* gene product and other higher eukaryotic enzymes. Evidently, the chromomethylase diverged from other known DNA-methyltransferases prior to the divergence of plants and animals. In addition, the chromomethylase lacks the extremely long N-terminal extension found in all other higher eukaryotic DNA-methyltransferases. Unlike the products of the *MET1* gene, that are readily detectable among the cDNA libraries of *A. thaliana* by PCR, the detection of *CMT* specific cDNAs appeared to be much more difficult. Based on comparison of the profiles of RT-PCR products obtained with poly(A)$^+$RNA preparations from flowers of the Kl-0 ecotype plants the spliced *CMT1* cDNA is approximately 100-fold less abundant than spliced *MET1* cDNA. This leads to an estimate of $\sim 10^{-7}$ of total mRNA for *CMT1* in floral tissues. In roots from Col-0 plants no *CMT1* cDNA was detected. *CMT1* mRNA is present in inflorescences and is at a much lower level or nonexistent in leaves, roots, growing seedlings and plants prior to formation of flower buds. *CMT1* inflorescence expression differs from that of *MET1*, which appears to be uniformly transcribed in meristematic tissues. Sequencing of PCR products covering the whole sequence of CMT1 demonstrated that the chromomethylase from buds and flowers is a 791 amino acid protein encoded in 20 exons. Molecular characterization of *CMT1* genomic and cDNA sequences from different ecotypes revealed an unexpected finding: at least, four of 13 *A. thaliana* ecotypes surveyed are evidently null for intact CMT1 protein, and another expresses mostly aberrantly processed mRNA. A G-to-T base substitution in the Metz coding sequence introduces a stop codon that terminates translation upstream of five of the six conserved methyltransferase blocks. In the Col-0 there is an A-to-G base substitution that introduces a splice acceptor site 8 bp upstream of the normal site, which is used 50% of the time in Col-0, resulting in a truncated protein lacking the downstream catalytic blocks. The cDNA analysis also shows the existence of, at least, one other alternatively processed form: skipping of exon 9 in ~1/2 of the mRNAs results in a truncated protein lacking nearly the entire catalytic domain. Thus, it appears that no more than about 1/4 of Col-0 mRNAs can encode an active protein, whereas only the correctly processed form was detected by the sequencing of cDNA from Kl-0, a representative of more common haplotype. The other three ecotypes with defective *CMT1* genes (No-0, Ler, and RLD) harbor a complete retrotransposon within exon 13 detected as a 4.7 kb genomic insertion named "*Evelknievel*". The presence of the retrotransposon has no effect on transcription and splicing upstream, as levels of inflorescence mRNAs assayed using upstream primers are similar to levels found in other ecotypes. Full-length cDNAs of about expected size have been amplified from No-0; sequencing reveals that the penultimate base of the 3'-long terminal repeat (LTR) is a splice donor site with the last base of the transposon splicing to the correct splice acceptor site of exon 14. As a result, the entire transposon (except for the first base of the 3'-LTR) is spliced out. The open reading frame is shifted resulting in synthesis of a truncated protein. The frameshift at this splice acceptor site caused by transposon insertion is nearly identical to that produced by alternative processing from the aberrant splice site in Col-0. The absence of the downstream methyltransferase conservative segments that form part of the catalytic domain is expected to inactivate the truncated proteins in these ecotypes. Blot hybridization analysis of genomic DNAs demonstrates that *CMT1* is a single-copy gene in *A. thaliana*. This nonessentiality of CMT1 could be explained as a redundant function if other chromomethylases exist in *Arabidopsis*. Indeed, two different sequences CMT2 and CMT3 evidently related to CMT1 were isolated from *A. thaliana* genomic DNA (McCallum et al. 2000). RT-PCR isolation and sequencing of the full coding regions of CMT2 and CMT3 cDNAs revealed that their intron/exon boundaries are similar to those of CMT1. Quantitative RT-PCR expression studies showed that CMT2 and CMT3 are ubiquitously expressed at moderate levels as expected for genes involved in gene silencing by DNA methylation. By PCR screening of EMS-induced mutants in the CMT1-null ecotype (N-0) four independent mutations of CMT3 were selected including one with a truncation that knocks out CMT3. The identification of three individual plants that are homozygous for the CMT3 knockout demonstrated that loss of function for, at least, two chromomethylases (CMT1 and CMT3) is compatible with viability. *CMT1* seems to be nonfunctional in all major strains of *A. thaliana* but *CMT3* has been shown to be functional (McCallum et al. 2000). As it was stated above, the *clark kent* are phenotypically stable in an antisense-*MET1* or in the *met1* mutant background, where they lack most CpG methylations but maintain the other methylation types (CpNpG and non-symmetric), showing that non-CpG methylation is critical for the maintenance of *SUP* gene silencing. To identify loci important for maintenance of methylation and silencing of *SUP*, a mutant screen for suppressors of a nonreverting *clark kent* allele, *clk-st*, created by introducing an additional *SUP* locus into *clark kent-3* plants was performed (Lindroth et al. 2001). The *clk-st* seeds were mutated with EMS, and individual M2 families were screened for mutations that derepress *SUP* gene silencing and result in plants with a wild-type floral phenotype. Of sixteen independent recessive mutants obtained five were chosen for initial study. Of these, four completely reverted the *clark kent* phenotype to yield wild-type flowers, and one displayed partial reversion. Each of five mutants failed to complement each other indicating that they are loss-of-function alleles of the same gene. One of these mutations was mapped to the bottom of chromosome I, near the *CHROMOMETHYLASE3* (*CMT3*) gene. In a cross to a nonsense allele of *CMT3* isolated previously (*cmt3-2)*, these mutants failed to complement, showing that all five suppressor mutants are indeed alleles of *CMT3* (designated as *cmt3-3*, *cmt3-4*, *cmt3-5*, *cmt3-6*, and *cmt3-7)*. The molecular lesions in the *cmt3* mutants were identified by sequencing 5021 base pairs (bp) of the *CMT3* gene from each homozygous mutant line. A single C/G to T/A transition mutation was found in each mutant, in every case altering the coding region of *CMT3*. The DNA polymorphisms created by the *cmt3-4*, *-5*, *-6*, and *-7* mutations were used to generate molecular markers, and these markers have been found to perfectly cosegregate with the suppressor mutant phenotypes. The *cmt3-5* and *cmt3-7* alleles contain stop codons terminating CMT3 after 95 or 27 amino acids, respectively, and, thus, they likely represent null alleles. The *cmt3-3*, *cmt3-4*, and *cmt3-6* alleles are missense mutations within the *CMT3* methyltransferase segment. Four additional *cmt3* alleles have been identified by sequencing the *CMT3* gene from each of remaining 11 mutants: *cmt3-9* and *cmt3-11* appeared to be phenotypically strong suppressors and contain nonsense mutations; *cmt3-8* and *cmt3-10* are phenotypically weak alleles and contain missense mutations in the methyltransferase segment. Thus, 9 of the 16 mutants isolated were alleles of *CMT3*. The effects of *CMT3* on methylation patterning were determined by bisulfite genomic sequencing. The methylation profiles of three genotypes were compared: line *clk-st*, *cmt3-7* in the *clk-st* background, and a *met1* mutant line that had developed a *clark kent* phenotype. Namely, the methylation patterns of *SUP* gene, long terminal repeat (LTR) of a pericentromeric *Athila* retrotransposon and 180-bp centromeric repeat sequence were analyzed. The *cmt3-7* mutant showed a nearly complete loss of CpNpG methylation in all sequences tested but it retained the majority of CpG methylation. In contrast, *met1* mutations showed a marked reduction in CpG methylation but had little effect on the level of CpNpG methylation. *cmt3-7* displayed variable effects on asymmetric methylation ranging from no reduction to nearly complete loss at the 5'-end of the *SUP* locus. In this region the asymmetric methylation may depend on the CpNpG methylation. Three additional *cmt3* alleles (*cmt3-4*, *cmt3-5*, and *cmt3-6*) showed a pattern of methylation similar to that of *cmt3-7*. The promoter of the *FWA* locus contains direct repeats that are methylated predominantly at CpG sites in wild-type plants, causing *FWA* expression to be silenced (Soppe et al. 2000). When this methylation is lost, either spontaneously or in the *ddm1* mutant, the *FWA* gene is overexpressed causing a dominant late-flowering phenotype (ibid). Similar CpG methylation patterns were found in line *clk-st* and in *cmt3-7* by direct sequencing of PCR products from bisulfite-treated genomic DNA. However, this CpG methylation was lost in a *met1* mutant line that had developed an *fwa* late-flowering mutant phenotype. Furthermore, no *fwa*-like late-flowering phenotypes have been observed in any of the *cmt3* alleles even after several generations of inbreeding. Thus, the *cmt3* mutations do not appear to affect the CpG methylation or gene silencing at the *FWA* locus. To determine whether loss of CpNpG methylation in the *cmt3* mutants is genome-wide, Southern blot analysis with methylation-sensitive restriction enzymes was performed. Both *cmt3-5* and *cmt3-7* showed an increased level of digestion at the *Athila* LTR sequences with the enzymes *Eco* RII (inhibited by methylation of the inner cytosine within its recognition site CC(A/T)GG) and *Msp* I (inhibited by methylation of the outer cytosine in its recognition site CCGG) but showed a level of digestion equal to the wild type with the enzymes *Hpa* II and *Hha* I, which are inhibited by CpG methylation in their recognition sites. Using similar restriction enzyme analyses, it was found that *cmt3* mutants exhibit decreased CpNpG methylation but not CpG methylation at the centromeric 180 bp repeat sequence and Ta3 retrotransposon sequence. Therefore, CMT3 seems to be a specific methylase for CpNpG methylation, the enzyme has site specificity different from that of the DNMT1/MET1 class of methyltransferases. Because *cmt3* mutants show a loss of CpNpG methylation in a background that is wild type for *MET1,* the MET1 cannot substitute for the function of CMT3 at these sites.

Screening of the maize cDNA and genomic libraries starting with conserved methyltransferase domains as the query sequence resulted in the recovery of a class of clones with significant similarity to chromomethylase genes of *Arabidopsis* (Papa et al. 2001). These sequences represent a small family of two genes, named *Zmet2* and *Zmet5*. The *Zmet2* and *Zmet5* nucleotide sequences are 90% identical over the coding regions. Amino acid sequence alignments of ZMET2 and ZMET5 proteins with the *Arabidopsis* chromomethylases CMT1, CMT2 and CMT3 showed conservation in motifs I, IV, VI, VIII, IX, and X. Phylogenetic analysis indicated that the ZMET2 and ZMET5 proteins are more closely related to CMT1 and CMT3 than to CMT2. Alignments of ZMET2 and CMT1 revealed 44% amino acid identity and 57% similarity. CMT1 and ZMET2 have 87% amino acid similarity across six conserved functional domains. Typically for chromomethylases the N-terminal domain of ZMET2 is smaller than those in the MET1 class of maintenance methyltransferases but it does contain putative nuclear localization signal. A chromodomain is present between motifs I and IV, and its amino acid sequence and position are conserved between ZMET2 and CMT1. The inferred ZMET2 protein has 912 amino acids and a predicted mass of 101 kD. In addition to chromodomain and a conserved methyltransferase domain, all chromomethylases contain a bromo adjacent homology (BAH) domain that has been implicated in linking DNA methylation, replication and transcriptional regulation in mammals (Callebaut et al. 1999). ZMET1 and MET1 both contain two BAH domains in the N-terminal regulatory region. Thus, common mechanisms may be controlling these two classes of methyltransferases or targeting them in the nucleus. This finding also supports the phylogenetic evidence suggesting that the chromomethylases and MET1-type enzymes have a common ancestor (Cao et al. 2000).

The F2 family of maize plants possessing a *Mutator* transposable element (*Mu*) insertion in *Zmet2* was identified by PCR with a pair of primers, of which one was specific for *Mu* and another for *Zmet2*. The respective allele, *zmet2-m1::Mu*, was sequenced. The *Mu* element appeared to be inserted into DNA-methyltransferase exon 18, which encodes motif IX. To determine the likely effect of the *Mu* insertion, the aberrant transcript resulting from this insertion event was sequenced. The aberrant transcript contains a stop codon after amino acid 833. The resulting protein lacks motif required for SAM binding, and is expected to lack an enzymatic function. Twelve individual plants from an F4-derived F5 family composed of three wild-type plants, seven plants heterozygous for the *zmet2-m1::Mu* allele and five plants homozygous for the *zmet2-m1::Mu* allele were analyzed by HPLC to assess the effect of the mutation on global methylation levels. A 12.6% decrease in the 5-methylcytosine content was observed in plants homozygous for *zmet2-m1::Mu* compared with siblings homozygous for wild-type *Zmet2*. Heterozygous plants also were hypomethylated though to a lesser extent: their reduction in methylation level was ~27% of the reduction in methylation observed in the homozygous *zmet2-m1::Mu* class. Southern blot analysis with restriction enzymes having different methylation sensitivity and the genomic bisulfite sequencing of the 180 bp knob sequence in the same DNA samples revealed significant reduction of CpNpG methylation in the *zmet2-m1::Mu* mutant, whereas methylation at CpG-sites and asymmetric cytosines was unaffected. Since CpNpG methylation accounts for about 30-40% of total cytosine methylation in genomic DNA of cereals (Gruenbaum et al. 1981; Kirnos et al. 1981) the 12.6% reduction in the total cytosine methylation observed in the homozygous *zmet2-m1::Mu* plants may be calculated to a 30-50% reduction in methylation at CpNpG sites. Complete reduction in methylation at a given methylated CpNpG site was observed rarely but nearly every site analyzed showed some reduction. This indicates that the partial reduction in CpNpG methylation is caused by random reduction of methylation at most or all CpNpG sites rather than complete reduction at some sites with no reduction at others. Whether DNA from certain cells is completely devoid of CpNpG methylation, whereas others have normal methylation level, or conversely, if CpNpG methylation is lost randomly in all cells, still remains to be seen. There are several potential explanations for an incomplete reduction in CpNpG methylation in homozygous *zmet2-m1::Mu* plants. The most likely explanation is that *Zmet5*, a homolog of *Zmet2*, has, at least, partial overlapping function and expression. This view is supported by high degree of conservation between *Zmet2* and *Zmet5* sequences and their expression profiles. A second possibility is that an enzyme unrelated to *Zmet2* is capable of maintaining CpNpG methylation with a reduced frequency. The cloned MET1 from pea displayed both CpG and CpA/TpG methyltransferase activities *in vitro* (Pradhan et al. 1998). In addition, the MET1 antisense *Arabidopsis* plants showed some reduction in methylation at CpNpG sites (Finnegan et al. 1996), supporting the possibility that this class of methyltransferases may have CpNpG methylation activity *in vivo*, though, of course, this effect may well be indirect. Finally, it is possible that the *zmet2-m1::Mu* insertion does not reduce the activity of the enzyme completely, although for the protein lacking a critical domain this letter possibility seems to be highly unlikely. To test the stability of the methylation levels over generations, the homozygous *zmet2-m1::Mu* mutant plants derived from heterozygous F3-derived plants were compared with homozygous *zmet2-m1::Mu* mutant plants derived from several generations of self-pollinated homozygous mutant plants. The percentage of methylated cytosines was consistent among all homozygous mutant progeny regardless of pedigree and did not decrease upon self-pollination of homozygous mutants. To determine the extent of remethylation, when the *zmet2-m1::Mu* mutation is removed by segregation, backcross plants with a homozygous *zmet2-m1::Mu* plant as a grandparent were produced. The inbred line, Mo17, was crossed to a homozygous *zmet2-m1::Mu* plant, and the resulting F1 plant was then backcrossed to the Mo17 parent line. Restriction enzyme analysis of backcross progeny indicated that all individuals without *Mu* insertion displayed substantial remethylation of repetitive centromeric sequences. Furthermore, analysis of the same DNA samples by HPLC indicated that genomic levels of cytosine methylation in homozygous normal backcross progeny were lower than those of Mo17, the nonmutant parent, but significantly higher than those of heterozygous backcross siblings. These data are consistent with the hypothesis that methylation levels are partially but not completely restored in the first generation of the homozygote wild-type progeny obtained from a homozygous mutant parent. Thus, either *ZMET2* has *de novo* methylation activity or a separate *de novo* methyltransferase functions early in development, whereas *Zmet2* maintains CpNpG methylation patterns.

All data reviewed indicate that chromomethylases function *in vivo* to maintain CpNpG symmetrical methylation patterns. This is consistent with two observations. First, CpNpG is found in most angiosperm and gymnosperm genomes but is very limited in frequency in organisms other than plants. Second, genes for chromomethylases have been found in plant species ranging from monocots to dicots but not in the genomes of any other organisms. Therefore, chromomethylases, which apparently evolved after the divergence of plants from other organisms, offer plant genomes a second means to propagate methylated cytosines. The conservation of the chromomethylase function across species as diverse as *Arabidopsis* and maize suggests that these genes seem to provide a function that offers an evolutionary advantage to the plant organisms.

The conserved domains of chromomethylases may provide insight into the purpose of these enzymes. In addition to the conservative methyltransferase domains, chromomethylases contain a chromodomain and a BAH domain. Chromodomains are found in several proteins involved in repression of transcription at the chromatin level (Cavalli and Paro, 1998). For Polycomb and HP1 proteins the chromodomain is critical for the proper targeting. The chromodomains of the *Drosophila melanogaster* dosage compensation proteins MOF and MSL-3 are involved in binding to noncoding RNA molecules (Akhtar et al. 2000). The interaction of chromodomains with RNA may be the mechanism for targeting proteins containing chromodomains to specific regions of chromosomes. Recently, Swi6, an HP1 homolog, was shown to interact directly with methylated Lys-9 of histone H3 (Rea et al. 2000; Nakayama et al. 2001). This suggests that chromomethylases may be targeted by their chromodomain to heterochromatic regions marked by H3 Lys-9 methylation.

A BAH domain also is found in the N-terminal portion of chromomethylases. It is interesting that all Dnmt1/Met1 DNA-methyltransferases and chromomethylases contain BAH domains. The Dnmt1/Met1 proteins contain two BAH domains, whereas chromomethylases contain only one BAH domain. This common feature may suggest a similar function or targeting for these two groups of methyltransferases. One function proposed for the BAH domain is to link DNA methylation to replication. These two classes of methyltransferases both may be involved in maintaining the symmetric methylation patterns, with the Dnmt1/Met1 class acting on hemimethylated CpG sites and the chromomethylases methylating the hemimethylated CpNpG sites soon after replication. This corresponds to earlier observations of distinct CpG and CpNpG methyltransferases found in plants (Pradhan, Adams, 1995), though the same authors reported the pea MET1 methylase has both CpG and CpNpG types of activity (Pradhan et al. 1998). *CMT* genes have thus far only been found in the plant kingdom, this agrees well with the observation that plants have a much higher incidence of CpNpG methylation than do other organisms such as mammals. Despite a nearly complete loss of genomic CpNpG methylation, null *cmt3* mutants are morphologically normal even after five generations of inbreeding. In contrast, *met1* mutants exhibit severe developmental abnormalities (Finnegan et al. 1996; Jacobsen et al. 2000). One explanation for this is that CpNpG and CpG methylations may act in a partially redundant fashion to silence most genes. Viability despite severe loss of genomic methylation makes *Arabidopsis* an ideal model system for elucidating the roles of DNA methylation in epigenetic and developmental processes.

3. DRM are the Plant *de novo* DNA-Methyltransferases

Dnmt3 class of DNA-methyltransferases was shown to act as *de novo* methyltransferases in animals (Okano et al. 1998, 1999; Lyko et al. 1999). A search for similar proteins in the *Arabidopsis* and maize databases has lead to identification of three full length cDNA clones in *Arabidopsis* and one in maize (Cao et al. 2000). The predicted proteins denoted as DRM1, 2 and ZMET3, respectively, have been found to be similar to each other along their entire length. They exhibit 28% amino acid identity in the N-terminal domains and 66% identity in the C-terminal catalytic domains. Of all DNA-methyltransferases known these proteins appeared to be most similar (an average of 28% amino acid identity along C-terminal catalytic domains) to animal Dnmt3 proteins. No significant similarity between the N-terminal domains of the plant and animal proteins was detected, and both DRM and ZMET3 lack the PWWP and cysteine-rich motifs present in the Dnmt3 methyltransferases (Xie et al. 1999; Xu et al. 1999). Quite unexpectedly, these plant proteins were found to have an unusual arrangement of the conserved catalytic motifs. Other methyltransferases, including Dnmt3, contain motifs I, II, III, IV, V, VI, IX, and X from the N terminus to the C terminus of the protein (motifs VII and VIII are not highly conserved and difficult to distinguish in many methyltransferases). However, both the *Arabidopsis* DRM and maize ZMET3 display an altered order of these motifs, VI, IX, X, I, II, III, IV, V, as if a rearrangement has taken place at a region of several amino acids between motifs X and I. As a matter of fact, the *Arabidopsis* proteins owe this rearrangement their peculiar name, the domains rearranged methyltransferases (DRMs). The maize sequence has been named ZMET3, as this represents the third class of methyltransferase isolated from maize. A search for other plant ESTs by using *DRM* and *ZMET3* as queries revealed the presence of a soybean 3'-cDNA sequence (accession no. A1736568) that displays high levels of identity to both *Arabidopsis* and maize sequences. This partial sequence predicts a polypeptide that encodes the methyltransferase catalytic motifs IX, X, I, II, III, IV, and V, which are of the same order seen in both *Arabidopsis* and maize. During search for genes, whose transcripts accumulate soon after wounding of tobacco leaves, a particular sequence was identified that was found later to be another member of plant DRM family, NtDRM1 (Wada et al. 2003). A full length tobacco cDNA of 2,540 bp was shown to encode a protein of 608 amino acids, containing, at least, six highly conserved motifs found in DNA-methyltransferases. Their arrangement was different from "canonical" order of I-IV-VIII-X in the C terminus and quite similar to members of DRM family previously found, VI-VIII-IX-X-I-IV. A high homology between NtDRM1 and

DRMs from *Arabidopsis* (DRM1 and DRM2) and maize (ZMET3) was observed. In the solved structure of a prokaryotic *Hha*I methyltransferase complex with target DNA the motifs I and X lie parallel to one another in the tertiary structure and they are physically associated (Cheng et al. 1993; Klimasauskas et al. 1994). The C terminus of motif X and the N terminus of motif I are very close together in the three-dimensional structure. Because these amino acids are directly adjacent to one another in the primary sequence of DRM members, it is conceivable that, despite the motif rearrangement, the overall fold of the plant proteins is similar to that of *Hha*I. Consistent with their putative function as DNA-methyltransferases, all DRM methylases studied (DRM1, DRM2, ZMET3 and NtDRM1) contain conserved nuclear targeting sequences in their N-terminal domains. Specific nuclear localization *in planta* was directly demonstrated for NtDRM1 (Wada et al. 2003). A series of the ubiquitin-association (UBA) domains (three in DRM2 and two in ZMET3 and NtDRM1) are also present. As UBA domains are not found in other classes of methyltransferases including the mammalian and fish Dnmt3 proteins, this possible association with the ubiquitin pathway may be restricted to the Dnmt3-like methyltransferases of plants. Southern blot analysis of *Arabidopsis*, maize and tobacco genomic DNA for the presence of related sequences detected several hybridizing bands in each, suggesting the presence of small gene families (Cao et al. 2000; Wada et al. 2003). Namely, three additional sequences with significant similarity were found in *Arabidopsis* genome. All three of these sequences appear to encode *DRM* pseudogenes.

The *DRM2* gene was mapped to the top of chromosome V between markers mi174 and mi322. *DRM1* maps to a nearby region, the estimated distance between *DRM1* and *DRM2* is about 230 kb, suggesting a map distance of approximately 1 centimorgan. The *DRM1* and *DRM2* genes, therefore, seem to have arisen from recent gene duplication. Their map positions do not correspond to known methylation mutants in *Arabidopsis*. The *DRM* probes also detected several weakly hybridizing clones, all of which mapped to two different regions on chromosome I and correspond exactly to the pseudogenes mentioned above. A related maize EST sequence seems to encode a protein that lacks a highly conserved PC site in motif IV of the catalytic domain. Genomic DNA of *Nicotiana tabacum* was found to contain four distinct sequences closely related to NtDRM1. Since *N. tabacum* is an amphidiploid, a pair of the DRM gene members seem to exist in each chromosome set originating from its ancestor lines, *N. sylvestris* and *N. tomentosiformis*.

RNA blot analysis was used to detect expression of *DRM2* in different tissues of *Arabidopsis* (Cao et al. 2000). A 2.5 kb transcript was easily detected on blots of total RNA from roots, leaves or inflorescences. This finding suggests that *DRM2* is expressed in most tissues. Probing of the same blot with *DRM1* did not detect any message suggesting that *DRM1* is expressed at a lower level than *DRM2*. Similar analysis in tobacco indicated that *NtDRM1* transcripts are ubiquitous in leaves, stems, flowers and roots (Wada et al. 2003). The expression levels were also high in all floral organs except for pistils. *NtDRM1* transcripts accumulate throughout all cell cycle in the synchronously cultured tobacco BY2 cells; this seems to be in a marked contrast to those of *NtMET1* gene encoding a maintenance DNA-methyltransferase expressed predominantly in the S-phase.

Alignments of the conservative catalytic motifs I-IV of different DNA-methyltransferases show that the plant DRM/ZMET3/NtDRM1and animal Dnmt3 enzymes form a distinct class of proteins that are closer to each other than to other types of methyltransferases including the Dnmt1/MET1 class, the CMT class, and the Dnmt2 class. This observation suggests that these different types of methyltransferases formed early in eukaryotic evolution before the divergence of plants and animals and, therefore, may share common functions. DRM2, ZMET3 and NtDRM1 like Dnmt3 may act as *de novo* methyltransferases.

A transgenic tobacco lines containing a selectable marker gene controlled by a derivative of the 35S promoter of the cauliflower mosaic virus (CaMV) devoid of the CG and CNG methylation acceptor sites have been used to study a possible role of DNA methylation at asymmetric sites (Dieguez et al. 1998). Silencing was triggered by crossing to the silencer locus of tobacco line 271. This line contains inactive methylated copies of the 35S promoter and is able to silence the homologous promoter copies at ectopic chromosomal positions. The mutated promoter lacking the CG/CNG methylation acceptor sites was found to be as susceptible to trans-silencing as the unmodified 35S promoter. Thus, methylation at CG and CNG sites is not a prerequisite for the initiation of the epigenetic gene inactivation. Interestingly, while methylation of cytosines at asymmetric sites was only slightly affected by silencing in the unmodified promoter, it was significantly increased in the absence of CG/CNG sequences. However, silencing without CG/CNG methylation was immediately relieved in the absence of silencer. Thus, CG/CNG methylation is probably essential for the maintenance of previously established epigenetic states. Nevertheless, the asymmetric methylation can exist in the absence of symmetric methylation and may contribute to gene silencing.

To study the possible role of the *DRM* genes in gene silencing by *de novo* DNA methylation, T-DNA insertion mutations in both *DRM1* and *DRM2* were produced and crossed together to create *drm1 drm2* double homozygous plants (Cao, Jacobsen, 2002). RT-PCR confirmed the expression of both *DRM1* and *DRM2* in wild-type plants but not in *drm1 drm2* double mutants, it means that the T-DNA insertions are likely to disrupt gene function. The *drm1 drm2* double homozygotes have morphology similar to the wild-type WS strain even after five generations of inbreeding. Using the methylation-sensitive restriction enzymes HpaII and MspI, which are inhibited by either CpG and/or CpNpG methylation in their recognition sites, no detectable loss of methylation at the repetitive centromeric repeat sequences was observed, suggesting that *drm* mutations do not affect maintenance methylation of these repeats. To test whether the *DRM* loci affect the *de novo* methylation associated with transgene silencing, the *FWA* gene was used. The promoter of *FWA* is normally methylated within two direct repeats; it is causing *FWA* expression to be silenced. In epigenetic *fwa* mutants, in which this methylation has been lost, *FWA* expression is ectopically activated in vegetative tissue causing a dominant late flowering phenotype (Soppe et al. 2000). These epigenetic *fwa* alleles are stable; the *FWA* direct repeats do not become spontaneously remethylated even after several generations of inbreeding. However, when an extra copy of the *FWA* gene is transformed into wild-type plants, the direct repeats become *de novo* methylated at a very high frequency and transgene expression is silenced.

Using *FWA* transformation as a *de novo* methylation assay, both the parental WS strain and the *drm* mutant strains were transformed. In wild-type WS the resulting transgenic plants displayed an early flowering phenotype similar to that of wild-type; therefore, the *FWA* transgene was efficiently silenced. Southern blot analysis showed that the *FWA* transgene was *de novo* methylated at CpG dinucleotides present within *Cfo*I restriction sites. However, *FWA* transformed into *drm1 drm2* double homozygotes produced plants with a late flowering phenotype, and the *de novo* methylation of the transgenes was blocked. Non-transformed *drm* mutant plants do not show a late flowering phenotype, and *drm* mutations do not affect preexisting methylation at *Cfo*I sites. Therefore, *DRM* is required for *de novo* methylation of *FWA* transgenes but not for maintenance of CpG methylation and silencing of the endogenous *FWA* gene.

The late flowering phenotype in *drm1 drm2 FWA* transformants was inheritable in both F2 and F3 generations. Furthermore, when late flowering *drm1 drm2 FWA* transformants were crossed to wild-type plants, the F1 plants retained a late flowering phenotype. Therefore, once *FWA* transgenes are hypomethylated (due to the presence of *drm* mutations), they retain the hypomethylated and active state even when exposed to wild-type *DRM* alleles in later generations. This suggests that *FWA* transgenes are most susceptible to the *DRM*-dependent *de novo* methylation either during the transformation process itself or during the first generation after transformation. These results are consistent with the observation that the originally isolated *fwa* hypomethylated epigenetic alleles are stable in the wild-type *DRM* backgrounds.

Using the same *FWA* transformation assay, the *drm1* and *drm2* single mutants were also tested. It was found that *drm2* but not *drm1* blocked transgene-associated *de novo* methylation and silencing. This is consistent with *DRM2* RNA being expressed at much higher levels than *DRM1* RNA and suggests that DRM2 is the predominant *de novo* DNA-methyltransferase in *Arabidopsis*.

To study the role of the *DRM* genes in the maintenance of preexisting methylation and silencing at the *SUP* locus, the *drm1 drm2* double mutant was crossed to two different epigenetic hypermethylated *sup* alleles (*clark kent* alleles), *clk-3* and *clk-st*. *clk-3* is an allele, in which the *SUP* gene has become densely hypermethylated and silenced but which spontaneously reverts to a wild-type unmethylated allele in 3% plants (Jacobsen, Meyerowitz, 1997). *clk-st* is a transgenic strain containing a 24 kilobase *SUP* inverted repeat transgene locus on chromosome III. In *clk-st*, both the inverted repeat *SUP* genes and the endogenous *SUP* gene are heavily methylated and silenced, causing a stable (nonreverting) epigenetic *clark kent* phenotype (Lindroth et al. 2001). *drm1 drm2 clk-3* and *drm1 drm2 clk-st* triple mutant plants retain a strong and heritable *clark kent* phenotype, showing that *drm* mutations do not suppress preexisting gene silencing at the *SUP* locus. Bisulfite genomic sequencing of the 5′-end of the *SUP* locus of these triple mutant strains showed that *drm1 drm2* mutants retained a high level of CpNpG methylation in both *clk-3* and *clk-st* backgrounds, they have significantly reduced but not eliminated *SUP* asymmetric methylation. CpG methylation is not adequately assayed in this region, as there is only one CpG site, which shows low and spurious levels of methylation. In summary, the *drm1 drm2* double mutants retained the majority of the preestablished DNA methylation character at *SUP*. To test whether the *drm* mutations block the *de novo* methylation of *SUP*, the silencing properties of the *clk-st* strain were utilized. The *SUP* inverted repeat transgene locus present in *clk-st* has been found to induce the *de novo* methylation and gene silencing of a previously unmethylated and active *SUP* endogene. This silencing phenomenon occurs after two or more generations of exposure of the *SUP* endogene to the *SUP* inverted repeat. To test for *de novo* methylation, the preexisting methylation present in *clk-st* was first erased by using a *cmt3-7* mutation, which has eliminated the majority of CpNpG and asymmetric *SUP* methylations causing reactivation of *SUP* expression. A *cmt3-7 clk-st* plant was crossed to a *drm1 drm2 clk-st* plant. The F1 plants from this cross displayed a wild-type *SUP* phenotype. In the F2 progeny a plant line was selected that retained a wild-type *SUP* phenotype and was homozygous for the wild-type *CMT3* allele, *clk-st* inverted repeat *SUP* locus and both *drm1* and *drm2*. Bisulfite sequencing of this line near to 5'-end of the *SUP* gene showed that it had a very low level of the CpNpG and asymmetric cytosine methylations, confirming that the *SUP* genes in this line had not yet undergone the *de novo* methylation. Moreover, all analyzed plants of self-pollinated F3 and F4 progeny displayed a wild-type *SUP* phenotype. The stable wild-type phenotype suggested that *drm1 drm2* double mutation blocked the *de novo* methylation and silencing of *SUP* that is normally induced by the inverted *SUP* repeat. This finding was confirmed by crossing the line described to a plant doubly heterozygous for *drm1* and *drm2*. F1 plants from this cross were genotyped for the *drm* mutations and then allowed to self-pollinate. Four of them were *drm1 drm2* double homozygotes, and the F2 progeny from these plants all retained a wild-type *SUP* floral phenotype. Bisulfite DNA sequencing confirmed that these plants showed a very low level of cytosine methylation. The remaining seven F1 plants were *drm1 drm2* double heterozygotes, and the F2 progeny from all of them segregated plants with a *clk* phenotype (96 *clk* plants of 993 total). Bisulfite sequencing of these several *clk* plants confirmed that CpNpG and asymmetric methylations were reestablished. Thus, the *de novo* methylation and silencing of *SUP* that is caused by the *clk-st* inverted repeat is dependent on the presence of wild-type *DRM* alleles. The *DRM* genes are important for the establishment but not the maintenance of gene silencing at *FWA* and *SUP* and are required for *de novo* methylation of cytosines in all known sequence contexts (CpG, CpNpG and asymmetric). While the direct repeat containing *FWA* gene was only susceptible to the *DRM*-dependent *de novo* methylation in the first generation after transformation, the *SUP* inverted repeat containing transgene locus was affected by *DRM* genes in many generations after integration. One interpretation of this finding may be that DRMs methylate direct and inverted repeats by different mechanisms.

The observation that *drm* mutants block the *de novo* methylation of *FWA* and *SUP* but do not cause a major loss of preexisting methylation of these genes after inbreeding suggests that DNA sequences do not normally lose their methylation during the plant life cycle. These data are consistent with results showing a lack of genome remethylation after exposure to demethylating mutations and support a view that the fundamental distinction between plant and animal DNA methylation is the lack of genome-wide resetting of methylation pattern (demethylation and *de novo* methylation) during plant development.

A possible role of *DRM* loci in the maintenance of asymmetric methylation was studied in a more detail using bisulfite genomic sequencing of two endogenous sequences known to be highly methylated at asymmetric sites, namely, *FWA* and *MEA-ISR* (Cao, Jacobsen, 2002a). The *FWA* locus, as it was already noted, encodes a homeodomain-containing protein, whose expression is silenced in the vegetative tissues of wild-type plants. This silencing is associated with methylation of two direct repeats in the 5'-region of the gene (Soppe et al. 2000). When this methylation is lost, either in spontaneous hypomethylated epigenetic mutants such as *fwa-1*, or in the methylation mutants *ddm1* and *met1*, the *FWA* gene is overexpressed causing a dominant late flowering phenotype. In wild type the *FWA* direct repeats contain 89% CpG, 18% CpNpG and 4% asymmetric DNA methylations. *MEA-ISR* is a 183 bp sequence present in seven direct repeats lying in an intergenic region between the imprinted *medea* (*MEA*) gene and the aldehyde oxidase gene near to the upper end of chromosome 1, approximately 500 kbp from the end. These repeats are also found in 12 other genomic locations, all of which are also subtelomeric, and this region was named *MEA-ISR* (Intergenic Subtelomeric Repeat). In wild-type strains this repeat has been found to contain a high proportion of methylated cytosines, namely 87% CpG, 47% CpNpG, and 18% asymmetric.

When a *drm1 drm2* double mutant strain was compared with the parental Wassilewskija (WS) strain by bisulfite genomic sequencing it was found that the *drm1 drm2* mutations eliminated all asymmetric methylation of both the *FWA* and *MEA-ISR* sequences. Therefore, the *DRM* loci are important for the methylation of asymmetric cytosines. The *drm1 drm2* mutations also eliminated the CpNpG methylation of both *FWA* and *MEA-ISR*. Furthermore, the *cmt3-7* mutation was found to reduce but not completely eliminate the CpNpG methylation in these loci. Therefore, both *DRM* and *CMT3* are required for proper maintenance of the CpNpG methylation patterns, and at some loci such as *FWA* and *MEA-ISR* the *drm1 drm2* double mutant is more efficient than *cmt3-7* with reducing the CpNpG methylation. The triple *drm1 drm2 cmt3-7* mutants lack all traces of asymmetric and CpNpG methylations at *FWA* and *MEA-ISR*. However, the CpG methylation levels were similar to the wild type. The *drm1 drm2* double mutants showed a complete loss of non-CpG methylation at both *FWA* and *MEA-ISR*. Interestingly, the *drm1 drm2* plants do not show a late flowering phenotype typical of plants, in which expression of the *FWA* gene has been reactivated. This suggests that non-CpG methylation does not play a major role in the *FWA* silencing.

Epigenetically silenced alleles of the *SUP* locus (the *clark kent* alleles) are densely methylated in all sequence contexts (Jacobsen, Meyerowitz, 1997; Lindroth et al. 2001; Kishimoto et al. 2001). Whereas the originally isolated *clk* alleles spontaneously revert to an unmethylated state in about 3% plants (Jacobsen, Meyerowitz, 1997), a transgenic allele called *clk-st* shows a stable nonreverting phenotype, making it more suitable for genetic studies (Lindroth et al. 2001). *clk-st* contains a single inverted repeat of the *SUP* locus, which can induce the *de novo* methylation of itself as well as of the endogenous *SUP* locus. The methylation profile consists of 16% CpG, 55% CpNpG and 16% asymmetric. The *cmt3-7* has been found to eliminate the CpNpG methylation in this region and reduce the asymmetric methylation by approximately 60%. To test the effect of the *drm* mutations on *SUP* methylation, *clk-st* was compared to a *drm1 drm2 clk-st* strain. Contrary to what was found with *FWA* and *MEA-ISR*, the *drm1 drm2* double mutants were found to reduce but not completely eliminate the asymmetric methylation in this case. However, the residual asymmetric methylation was not detected in the *drm1 drm2 cmt3-7 clk-st* strains. Therefore, *DRM* and *CMT3* act redundantly to control *SUP* asymmetric methylation. Because the *drm1 drm2 cmt3-7* triple mutants eliminate all asymmetric methylation at *SUP*, an analysis of the methylation remaining in the *cmt3-7* mutants should reflect the residual activity of *DRM* and that remaining in *drm1 drm2* should reflect the residual activity of *CMT3*. The positions of the asymmetric methylation of *SUP* remaining in either *cmt3-7* plants or *drm1 drm2* plants appeared to be largely overlapping. An analysis of the sequence context of the asymmetric methylation in these mutants suggests that both CMT3 and DRM prefer to methylate sites that follow the cytosine residues. However, both CMT3 and DRM showed a bias against sites that immediately precede cytosines; CpA and CpT methylations were much more frequent than CpC methylation. It means that CMT3 and DRM can methylate the same asymmetric sites, and both show roughly the same preference for particular DNA sequence contexts. The effects of the *drm* mutations on CpNpG methylation at *SUP* are also different from that seen at *FWA* and *MEA-ISR*. *drm1 drm2* only reduce CpNpG methylation at the *SUP* locus by about 30%, whereas *cmt3-7* completely abolishes this methylation. Thus, the effect of *drm1 drm2* on CpNpG methylation is locus-specific completely eliminating the CpNpG methylation of *FWA* and *MEA-ISR* but only reducing it moderately at *SUP*. The effects of the various single, double and triple DNA-methyltransferase mutants on two pericentromeric sequences, the *Ta3* retrotransposon, and the 180 bp centromeric repeat sequences were tested using Southern blot analysis with *Hpa*II and *Msp*I. Neither the *drm* single mutants nor the *drm1 drm2* double mutants affected the pattern of enzymatic digestion, suggesting that the *DRM* genes do not play an essential role in maintaining CpG or CpNpG methylations at these sequences. In contrast, respective DNA from *cmt3-7* and *drm1 drm2 cmt3-7* triple mutants showed nearly complete digestion by *Msp*I but not *Hpa*II showing that *CMT3* is responsible for the maintenance of CpNpG methylation at both *Ta3* and the centromeric repeats.

The bisulfite sequencing and Southern blot data show that methylation of CpG sites are unaffected in the *drm1 drm2 cmt3-7* triple mutant strains at all sequences tested (Lindroth et al. 2001). In contrast, as it was already noted above, in every sequence tested the *met1* mutant showed a strong reduction in the CpG methylation. Therefore, like its mammalian counterpart Dnmt1, MET1 appears to be the primary methyltransferase for the maintenance of CpG methylation. However, consistently with earlier studies (Finnegan et al. 1996; Ronemus et al. 1996; Bartee, Bender, 2001) the *met1* mutants were found to greatly reduce CpNpG methylation at *FWA*, *MEA-ISR*, *Ta3* and the centromeric repeat sequences (Cao, Jacobsen, 2002a) and eliminate the asymmetric methylation of *FWA* and *MEA-ISR* as detected by bisulfite DNA sequencing. As *MET1* cannot substitute for the maintenance of the CpNpG and asymmetric methylations in *drm1 drm2 cmt3-7* strains, it seems most likely that the losses of CpNpG and asymmetric methylations in *met1* are not directly caused by promiscuous enzymatic activity of the MET1 enzyme but are the secondary effects caused by the primary loss of CpG methylation.

We have studied effects of MET1 inactivation in a number of transgenic lines of *Arabidopsis* containing the antisense constructs of *MET1* under the control of copper-inducible promoters (Ashapkin et al. 2002). Variable patterns of *DRM2* gene methylation at cytosine residues were found in wild-type and transgene plants. Moreover, induction of antisense-*MET1* expression with copper ions has been found to lead to further alterations of *DRM2* methylation patterns. Interestingly, along with cytosine methylation the DRM2 gene was found to be methylated at some adenine residues (Gm^{6}ATC); the adenine methylation degree of DRM2 gene was variable between wild-type and transgenic plants. General conclusion from this work was that MET1 activity somehow affects methylation (and probably expression) of *DRM2* gene. As a matter of fact, we found similar MET1-dependent variations in the methylation patterns of all *Arabidopsis* DNA-methyltransferase genes including CMT as well as the gene coding for a putative adenine DNA-methyltransferase and *MET1* itself (Ashapkin et al., to be published elsewhere).

Though neither *cmt3* mutants nor the *drm1 drm2* double mutants show morphological differences from wild type, they had a pleiotropic phenotype including developmental retardation, reduced plant size and partial sterility (Cao, Jacobsen, 2002a). However, defects commonly seen in the *ddm1* and *met1* methylation mutants such as *clavata*-like flowers, *apetala2*-like flowers, *agamous*-like flowers, *sup*-like flowers or extremely late flowering were never observed in the *drm1 drm2 cmt3-7* plants. Thus, *DRM* and *CMT3* act in a redundant fashion to control some elements of plant growth and development, which may be different from those affected in *ddm1* and *met1* mutants.

Thus, the *DRM* and *CMT3* genes encode DNA-methyltransferase enzymes that show overlapping roles in the control of the asymmetric and CpNpG methylations. However, the activities of these methyltransferases are highly dependent on the locus under study, giving a surprising number of different patterns of dependence on either *DRM*, *CMT3*, or both. It seems quite likely that several factors could be involved in targeting the particular sequences to different DNA-methyltransferases, including the DNA sequences themselves, chromatin modifications present at specific loci, and, last but not least, the cross talk between different kinds of DNA methylations.

In a recent paper concerning the study of *Nicotiana tabacum* DRM methylase (NtDRM1) Wada and coauthors (2003) describe results of direct measurement of DNA methylation activity of the enzyme with different DNA substrates. The NtDRM1 gene was fused in-frame to the *GST* gene and expressed in a baculovirus-mediated insect cell expression system (Sf9) that lacks any endogenous DNA-methyltransferase activity. The activity of purified recombinant NtDRM1 in an *in vitro* assay with poly(dI-dC)-poly(dI-dC) as a methylation substrate was found to be by an order of magnitude higher than that of Dnmt3a, proven to be an animal *de novo* DNA-methyltransferase. The site (sequence) specificity of the NtDRM1-mediated methylation was determined by using synthetic oligonucleotide substrates with different sequences but of the same base composition. Namely, three palindromic 24-mer oligonucleotides were used, of which one contained 6 CpG sites, another – 6 CpNpG sites, and still another – 6 CpNpN sites, where N was A or T. The highest methylation activity was seen with the CpNpN substrate, ^{3}H incorporation from [methyl-^{3}H]-labeled SAM (S-adenosyl-L-methionine) increasing almost linearly within incubation periods up to 1.5 h, whereas methyl transfer efficiency was lower with CpNpG and almost absent with the CpG substrate. Thus, NtDRM1 seems to be a *de novo* methylase for non-CpG DNA sequences. The possible maintenance methylation activity of NtDRM1 was studied with double-stranded CpG or CpNpG containing oligonucleotide substrates methylated at all cytosine residues in one strand. The general finding was that NtDRM1 does methylate such hemi-methylated substrates but the activity of this methylation is significantly lower as compared to methylation of respective fully unmethylated oligonucleotides, thus confirming the view of DRM members as non-maintenance *de novo* methylases. The sequence specificity of NtDRM1 was directly determined by the bisulfite mapping of cytosines in DNA methylated by purified NtDRM1. The methylation frequencies were estimated for 15 independent clones of methylated sequence. The enzyme showed the highest methylation of cytosines in CpNpN, followed by CpNpG and least in CpG. The percentages of methylated sites in the total in 15 clones were 75% for CpNpN, 60% for CpNpG and, in sharp contrast, 10% only for CpG. It appears that the third base, including guanine, does not appreciably influence the methylation activity. The only evident exception seems to be T located at the 3'-end of CpCpT and CpTpT that was found to significantly reduce the methylation frequency. The nucleotide located at the 5'-end of the target cytosine containing sequences showed no significant apparent effects on the enzyme site specificity. These observations suggest that NtDRM1 essentially recognizes and methylates all cytosines, although some particular combinations such as CpG, CpTpT and CpCpT appear to be less favorable. To examine whether or not NtDRM1 functions as a *de novo* methylase *in vivo*, DNA from insect Sf9 cells was analyzed by HPLC. Samples from Sf9 cells expressing NtDRM1 were found to contain m^5C at 1.8% of total cytosines that is somewhat less compared to wild-type Sf9 cell DNA *in vitro* methylated (2.7%) by purified NtDRM1 but far more as compared to intact DNA samples from wild-type Sf9 cells that do not contain m^5C at all. Thus, in fact, NtDRM1 appears to be a real *de novo* cytosine DNA-methyltransferase *in vivo*.

All these data are consistent with a general view that plant DRM members are *de novo* methylases of non-CpG DNA sequences. The molecular basis for preferential methylation of non-CpG sites deserves more attention. Results with direct methylation mapping suggest that NtDRM1 does not necessarily favor CpNpN and its CpNpG subgroup but rather evades CpG (as well as some other!) sequences resulting in an apparent preference. Such rejection of a particular sequence has never been reported before for DNA-methyltransferases of higher eukaryotes. It would be interesting to learn whether some peculiarities in the DRM enzyme structure are associated with specific rejection of CpG sites. One evident possibility is that the rearranged catalytic domain is responsible for such behavior, since only plants exhibit active methylation in non-CpG sites and have DRMs.

ARE THERE SIGNALS FOR THE *DE NOVO* DNA METHYLATION?

As an attempt to understand the nature of *cis*-acting signals that may cause genes to be hypermethylated, the genomic sequences of *SUP* and *AG* were analyzed (Jacobsen et al. 2000). These sequences appeared to be quite typical with respect to GC content and in overall frequency of possible target dinucleotides and trinucleotides. However, in the most densely methylated region of *SUP* (beginning at nucleotide –132) a pyrimidine-rich sequence containing many CT dinucleotides is present (5′-ATCACACACAC$(CT)_4$CAT$(CT)_2$ATAT $(CT)_8$A-3′). Methylation was detected at every C in this sequence. Other pyrimidine-rich sequences were also found in both promoter and intron regions that were methylated in *AG*. These are 5′-$(CT)_4$TTTTTTCTTCATTTCC-3′ and 5′-$(CT)_3$TTTTCTT$(CT)_2$TCTTT$(CT)_2$ TACTTTCCTTTCTTAT$(CT)_2$AG$(CT)_3$TT$(CT)_3$C-3′ sequences, respectively. 874 similar sequences were found in a total of 97,935 sequence entries representing 81% of the complete *Arabidopsis* genome. To test whether such sequences are methylated in the antisense-*METI* lines, two additional sequences were analyzed by bisulfite sequencing, one in the promoter region of *CARPEL FACTORY* containing the sequence 5′-$(CT)_{22}$-3′ and another near to the transcription start region of the *LEAFY* gene containing the sequence 5′-T$(CT)_6$ATCA$(CT)_3$TTTT$(CT)_3$TTCTTTA$(CT)_2$-3′. Methylation was not detected at these sequences. Thus, not all sequences that are rich in CT dinucleotides become hypermethylated in the antisense-*METI* plants, suggesting that some other structural aspect of the *SUP* and *AG* sequences is important. One possibility is that, if these pyrimidine-rich sequences are involved in targeting of *SUP* and *AG* for hypermethylation, such targeting might involve unusual DNA structures. Indeed, the pyrimidine-rich sequences in *SUP* and *AG* are predicted to form hairpin structures, whereas those in the unmethylated *CARPEL FACTORY* and *LEAFY* genes are not (Jacobsen et al. 2000). In this light it is interesting to note that mammalian DNA-methyltransferase has been shown to preferentially methylate hairpin structures *in vitro* (Smith et al. 1994). Furthermore, inverted repeats appear to be particularly good targets for *de novo* methylation in *Arabidopsis* (Luff et al. 1999). In the standard strain Columbia (Col) there are three nearly identical *PAI* genes in three unlinked genomic sites, and these genes are unmethylated (Bender, Fink, 1995), whereas in the Wassilewskija strain (WS) there was a duplication at one of three *PAI* loci that have yielded a tail-to-tail inverted repeat of two *PAI* genes, *PAI1-PAI4,* and in this strain there is dense methylation of all four *PAI* sequences. When the inverted repeat from WS was combined with unmethylated Col *PAI2* and *PAI3* genes by genetic crossing, the unmethylated Col genes became *de novo* methylated within a few generations of inbreeding. These results suggest that the inverted repeat provides the primary signal for *de novo* methylation of related sequences elsewhere in the genome. F3 lines displayed substantial methylation of *PAI2* that is almost identical to *PAI1* but there was no methylation of *PAI3* that is a divergent member of *PAI* gene family with approximately 90% identity to *PAI1* and *PAI2*. In representative inbred lines, *PAI2* became as densely methylated as in WS by the F4 generation. *PAI3* was methylated more stochastically between F4 and F6 generations with only some lines achieving full methylation. The effect of the WS *PAI1–PAI4* locus on the allelic Col singlet *PAI1* gene could be tested by combining two alleles in a heterozygote. DNA prepared from F1 heterozygotes showed a small share of methylated Col *PAI1*, when WS but not Col was the female parent in the cross. Therefore, the *de novo* Col *PAI1* methylation can occur at a low efficiency during F1 generation but factors that mediate this process behave differently, when passed through the female versus the male gametes. This methylation became substantial by F4 generation of self-pollinated heterozygotes. Interestingly, in such selfed heterozygous (WS *PAI1–PAI4*/Col *PAI1*) lines *PAI2* also became methylated by the fourth generation, whereas *PAI3* was not methylated even after five generations. It is more interesting, when WS *PAI1–PAI4* inverted repeat was segregated away from methylated WS *PAI2* and *PAI3* genes by crossing with Col, their methylation was maintained for, at least, five generations but with methylation density reduced. Sequenced methylation patterns for *PAI2* from a representative hybrid line showed that the residual methylation was almost entirely at symmetric cytosines. Neither methylated *PAI2* nor methylated *PAI3* singlet locus triggered *de novo* methylation of unmethylated *PAI* sequences. Similarly, homozygous P1Hyb lines with a methylated Col *PAI1* singlet gene (the F4 generation lines) showed no methylation changes upon inbreeding. In plants homozygous for Col *PAI1* and *PAI3* but heterozygous for WS *PAI2*/Col *PAI2* no *de novo* methylation of Col *PAI2* was detected after two or three generations of inbreeding before segregating away the WS *PAI2* allele. To more precisely map the determinants at the *PAI1–PAI4* locus that trigger methylation, a promoterless inverted repeat *pai1–pai4* transgene was constructed and introduced as a single-copy insert in the *Arabidopsis* genome (Luff et al. 1999). In all independent random insertion lines examined (both Col and WS) this homozygous *pai1–pai4* construct displayed *de novo* methylation over the symmetrical portions of the sequence within twice inbred generations. The methylation became progressively denser in subsequent generations without a spread to flanking regions. No fortuitous transgene RNA expression was detected by an RT-PCR assay. Thus, the inverted repeat DNA structure *per se* rather than the RNA or protein product of the endogenous locus provides a signal for its own methylation. In no single-copy line did the promoterless transgene trigger methylation of endogenous *PAI* genes. However, in some multicopy lines both transgenes and endogenous *PAI* genes were methylated by the second homozygous generation. Therefore, it is likely that the promoterless transgene carries all the sequence determinants for self- and *trans*-methylation but when present as a single copy, it might require several additional generations of inbreeding and/or an appropriate insertion site to trigger *trans*-methylation. Based on these observations, it has been proposed that the *PAI1–PAI4* inverted repeat might form a hairpin or cruciform that serves as an ideal substrate for methylation just along the regions of *PAI* identity. This unusual structure might also trap unlinked identical sequences to promote their *trans*-methylation. The observation that *PAI* methylation primarily affects cytosines at symmetric sites in the absence of the *PAI1–PAI4* inverted repeat but both symmetric and asymmetric sites in the presence of the inverted repeat suggests that *PAI* genes in the inverted repeat context might stimulate the activity of some additional DNA-methyltransferase(s).

The observation that wild-type levels of DDM1 or MET1 activities are not required for the establishment or maintenance of *SUP* hypermethylation contrasts sharply with results showing that methylation of the multi-copy *PAI* loci is abolished in a *ddm1* mutant background (Jeddeloh et al. 1998). This suggests that the mechanism that maintains methylation at single-copy genes like *SUP* and *AG* may be different from the mechanism that maintains methylation at repetitive sequences like *PAI* genes.

Is DNA Methylation Itself Regulated by DNA Methylation?

Mutations in the *SUPERMAN* gene affect flower development in *Arabidopsis*. Seven heritable but unstable *sup* epi-alleles (the *clark kent* alleles) were found with phenotypes similar to but weaker than that of the known *sup* mutants (Jacobsen, Meyerowitz, 1997). The *sup*-5 allele, which has a nearly complete deletion of the *SUP* gene produces an increased number of stamens (usually 12-13) and carpels (usually - 3) on the first ten flowers formed on the plant. The stronger *clark kent* allele (*clk*-3) has an average of 8 stamens and 3-4 carpels, while the weaker *clark kent* allele (*clk*-1) has an average of 6-7 stamens and 3 carpels. The genetic complementation tests showed *clark kent* alleles to be allelic forms of *SUP* gene. *In situ* hybridization experiments show that *SUP* RNA expression is reduced in *clk*-3. In wild type the expression of *SUP* RNA occurs early during floral meristem development in the incipient stamen primordia. In *clk*-3 homozygotes, however, this expression was reduced in some floral meristems and undetectable in others. Despite the evidence that *clk* and *sup* are allelic, the sequencing of the *SUP* coding region from a number of *clark kent* alleles failed to reveal any nucleic acid sequence differences from the wild-type. In addition, the cloned *SUP* gene from a *clk*-3 genomic library complements the *clk*-3 and *sup*-5 mutants in transgenic plants, as if cloning the *clk*-3 allele restores it to wild type. Together these results suggest that the *clk* alleles represent an alternative epigenetic state of the *SUP* gene. Methylation patterns within the SUP gene were analyzed in different genotypes using bisulfite genomic sequencing. While there was no cytosine methylation detected in the wild-type or in a *sup* nonsense allele (*sup*-1), the extensive methylation was found in the *clk* alleles covering the start of transcription and most of the transcribed region. The *clk*-3 allele contained 6 more 5-methylcytosines (a total of 211 detected) than the weaker *clk*-1 allele, possibly providing an explanation for the slight difference in phenotypic strength of these alleles. In a *clk*-3 phenotypic revertant only 14 of the original 211 5-methylcytosines remained, this correlated with restoration of the wild-type RNA expression level. The methylation pattern in *clk* is dense and essentially non-sequence specific: both symmetric (CG and CNG) and non-symmetric cytosines are methylated. However, the pattern of methylation is nearly identical in different *clk* alleles. For instance, *clk*-1 and *clk*-3 share 204 methylated cytosines; 7 sites are methylated in *clk*-3 only, and one site is methylated in *clk*-1 only. Also, in the most densely methylated region the *clk*-2, *clk*-5, and *clk*-6 alleles have the hypermethylation patterns very similar to those in *clk*-1 and *clk*-3. The reproducibility of these patterns in independently isolated *clk* lines suggests that the mechanism, by which these sequences become methylated, is rather specific. A number of the *Arabidopsis* antisense-*MET1* lines exhibit phenotypes resembling *sup* mutants. In these lines the hypermethylation pattern of *SUP* gene was similar to that seen in the *clk* lines. 186 methylcytosines were detected, all but three of which were in the same positions as those in *clk*-3. Thus, while overall methylation in these antisense cytosine methyltransferase plant lines is decreased by about 90%, the *SUP* gene has become hypermethylated. These results challenge the original interpretation that various phenotypes in these plants are always caused by demethylation of specific genes. To test whether the *clk* lines have the general demethylation defects similar to those in the antisense cytosine methyltransferase lines, the methylation status of the 180 bp centromeric repeat and the rDNA loci (both are significantly hypomethylated in the antisense cytosine methyltransferase lines) were analyzed by Southern blot method. Five *clk* lines were normally methylated in these repetitive genes, therefore, their defects are different from those in the antisense cytosine methyltransferase lines. One possible interpretation of these data is that the antisense cytosine methyltransferase lines cause misregulation of a component of the methylation pathway other, than methyltransferase itself, that results in hypermethylation of some genomic regions. If this hypothesized component was mutated in the *clk* lines, this might cause only a portion of the antisense cytosine methyltransferase phenotype, namely, hypermethylation of *SUP*. To test whether maintenance of the dense cytosine methylation present at the *SUP* locus might require wild-type activity of *DDM1* or *MET1*, the *clk*-3 plants containing a closely linked (10 cM) *gl1*-1 mutation causing a readily selectable epidermal hairless phenotype were crossed to the *ddm1* or *met1* plant lines and the double mutant lines with reduced methylation at the centromeric repeats (diagnostic for *ddm1* or *met1* homozygosity) and hairless epidermis (diagnostic for *clk*-3) were selected (Jacobsen et al. 2000). Most of these plants displayed *clk -3*-like flowers. Bisulfite genomic sequencing confirmed that these plants were hypermethylated at *SUP*. Repeated self-pollination of these plants showed that *SUP* hypermethylation was stable in the *ddm1* and *met1* homozygous backgrounds for, at least, three generations. Thus, neither *DDM1* nor *MET1* activity is required for maintenance of *SUP* hypermethylation. To test whether *SUP* hypermethylation and silencing can occur spontaneously in the hypomethylation mutants, several independent *ddm1* and *met1* homozygote lines were allowed to self-pollinate for few generations. Plants with *sup*-like phenotypes coupled with *SUP* hypermethylation were found in most of the lines tested. Some of them have occurred in the second generation after self-pollination, while others only after three or four selfed generations. New *SUP*-hypermethylation (sup^{hm}) alleles appeared in the hypomethylation homozygotes in a sporadic fashion: generally, 1-2 plants of the 50 analyzed show *sup*-like phenotypes. This non-Mendelian inheritance suggests that the hypomethylation homozygotes are chimeric for *SUP* hypermethylation. A clear demonstration of this was done by crossing plants of a *met1* homozygous line to a *sup* deletion mutant line. All plants of the *met1* line had wild-type flowers in the second generation of self-pollination. When a first-generation homozygous plant was used as a male in a complementation cross to *sup* deletion mutant line, nevertheless, 1 of 12 F1 plants showed a clear *sup* phenotype coupled with hypermethylation of *SUP*. Thus, *met1* plant was a chimera showing *SUP* hypermethylation in only small proportion of its pollen, this is consistent with hypermethylation arising during meiosis or in small somatic sectors of the floral tissue. The hypermethylation patterns of the *SUP* locus were analyzed in a number of independent *ddm1* and *met1* homozygote lines. It appeared to be nearly identical in all these lines and quite similar to that found in *clk* lines and in an antisense-*METI* line. Thus, regardless of the cause of hypermethylation at *SUP,* the methylation pattern within this region of the gene is consistent.

Another striking phenotype typically present in the antisense-*METI* lines is characterized by flowers that resemble those of the floral homeotic mutant *agamous* (*ag*). In *ag* mutants the stamens are converted to petals and the ovary to a new internal flower. *AG* encodes a MADS-box protein and its RNA is expressed in the incipient and developing stamens and carpels. The *ag* phenotype of the antisense-*METI* plants is highly variable with *AG*-like (wild-type for *AG*) and *ag*-like (similar to *ag* mutants) flowers occurring on the same plant. By *in situ* hybridization these *ag*-like flowers were shown to have reduced levels of *AG* RNA. Besides, by bisulfite genomic sequencing it was found that *AG* is methylated in the antisense-*METI* plants but not in the wild type ones. Extensive methylation was detected in two regions of *AG*, the promoter (16 methylated Cs) and the large second intron (37 methylated Cs). Methylation status of these two regions in either *ag*-like or *AG*-like flowers taken from a single antisense-*METI* plant was separately assayed. Interestingly, the methylation in the promoter region was observed in both types of flowers, whereas the methylation in the intron occurred exclusively in the *ag*-like flowers. The composition of methylated sites at *AG* in the antisense-*METI* plants was similar to that seen previously at *SUP*. The methylation was found mostly at non-symmetric (other than CpG and CpNpG) sites. Despite the low sequence-context specificity there was some bias in a favor of methylation at Cp(A/T)pG trinucleotides and a bias against methylation at CpC dinucleotides. Methylation at CpG sites was not important for silencing of *AG*, as there were no CpG sites found in either the methylated promoter region or methylated intron region. As *METI* is believed to be important for CpG methylation, these data suggest that in the antisense-*METI* background some other methyltransferase with a different specificity rather than hyperactivity of residual *METI* catalyzes the hypermethylation of *AG*. Methylation at *AG* appears to be less dense than at *SUP*. In addition, none of the cytosines at the locus are fully methylated, as it is at the *SUP* locus. This may relate to the fact that the *AG*-hypermethylation (ag^{hm}) phenotype is less stable than the sup^{hm} one; the antisense-*METI* plants show a sup^{hm} phenotype throughout the entire body of the plant but only some sectors with an ag^{hm} phenotype. The sectors of ag^{hm} flowers always develop in plants that already have the sup^{hm} phenotype, it suggests that hypermethylation of *SUP* occurs first and hypermethylation of *AG* may or may not follow.

SUP was found to be hypermethylated at a high frequency in antisense-*METI* and in both *ddm1* and *met1* lines and *AG* became hypermethylated in the antisense-*METI* line. This suggests that hypermethylation may be a common event in mutants showing overall hypomethylation. As some other epimutations seen in the various hypomethylation mutants are recessive, it seems that, at least, some of these may also be due to excessive methylation of specific genes. This ectopic hypermethylation suggests that some aspect of the methylation pattern fidelity is compromised when overall genomic methylation is decreased. One possible model to explain this is that factors that control the fidelity of genomic methylation are themselves regulated by DNA methylation. A second possibility is that residual DNA-methyltransferase activity in these hypomethylation mutants is in some way hyperactivated when overall methylation is too low, resulting in ectopic methylation of some genes. It is not clear which DNA-methyltransferases might be important for the establishment and/or maintenance of the methylation patterns found at *SUP* and *AG*. An other interesting aspect of the possible challenging regulation of DNA methylation by DNA methylation in plants may be the fact that the introducing of an antisence cytosine methyltransferase (*METI*) construct in *Arabidopsis* plants changes essentially the adenine methylation pattern of *DRM2* (Ashapkin et al. 2002).

H3 Histone Methylation or How DNA Methylation Patterns are Established and Maintained?

As it was already noted, some of the developmental defects seen in *met1* and *ddm1* mutants, though clearly result from loss of the CpG DNA methylation at the particular genes, nevertheless, they segregate as dominant Mendelian traits independent of their "parental" *met1* and *ddm1* mutations. This is because the loss of CpG DNA methylation is not readily regained, when these mutants are crossed to wild type plants. Loss of non-CpG methylation in plants with combined mutations in the *DRM* and *CMT3* genes also causes a suite of the developmental defects but these are always fully recessive and, unlike phenotypes caused by *met1* and *ddm1,* are not inherited independently of the *drm* and *cmt3* mutations. This disparity seems to be a consequence of a basic difference in the mode of maintenance of two types of DNA methylations in the plant cell. The maintenance activity of MET1 can replicate the CpG DNA methylation even when the initial trigger for DNA methylation is genetically removed (Jones et al. 2001; Aufsatz et al. 2002a, 2004). This may be explained in part by the fact that Dnmt1-type DNA-methyltransferases have a strong preference for hemimethylated substrates such as those left by DNA replication of a CpG dinucleotide that was initially methylated on both strands.

Non-CpG DNA methylation appears to require for its maintenance the active signals to continually target the regions of DNA for methylation (Chan et al. 2005). In the case of non-CpG methylation this signal seems to come from histones associated with DNA. As it was already noted, some EMS-induced suppressors of a nonreverting *clark kent* allele, *clk-st,* were found to be mutant alleles of gene for CMT3 methylase. Another group of supressor mutants were identified as alleles of a new locus, named *KRYPTONITE* (*KYP*) (Jackson et al. 2002). These latter mutations are recessive and do not exibit any notable phenotypic effects other than suppression of the *clark kent* alleles even after extensive inbreeding. The *KYP* locus was cloned and found to code for a 624-amino acid SET-domain containing protein similar to Su(var)3-9 class of histone H3 Lys9 methyltransferases (Baumbusch et al. 2001). Generally, SET domain is a common feature of a vast group of evolutionary conservative proteins that are believed to play a key role in epigenetic control of gene expression by methylating the different Lys residues in histones H3 and H4 (Lachner et al. 2003; Cheung, Lau, 2005). These proteins are divided into four classes as typified by their *Drosophila* member genes E(Z), TRX, ASH1 and SU(VAR)3-9. The first plant genes identified encoding the SET domain proteins were *E(Z)* homologs, the *CURLY LEAF* (*CLF*) involved in the control of leaf and flower morphology and flowering time (Goodrich et al. 1997), and *MEDEA* (*MEA*), an inhibitor of the endosperm development in the absence of fertilization, also implicated in imprinting of paternal genes (Grossniklaus et al. 1998; Vielle-Calzada et al. 1999; Luo et al. 2000). In a complete sequence of the *Arabidopsis* genome more than 30 such genes were detected, they can be grouped based on the characteristics of the SET domains and cysteine-rich regions into four classes mentioned, E(Z), TRX, ASH1 and SU(VAR)3-9. A study by RT-PCR indicated their spatially and temporally differential expression patterns during plant development (Baumbusch et al. 2001). The high number of genes and their diverse expression patterns may reflect a high complexity of the epigenetic control of gene activity during plant development. *KYP* is the forth of total nine Su(var)3-9 class genes, *SUVH1-SUVH9*. It possess all motifs previously found to be critical for histone methylating activity in SUV39H1 proteins of mammals and yeast including both specific residues within SET domain and two flanking cysteine-rich motifs (Rea et al. 2000; Nakayama et al. 2001). Interestingly, all *Arabidopsis* SUVH members contain an arginine residue instead of first histidine in the conservative SET motif HΦNHSC, a change shown in mammalian SUV39H1 proteins to increase histone methylating activity by 20-fold or more (Rea et al. 2000). The purified KYP protein was indeed shown to methylate histone H3 (but not other histones) at Lys9 position. Since all three *clark kent* suppressor *kyp* alleles contain mutations reducing or eliminating function of SET domain, H3 Lys9 methylation appears to be necessary for maintaining the silent state of *SUPERMAN* gene. The methylation patterns of *SUP* in *kyp-1*, *cmt3-7* and *met1* mutants were compared to those in *clk-st* by bisulfite genomic sequencing (Jackson et al. 2002). As it was noted above, the methylation pattern in *clk-st* is dense and essentially non-sequence specific; namely, about 16% of CpG and non-symmetric cytosines and 55% CpNpG cytosines are methylated. Quite expectedly and in accordance with earlier studies the *met1* mutation has been found to eliminate the major part of CpG methylation but practically with no effects on non-CpG methylation. The *cmt3-7* mutation had relatively small effect on CpG methylation; it completely eliminated CpNpG methylation and significantly diminished non-symmetric methylation. The *kyp-1* mutation had moderate effect on CpG methylation (weaker than *met1* but stronger than *cmt3-7*) and caused nearly complete elimination of both types of non-CpG methylations. In general, the methylation patterns in *kyp* mutant were closer to those in *cmt3* than in *met1* mutants. In accordance with this view the *kyp* mutants (similarly to *cmt3* mutants) neither developed late-flowering phenotype nor affected methylation of silenced *FWA* locus known to be methylated mainly at CpG sites. As a matter of fact, both *cmt3* and *kyp* do cause some undermethylation of *FWA* at CpNpG sites but this undermethylation seems to be without effect on the silent state of the gene. Similar undermethylation at CpNpG but not CpG sites has been found in other sequences tested, namely, 180 bp centromere repeat, LTR sequence of *Athila* retrotransposon and single-copy *Ta3* retrotransposon. Though generally the effects of *kyp* mutations on the DNA methylation patterns resemble those of *cmt3* mutation, some subtle differences do exist. Namely, *kyp* effects on CpNpG methylation seem to be weaker than these of *cmt3* at most loci studied, whereas it's effects on CpG and non-symmetric methylation are greater than these of *cmt3*, at least, at the *SUP* locus. Both *cmt3* and *kyp* induce some small level of expression of two normally silenced retrotransposon sequences, namely, TSI sequence of *Athilla* and *Ta3*. Since CMT3 is unique among cytosine DNA-methyltransferases in possessing a HP1-related chromodomain, the similar effects of *kyp* and *cmt3* on gene methylation and silencing suggested that CMT3 could bind directly to Lys9-methylated N-tail of histone H3, thereby, targeting respective sites of genomic DNA for the CpNpG-specific methylation. This straightforward model has not found support in the *in vitro* binding studies. Neither CMT3 itself no it's isolated chromodomain peptide were found to specifically bind to a matrix containing Lys9-methylated histone H3 peptides, whereas such specific binding was readily observed with mouse HP1β chromodomain. On the other hand, an *Arabidopsis* homolog of HP1, LHP1, specifically binds to both CMT3 and Lys9-methylated histone H3. The Lys9-methylated N-tail of H3 histone, therefore, may well serve as the signal targeting specific CpNpG sites for methylation by LHP1-mediated binding to CMT3.

The 180 bp centromeric repeats in the *Arabidopsis* genome are the tandem arrays that span the core centromeres of all five chromosomes. They have been shown to be heterochromatic based on decreased recombination frequencies, increased condensation and high levels of CpG and CpNpG methylations. Transcription of these repeats was never detected in any methylation mutant known (*ddm1*, *met1*, *cmt3*, *cmt3 met1* and *kyp*). Therefore, they seem to be the excellent loci, which can be used for investigation of the relationship between histone methylation and DNA methylation without the complication of transcriptional effects (Johnson et al. 2002). As it was already mentioned, in the wild-type *Arabidopsis* the 180 bp repeats are highly methylated at CpG sites (~71%) and moderately methylated at CpNpG sites (~38%) (Lindroth et al. 2001). Homozygous mutations *cmt3* and *kyp* have little effects on CpG methylation and remove essentially all (*cmt3*) or a considerable part (*kyp*) of CpNpG methylation. The *ddm1*, *met1*, and *cmt3 met1* mutants eliminate almost all CpG and part of CpNpG methylation. The relationship between DNA methylation and histone modifications in heterochromatin was studied by chromatin immunoprecipitation (ChIP) assays. A strong enrichment in 180 bp repeat sequence was found in the chromatin fraction containing Lys9-methylated histone H3. Thus, H3-Lys9 methylation seems to mark heterochromatin in plants as it does in fungi and animals. Much of this methylation is dependent on the *KRYPTONITE*, since *kyp* mutants have greatly reduced levels of H3-Lys9 methylation. Examination of the DNA-methyltransferase mutants (*cmt3*, *met1*, and double mutant *cmt3 met1*) revealed little effects on the level of H3-Lys9 methylation. Therefore, the decrease in either CpNpG or CpG methylations *per se* does not directly affect H3-Lys9 methylation. Conversely, *ddm1* mutation caused a significant reduction in H3-Lys9 methylation, suggesting that the chromatin remodeling is important for the maintenance of H3-Lys9 methylation. The *ddm1* mutations have profound effects both on all types of DNA methylations and on H3-Lys9 methylation, whereas *cmt3 met1* double mutants drastically reduce DNA methylation but have little effect on H3-Lys9 methylation, and *kyp* eliminates most H3-Lys9 methylation but only has an intermediate effect on CpNpG methylation with no effect on CpG methylation. So, the DDM1 seems to play the independent roles in both histone methylation and DNA methylation. *Ta3* is a single copy *copia*-like retrotransposon located in the pericentromeric region of *Arabidopsis* chromosome 1. As it was already noted, *Ta3* sequence is highly methylated and silenced in *Arabidopsis* and yet is transcribed in lines homozygous for *cmt3* mutations (Lindroth et al. 2001). In the *Arabidopsis* lines carrying either *cmt3* or *kyp* mutations the CpG methylation is practically not affected, whereas CpNpG methylation is considerably reduced (Jackson et al. 2002). Both *ddm1* and *cmt3 met1* eliminate almost all CpG and CpNpG methylations, whereas the *met1* single mutant eliminates almost all CpG methylation but has an intermediate effect on CpNpG methylation (Johnson et al. 2002).

A H3-Lys9-methylated chromatin fraction is greately enriched with silent *Ta3* sequence in wild-type plants. In the *ddm1* and *kyp* mutant lines, the H3-Lys9 methylation was reduced to background level. The transcriptional activity of *Ta3* in *ddm1* mutants was derepressed to an easily detectable level, whereas in *kyp* lines it was barely detectable. Thus, reduction in H3-Lys9 methylation *per se* (without accompanying reduction in DNA methylation) is not enough to significantly reactivate gene expression. The *cmt3* lines show a significant loss of H3-Lys9 methylation (to ~37% of the wild-type levels), whereas the *met1* lines are essentially unaffected. In the *cmt3 met1* double mutant lines the H3-Lys9 methylation is greatly reduced to just above background levels. Thus, at the *Ta3* locus the histone H3-Lys9 methylation is affected by DNA methylation. The magnitude of H3-Lys9 hypomethylation in this case seems to correlate with the level of the transcription reactivation. As it was shown by RT-PCR measurements, *cmt3* activates *Ta3* transcription to approximately 1/3 of the level observed in *ddm1* lines, whereas *met1* has a weaker effect (~1/10 of the *ddm1* levels). This suggests that CpNpG methylation plays a more important role in *Ta3* silencing than CpG methylation; it may be a direct consequence of the CpNpG sites location in the promoter region. The *cmt3 met1* double mutant reactivated the transcription better than either single mutant (~90% of *ddm1* levels). Thus, while CpNpG methylation is primarily responsible for gene silencing, CpG methylation also plays some role in it. Comparison of the *Ta3* RNA levels in each line and of shares of Lys9-methylated H3 associated with *Ta3* reveals an inverse relationship between transcription and H3-Lys9 methylation. One may suggest that DNA hypomethylation leads to a loss of H3-Lys9 methylation only when coupled with transcription. The *clark kent* (*clk*) alleles in *Arabidopsis* were studied as representative genes that are highly methylated and silenced but reside in the midst of euchromatin. Since the *kyp* and *cmt3* mutant lines were isolated as supressors of the *clk-st* (*clark kent-stable*) phenotype, they were directly compared to parental *clk-st* line by ChIP assay with antibodies to dimethylated H3-Lys9. The *SUP* locus was highly enriched with Lys9-methylated H3 in *clk-st* line, whereas in the *kyp* line the degree of such methylation was at the background level. Unexpectedly, in the *cmt3* lines that show a dramatic loss of CpNpG DNA methylation and complete reversal of *SUP* gene silencing, the negligible loss of H3-Lys9 methylation was observed. Thus, DNA methylation but not H3-Lys9 methylation is primarily responsible for *SUP* gene silencing.

Since *met1* allele used in these studies was a point mutation with partial loss of MET1 function, the possible role of CpG methylation in directing H3-Lys9 methylation has been readdressed using a *met1* null-mutant strain of *Arabidopsis* with CpG methylation completely eliminated (Tariq et al. 2003). The methylation status of histone H3 in the nontranscribed *Ta2* retrotransposon that resides in a pericentromeric region of chromosome 1 was determined. *Ta2* contains hypermethylated DNA that is packaged into nucleosomes with $H3Lys9^{Me}$ (Johnson et al. 2002). $H3Ly9^{Me}$ abundant at *Ta2* in the wild type plants was almost undetectable in the *met1* null mutant. Furthermore, in *met1, Ta2* consistently gained $H3Lys4^{Me}$, a marker for transcriptionally active genes. Indeed, in the *met1* mutant, transcription of *Ta2* was clearly activated. Depletion of CpG methylation at *Ta2*, therefore, likely changed the methylation status of histone H3. This change in H3 methylation could well be an indirect effect of transcriptional activation. To resolve this uncertainity, the levels of $H3Lys9^{Me}$ and $H3Lys4^{Me}$ in three additional target genes residing within the heterochromatic knob on chromosome 4 (At4g03760, T5L23.26, and At4g03870) were examined. Like *Ta2*, these loci in the wild type plants are associated with high levels of $H3Lys9^{Me}$, which was drastically reduced and replaced by $H3K4^{Me}$ in the *met1* mutant. As concerning their transcription in the *met1* plants, it was detected for *Ta2*, At4g03760, and T5L23.26 but not At4g03870. Thus, the loss of $H3Lys9^{Me}$ in the *met1* mutant could occur in the absence of transcription and seems not to be an indirect effect of transcriptional activation. This observation was further supported by a more thorough study using a subpopulation of the 180 bp centromeric repeats that are transcriptionally inert in both wild type and DNA methylation-deficient plants. As shown by a bisulfite analysis assay, CpG methylation of these repeats is completely erased in *met1* null mutant, whereas methylation at CpNpG and CpNpN sites is reduced to 57.6% and 73% of the wild type levels, respectively (Saze et al. 2003). The ChIP assays showed a strong enrichment for $H3Lys9^{Me}$ at these repeats in wild type plants and a 17-fold depletion in the *met1* mutant. Thus, methylation at H3Lys9 is indeed directed by DNA methylation at CpG sites, whereas non-CpG DNA methylation is directed by H3Lys9 methylation.

The lysine residues of histones can be either, monomethylated, dimethylated or trimethylated, and recent evidence suggests that various methylation states may have different functional significance. In *Arabidopsis* the high levels of Lys9-monomethylated and dimethylated histone H3 forms are readily detectable, whereas trimethylated form is practically absent (Jackson et al. 2004). Both monomethyl and dimethyl histone H3-Lys9 are concentrated in heterochromatin. In *kyp* mutants, dimethyl H3-Lys9 is nearly completely lost, whereas monomethyl H3-Lys9 levels are only slightly reduced. Recombinant KYP can add one or two but not three methyl groups to the Lys9 residue of histone H3. Another KYP-related protein SUVH6 seems to have similar H3-Lys9 methylation activity. Thus, at least, two members of Su(var)3-9 family are active in *Arabidopsis,* and dimethylation of histone H3 Lys9 seems to be critical for gene silencing and CpNpG DNA methylation.

The dependence of DNA methylation in plants on histone H3 Lys9 methylation is not without the precedent. Previous studies with *Neurospora* have shown that the loss of DIM5 H3 methyltransferase as well as mutations of Lys9 in histone H3 result in a complete loss of DNA methylation *in vivo* (Tamaru, Selker, 2001). Moreover, it was shown that trimethylated but not dimethylated H3 Lys9 marks the chromatin regions for cytosine methylation, and the DIM-5 specifically creates this mark (Tamaru et al. 2003). An important difference between *Neurospora* and plants in this respect still exists. There is a single DNA-methyltransferase in *Neurospora* responsible for all known types of cytosine methylation, namely Dim-2 (Kouzminova and Selker, 2001), and a single H3 Lys9 methylase Dim-5 seems to be absolutely required for its function. On the other hand, only CpNpG methylation seems to be dependent on histone H3 Lys9 methylation in plants. Moreover, there are eight *KYP* homologs in genome of *Arabidopsis*, and two of them are closely related to *KYP* (Baumbusch et al. 2001). It seems not quite improbable if some of them also function in regulating DNA methylation. In a number of studies on various organisms both histone H3 Lys9 and Lys27 methylations have been correlated with the transcription silencing, whereas H3 Lys4 methylation is widely believed to be a chromatin mark for active genes (reviewed in Lachner et al. 2003; Cheung, Lau, 2005). On the histone H3 tail, Lys9 and Lys27 are both reside within a highly related sequence motif ARKS. Nevertheles, their binding to the chromodomain proteins has been found to be quite discriminative both *in vitro* and *in vivo* (Fischle et al. 2003). Namely, Lys9-methylated histone H3 specifically binds HP1, whereas Lys27-methylated H3 specifically binds Pc (Polycomb). Moreover, in *Drosophila* S2 cells the methyl-Lys27 and Pc are both excluded from areas that are enriched in methyl-Lys9 and HP1. Swapping of the chromodomain regions between Pc and HP1 is sufficient for exchanging the nuclear localization patterns of these factors, indicating an essential role for their chromodomains in both binding and discrimination of target sites. Comparison of three-dimensional structures of respective complexes (Pc chromodomain-H3 peptide bearing trimethyl-Lys 27 vs HP1 chromodomain-H3 peptide bearing trimethyl-Lys 9) at 1.8 Å resolution showed that Pc chromodomain distinguishes its methylation target on the H3 tail via an extended recognition groove that binds to five additional amino acid residues preceding the ARKS motif.

Histone H3 is di- and trimethylated at Lys4 residue in active euchromatic regions but not in silent heterochromatic sites. Sites showing trimethylation correlate with transcription starts, while those showing mainly dimethylation occur elsewhere in the vicinity of active genes (Schubeler et al. 2004; Bernstein et al. 2005). In *Saccharomyces cerevisiae* the Set1 protein is a histone methylase that can catalyze di- and trimethylation of Lys4 and stimulate the activity of many genes (Santos-Rosa et al. 2002). H3 Lys4 dimethylation occurs at both inactive and active euchromatic genes, whereas trimethylation is located exclusively at the active genes. Furthermore, the methylation state of H3 Lys4 contributes to its association with chromatin remodelling protein ATPase Isw1p (Santos-Rosa et al. 2003). The binding of Isw1p to chromatin shows preference for di- and trimethylated Lys4 residues in H3. This binding does not appear to be direct but it requires some other proteins present in the yeast extract. Analysis of a methionine pathway gene, *MET16,* in yeast has shown that both Isw1p ATPase and Set1p are necessary for an efficient expression of the gene under inducing (methionine-less) conditions. Both enzymatic activities are also required to generate the activation-specific chromatin changes at the 5′-end of *MET16* (an appearance of a strong TATA box hypersensitive site and of a protein-free DNA stretch downstream) and to correctly position the RNA polymerase II transcription complex and a termination factor Rna15p over the coding region. The chromatin remodelling protein Chd1 (chromo-ATPase/helicase-DNA binding domain 1) is a component of SAGA and SLIK, two highly conservative yeast multi-subunit histone acetyltransferase (HAT) complexes, which preferentially acetylate histones H3 and H2B and deubiquitinate histone H2B. It happened to be the first chromodomain-containing protein that recognizes Lys4-methylated histone H3 (Pray-Grant et al. 2005). In a series of the pull-down assays with differently methylated histone peptides (H3 trimethylated at Lys4, Lys9, or both, and two control peptides of unmethylated H3 and H4 tails) a number of proteins specifically bound to the unmethylated H3 were purified from HeLa nuclear extract (Zegerman et al. 2002). The same set of proteins was also purified by affinity chromatography with the Lys9-methylated H3 peptide but not Lys4-methylated, Lys4-Lys9 double methylated H3 peptides or unmethylated H4 peptide. Thus, only Lys4 methylation of histone H3 disrupts the binding of the complex. The subunits of the complex identified by mass spectrometry include the histone deacetylases HDAC1/2, the ATPase chromatin-remodeling enzyme Mi-2^{3}, the RB-associated proteins Rbap48/46, the metastasis-associated antigens MTA1/2 and the methyl CpG binding domain MBD3. All these components are found in the NuRD repressor complex that is targeted to specific promoters by DNA binding transcriptional repressors (Zhang and Reinberg, 2001). The binding of NuRD to histone H3 and the deacetylase-mediated promotion of H3 Lys9 methylation may be a secondary step reinforcing the transcription repression. The prevention of NuRD binding to histone H3 due to Lys4 methylation may represent a mechanism, by which this type of H3 methylation counteracts the transcription repression.

A gene, coding for *Arabidopsis* protein similar to animal TRX family of SET-domain proteins, *ATX1* (*Arabidopsis TRITHORAX1)* is expressed during floral organ initiation and seems to play a major role in development, spatial arrangement and morphology of all floral organs. In a striking contrast to other known SET-domain containing proteins acting as negative regulators of plant homeotic genes, the product of *ATX1* positively and specifically affects the expression of several flower homeotic genes (Alvares-Venegas et al. 2003). An *atx1-1* "knockout" mutant with a T-DNA insertion between exons 10 and 11 of *ATX1* gene was identified. This insertion truncates the mRNA leading to deletion of both PHD fingers and SET domain both essential for protein activity. Mutant plants homozygous for *atx1-1* display a bunch of modified morphological phenotypes (small size, later flowering, numerous floral abnormalities). These phenotypes were not observed in plants heterozygous for the T-DNA insertion, indicating that *atx1-1* is a recessive mutation. The flowers of homozygous mutants display abnormalities in all parts. They develop aberrant locules and do not produce pollen, many stamens remain short, the pistil may lack stigmatic papillae and be curved or otherwise aberrantly shaped, the carpels may be partially or completely unfused, and petals may be of variable size, shape, and number. Most often the defect observed in sepals is an asymmetric placement and folding. Taken together, these observations suggest that the *atx1-1* mutation affects floral organ identity. The *in situ* hybridization experiments have shown that *ATX1* is widely expressed in the cells of the inflorescence meristem at stage 2, when flower primordia become visible. Later the *ATX1* transcripts are mainly detectable in petal primordia (stage 4), in the lobed stamens and in the gyneocium cylinder (stage 8). Still later (stage 10) *ATX1* transcripts are only detected in the gyneocium in association with ovule primordia initiation, whereas no transcripts are detected in mainly developed sepals, petals and stamens. Thus, *ATX1* expression seems to be specifically associated with the initiation of flower organs. In *Arabidopsis* a number of homeotic genes are known to determine flower organ identity, namely, a pair of the class A genes *APETALA* (*AP1* and *AP2*), class B genes *PISTILLATA* (*PI*) and (*AP3*), a class C gene *AGAMOUS* (*AG*), and the *SEPALLATA* (*SEP1*, *2* and *3*) genes (Pelaz et al. 2000; Honma and Goto, 2001). The expression of these genes was affected to different degrees in the homozygous *atx1-1* mutant. In mutant buds the lower expression levels were observed for the *AP1*, *AP2*, *PI* and *AG* genes, whereas in fertilized flowers their expression levels were not significantly affected. Apparently, wild-type *ATX1* activity is required to maintain normal mRNA levels of these genes at the early stages of flower development. The expression levels of other MADS-box genes (*AP3* and *SEP1*) as well as that of *AINTEGUMENTA* (*ANT*) belonging to the plant-specific AP2 family of transcription factors are not significantly affected by the loss of *ATX1* activity. Thus, *ATX1* maintains the active state of floral homeotic genes in a gene-specific manner. To test whether ATX1, like some other proteins containing SET-domain, could be a histone methylase, the respective parts of its polypeptide chain were expressed as GST fusion proteins in bacteria and the histone-methylation activity of purified polypeptides *in vitro* was tested. ATX1-SET domain has been found to possess a lysine methylase activity, albeit at much lower levels than KYP-SET. In contrast to KYP-SET, which mainly methylates Lys9 residue of histone H3, the ATX1-SET methylates Lys4. The role as an epigenetic regulator probably explains the great variability of *atx1-1* mutation phenotypic effects.

The histone H3 methylation patterns were studied in heterochromatic region obtained from the knob of chromosome IV known to contain multiple tandem arrays of long repeats and many transposons, of which Athila is a most prominent, as well as small number of expressed genes (Fransz et al. 2000). Chromatin immunoprecipitation (ChIP) followed by PCR amplification with a vast assortment of primer pairs specific for respective genes, upstream regions and transposons has shown that in the wild type (WT) plants all 15 transposons studied were associated with Lys 9-methylated H3 (H3mK9), whereas four of six known expressed genes were associated with Lys 4-methylated H3 (H3mK4) (Gendrel et al. 2002). In *ddm1* mutant, a dramatic shift in the pattern of histone methylation was observed. Thirty-two of 41 sequences, whose association with methylated histones could be detected, either lost H3mK9 or gained H3mK4, or both. These included both known and hypothetical genes, transposons and upstream regions; thus, heterochromatin underwent a major restructuring in *ddm1*. RT-PCR study of total RNA extracted from WT plants and *ddm1* seedlings showed an expression of many sequences to be up-regulated. Those include *Athila*, *Cinful*, *del-like* and novel retrotransposons as well as the *MULE* and *CACTA* class II transposable elements. In addition, one silent gene (encoding a phosphate translocator), and three other genes were up-regulated in *ddm1* along with, at least, 10 hypothetical genes. Ten-fold or more overexpression in *ddm1* was observed for 26 sequences, and in 20 cases this was correlated with a decreased association with H3mK9, an increased association with H3mK4, or both. On the other hand, 16 sequences were expressed at similar levels in the WT plants and *ddm1* irrespective of their histone methylation patterns. No methylated histone association has been detected for 11 sequences, of which 9 were transcribed and 3 were found to be up-regulated in *ddm1*. Thus, although there was a clear correlation, changes in histone methylation patterns were not obligatory in up-regulated sequences. Overall levels of H3mK9 and H3mK4 in the genome did not change appreciably in *ddm1*. Since *DDM1* is widely believed to be a part of a SWI2/SNF2-like chromatin remodelling complex (Jeddeloh et al. 1999), such remodelling seems to be important for correct distribution of H3mK9 and H3mK4 between heterochromatic and euchromatic domains. Whether the drastic reduction in DNA methylation seen in *ddm1* mutants is a consequence or cause of H3mK9-H3mK4 misdistribution remains to be established. In any case the cross-talk between histone code and cytosine methylation observed provides, at least, a tentative answer to the long-standing question of how proper DNA methylation patterns may be established and maintained.

Generally, the functional combination of two methylation systems (histones and DNA) could be the means to stabilize a silent state of respective chromatin regions. Therefore, it safe-guards the gene expression programs and protects genome integrity (Lachner et al. 2003). How exactly DNA methylation is coupled to histone methylation is far from clear. One can imagine that methylated DNA attracts methyl-CpG-binding proteins, which in their turn recruit the histone deacetylase complexes to deacetylate histone tails so that the tails become suitable for serving as substrates for H3 Lys9-methylation. Alternatively, it is also possible that the chromodomain-containing proteins bind to methylated histone tails and recruit cytosine DNA-methyltransferases to methylate adjacent DNA sequences. This latter possibility seems to be realized in the case of the H3 Lys9-methylation dependent DNA methylation by plant CMT3 methyltransferase.

Is dsRNA an Another Way of Establishing DNA Methylation Patterns?

During last several years the regulatory RNAs have been linked to various gene silencing phenomena in plants, animals and fungi (Matzke et al. 2004). Different types of regulatory RNA were shown to act in distinct ways to induce gene silencing. The short RNAs (21-26 nucleotides, nt), which are derived via cleavage of double-stranded RNA (dsRNA) precursors, serve as the specificity determinants for enzyme complexes that degrade, modify or inhibit the function of homologous nucleic acids. Gene silencing phenomena that are induced by nucleotide sequence-specific interactions mediated by RNA are termed collectively 'RNA silencing' (Voinett, 2002; Mlotshwa et al. 2002). The most familiar form of RNA silencing in plants occurs in the cytoplasm and has been termed posttranscriptional gene silencing (PTGS). This evolutionarily conservative process involves a perfect dsRNA that is processed by an RNase III activity termed Dicer into 21 nt short interfering RNAs (siRNAs) of both polarities. Following ATP-dependent unwinding of the siRNA duplex, the antisense siRNA guides a ribonuclease complex RISC (RNA-induced silencing complex) to the cognate mRNA and targets it for degradation. RNAi and related phenomena were initially observed following microinjection of dsRNA into cells or when transcription of transgenes, transposons or viruses yielded dsRNA. RNAi/PTGS/quelling plays a major role in defending organisms against foreign or invasive sequences. Host defense is not the only function of RNA silencing, however, the examination of native short RNA populations is revealing their pervasive involvement in the regulation of endogenous genes that are important for development.

Hamilton and Baulcombe (1999) were first to demonstrate the direct involvement of special kind of RNA in plant gene silencing. They have carried out analyses specifically devoted to detection of low molecular weight antisense RNA in four models of post-transcriptional gene silencing (PTGS) in plants. First one was "cosuppression", i.e. transgene-induced PTGS of a homologous endogenous gene. To this end, the tomato plants were transformed with cDNA coding for 1-aminocyclopropane-1-carboxylate oxidase (ACO) under 35*S* promoter. Of all lines tested two exhibited PTGS of the endogenous ACO mRNA. Low molecular weight RNAs from both silenced and non-silenced lines were separated by denaturing PAAG electrophoresis, blotted and hybridized to ACO sense and antisense RNA probes. Discrete RNAs of 25 nt, both sense and antisense, were present in both PTGS lines but absent in non-silenced lines. The second model analyzed was PTGS of a 35*S*-β-glucuronidase (GUS) transgene in tobacco. One of the lines tested (6b5) had a strong PTGS, another one (T4) – weaker PTGS, still another one (6b5x271) had no PTGS of transgene, since it was transcriptionally suppressed by a 35S promoter supressor present in one of its parental lines (271). Hybridization with a GUS-specific probe revealed that a 25 nt GUS antisense RNA is present in both PTGS lines but absent in non-PTGS line 6b5×271. The amount of antisense RNA was much higher in strong-PTGS line 6b5 compared with weaker PTGS line T4. In *Nicotiana benthamiana* plants expressing a 35S-GFP transgene the PTGS was initiated by infiltration of a single leaf with *Agrobacterium tumefaciens* containing the same transgene in a binary plant transformation vector. Following 2-3 weeks after this infiltration, the GFP fluorescence disappeared due to systemic spread of PTGS. A 25 nt GFP antisense RNA was detected in tissues exhibiting PTGS but not in equivalent leaves of plants that had not been infiltrated or had been infiltrated with *A. tumefaciens* without 35S-GFP transgene. The 25 nt RNA complementary to the positive (genomic) strand of potato virus X (PVX) was detected 4 days after inoculation of *N. benthamiana* with this virus. Thus, 25 nt antisense RNA complementary to targeted mRNAs accumulates in four types of PTGS. As a matter of fact, such RNAs were detected in some other PTGS but never in the absence of PTGS. These 25 nt RNAs are not degradation products of the target mRNAs because they have antisense polarity. More likely, these RNAs are produced by transcription of an RNA template. This is consistent with the presence of the 25 nt PVX RNA in PVX-infected cells that do not contain DNA template. In plants, heavy *de novo* methylation and silencing of multiple transgene copies integrated at the same locus have been proposed to occur through a DNA-DNA pairing process (Matzke et al. 1995). To account for this *de novo* methylation, the involvement of a DNA pairing-dependent process termed 'epigene conversion' was suggested. In the epigene conversion the strand interactions between methylated and unmethylated homologs produces hemimethylated intermediates, which are the favorite substrates for the maintenance DNA-methyltransferase.

A study of DNA methylation of the transgenic viroid cDNA sequences after their integration into the plant genome as a tandemly repeated copies have shown that their transcription may play a leading role in targeting the complementary DNA sequences for *de novo* methylation (Pelissier et al, 1999). Transgenic tobacco lines used contained the potato spindle tuber viroid (PSTVd) cDNA sequences, either viroid replication initiation competent (VRI+) or incompetent (VRI-) ones. Bisulfite genomic sequencing analysis showed that regions representing the 5'-junction between p35S promoter and PSTVd sequences and the 3'-junction between the PSTVd and pAnos sequences were virtually free of cytosine methylation in viroid-free plants transgenic for a dimeric VRI- PSTVd construct. Quite to the contrary, these sequences became by more than 90% of all C residues methylated in viroid-infected progeny of these plants. Of the m^5C residues 43.8% were found in a non-symmetrical sequence context, while 29.3% were detected at CpG and 26.9% at CpNpG sites. This distribution closely reflected the relative representation of these sites in this region (42.7%, 31.2% and 26.1%, respectively). Upper strand methylation ranged from 60 to 100% and lower strand from 72.5 to 100% for individual DNA molecules. Strikingly, 22 of 28 DNA molecules displayed > 90% methylation and included 11 molecules that were completely methylated. Cytosine methylation at symmetrical and non-symmetrical sites was also detected in the non-viroid-specific section of the p35S-PSTVd junction and it appeared to be restricted to the region immediately adjacent to the viroid sequence (positions -1 to -21). In this PSTVd-flanking region the overall level of methylation rapidly decreased from 75% (-1 to -5) to 29.4% (-6 to -21) with increased distance from the PSTVd sequence. In the proximal -1 to -21 p35S region, this corresponded to an average degree of C methylation of 38.9%, which was significantly lower than that detected for viroid sequence (94.7%). In the region further upstream (-22 to -128) the sparse methylation was detected. The methylation pattern at the 3'-junction was essentially similar, though the level of cytosine methylation was reduced compared with that at the 5'-junction: 64% symmetrical and 69.2% non-symmetrical C residues were methylated. The overall level of methylation decreased to 32.8% in the pAnos region immediately adjacent to the PSTVd sequence (positions +1 to +20). In the +21 to +117 region the partial methylation was found mainly in symmetrical positions. In the viroid-free plants no methylation was found in the corresponding 3'-junction sequences. Together with the situation observed in the promoter region, these results strongly suggest that in viroid-infected plants the PSTVd RNA-directed DNA methylation occurs in a region almost entirely restricted to the PSTVd cDNA trangene sequences.

In plants transgenic for a VRI+ PSTVd construct the DNA methylation patterns were essentially the same as those detected in the viroid-infected VRI-plants. Methylation was mainly limited to the viroid sequences with an overall methylation frequency of 92% in the viroid region and of 39% in the -1 to -21 proximal p35S region. The general pattern of methylation at the 3'-junction was also similar to that observed in the viroid-infected VRI-plants but significantly increased. In the viroid sequence, nearly all C residues (99%) were methylated whatever their genomic context. The pAnos region immediately flanking the PSTVd sequence (positions +1 to +20) showed an overall methylation level of 71.9%, which was significantly higher than that (32.8%) for the same area in the viroid-infected VRI-plants. The +21 to +117 region was 16.7% methylated, which was ~7-fold higher than the respective value observed in the pAnos region of the viroid-infected VRI-plants, and 54% m^5C residues were found in CpG or CpNpG sites, which contain 48.6% of all C residues. Only sparse methylation was detected in the DNA regions downstream of position +117. All these data provide a strong argument for the *de novo* methylation directed by unusual structures that could arise by pairing of RNA molecules with their genomic counterparts. This process should be termed RNA-directed and not RNA-mediated DNA methylation (RdDM). This is to emphasize that only DNA sequences complementary to the directing RNA are specifically methylated. Most, if not all, cytosines within the putative RNA-DNA triplex region are methylated irrespective of their sequence context. The recognition of specific structures in DNA that are formed during the RdDM process may strongly stimulate the activity of *de novo* DNA-methyltransferase(s). Since PSTVd replication involves generation of both plus and minus RNA strands, it is not known whether the RNA-DNA duplex or a triple helix structure is recognized by *de novo* DNA-methyltransferase(s). Along with heavy methylation of the viroid sequences, most of the individual DNA strands displayed a significant but lower level of methylation within the 5'- and 3'- PSTVd-flanking regions. The extent of this methylation is mainly restricted to the -1 to -21 promoter region and +1 to +40 pAnos region. It is conceivable that the *de novo* DNA-methyltransferase, which is directed to the place of RNA-DNA interactions may spread onto the adjacent sequences before it is released from the template. To define the minimal DNA target sequence for an efficient RdDM, the non-infectious subfragments of the viroid cDNA have been introduced into genomic DNA of tobacco plants (Pélissier and Wassenegger, 2000). The 60 bp long and larger fragments were found to be specifically and heavily methylated as full-length copies of the PSTVd cDNA (overall methylation levels of different 60 bp fragments varied between 63% and 76%) in nearly all tobacco leaf cells. In contrast, use of a 30 bp fragment leaded to a significantly decreased though still substantial methylation level (about 16%), about 45% of the leaf cells displayed no methylation at all. Since a spread of methylation into adjacent sequences is mainly restricted to the first 30-50 bp directly flanking the RNA-targeted DNA, plant DNA-methyltransferase(s) in question seems to specifically recognize the RNA-DNA hybrid structure, even if the region of complementarity is limited to a length of only 30 nt. As soon as the DNA-methyltransferase slips to flanking RNA-free region, it either leaves the template or its methylation activity ceases. Both strands of the target DNA sequence are heavily methylated at symmetric and asymmetric sites suggesting a mechanism operating on both strands simultaneously.

Whether transcriptional silencing and methylation of target gene promoters could really result from a *trans*-acting homologous RNA, was tested on tobacco plants expressing an unmethylated NOSpro-*nptII* target gene (Mette et al. 1999). A chimeric gene consisting of a nopaline synthase promoter (NOSpro) under the control of 35S promoter (35Spro) was constructed and used for plant transformation. The expression of the 35Spro-NOSpro transgene in plants could be altered in two ways. First, the 35Spro was flanked by *lox* sites to allow its excision by Cre recombinase. Second, the 35Spro could be inactivated by crossing to a 35S-silencing tobacco line 271. Transformed plants were analyzed for NOSpro RNA synthesis, activity and methylation of the NOSpro-*nptII* gene. A full-length polyadenylated NOSpro RNA produced in most plant lines did not lead to inactivation or methylation of the target locus. In contrast, in one line (9NP) the transgene 35Spro-NOSpro locus appeared to be somehow rearranged (to produce smaller than expected and non-polyadenylated transcripts) and caused methylation and inactivation of the NOSpro-*nptII* target gene. Interestingly, both methylation and silencing of NOSpro-*nptII* gene were readily reversed, when two loci were segregated in progeny. To determine whether elimination of NOSpro transcription would alleviate silencing, the 9NP plants were crossed with two other plant lines, one expressing the Cre recombinase, to remove the 35Spro driving transcription of NOSpro sequences, and the second containing the 35Spro silencing locus, *271*, to abolish transcription of the NOSpro. Significant silencing of the NOSpro-*nptII* gene was still observed in progeny of the cross with Cre line. In contrast, offspring of the cross to 271 line exhibited reversal of NOSpro-*nptII* silencing and reduction of its methylation. Complete sequencing of the rearranged 35Spro-NOSpro locus in 9NP plants showed it to contain two copies of the 35Spro-NOSpro gene that lacked NOSter sequences and were arranged as an inverted repeat (IR) with NOSpro sequences in the center. Only one of two 35Spro copies was complete and flanked by *lox* sites; the second copy was truncated and associated with only one *lox* site. In the progeny from the cross with the Cre plants, therefore, only the intact copy of the 35Spro was removed, whereas the remaining incomplete copy was still sufficient to transcribe NOSpro sequences and induce methylation and silencing of the target NOSpro-*nptII* gene. Both copies of 35Spro were inhibited upon crossing to the *271* line plants, this resulted in decreased or abolished NOSpro RNA synthesis, which was accompanied by reduced silencing and methylation of the target NOSpro-*nptII* gene. The most direct way how aberrant NOSpro transcripts could mediate methylation of the NOSpro-driven target genes seems to be a direct interaction of a diffusible NOSpro RNA with target promoters. A second possibility involving the DNA–DNA association between silencing and target NOSpro sequences seems to be ruled out by the observation that the NOSpro IR is still present at the *9NP* locus and retains its methylation status upon crossing to the 271 line, when the NOSpro-*nptII* target gene loses methylation and reactivates. Therefore, neither the NOSpro IR presence, nor its methylation, is sufficient for the *trans*-silencing ability of the *9NP* locus. The DNA pairing-mediated *de novo* methylation was postulated as a means to silence the multiple transgene copies integrated at the same locus, a mechanism that could prevent over-expression by controlling gene copy numbers. However, it is unknown, whether such homologous DNA pairing really occurs. Multiple transgene copies can be introduced into plant genomes and are actively expressed in most of the transgenic plants. In some relatively small subpopulation of transformants the *de novo* methylation and silencing of all transgene copies is triggered. On the other hand, RdDM and gene silencing seem to occur always, when significant quantities of aberrant RNAs are produced. Since gene methylation usually reinforces silencing, RdDM may represent a powerful mechanism for specific down-regulation of over-expressed genes. Because the aberrant RNA in 9NP line was synthesized from an inverted DNA repeat (IR) containing NOSpro sequences, one may well suggest that it must be double stranded to exert silencing effect. To test whether this RNA is indeed double stranded, RNA samples from silenced and non-silenced plants were treated with RNase I, an enzyme known to degrade single-stranded RNA, but not dsRNA (Mette et al. 2000). In non-silenced line the NOSpro RNA was rapidly degraded by this enzyme. In contrast, an RNase I resistant dsRNA of the expected size was present in plants from the silenced line 9NP. Such NOSpro dsRNA was not detectable in the original target line or in the presence of the 35S promoter suppressor *271* locus. If NOSpro dsRNA is really sufficient to induce silencing and methylation of the target NOSpro, then any NOSpro IR that is transcribed regardless of its location in the genome should act similarly to *9NP* locus. To test this, NOSpro IRs were created *in planta* at different genomic locations by site-specific recombination. A construct that contained a 35Spro-driven NOSpro direct repeat (DR), in which the second copy was flanked by two *lox* sites in an inverse orientation, was introduced into a tobacco line homozygous for the target NOSpro-*nptII* gene. The DNA blot analysis of three independent lines demonstrated the presence of the intact NOSpro DR that was unmethylated and actively transcribed. The presence of the transcribed NOSpro DR induced little or no methylation of the target NOSpro-*nptII* gene and did not affect its expression. After crossing to a plant line expressing the Cre recombinase that converted the NOSpro DR into a NOSpro IR, however, 50% of the progeny exibit methylation and silencing of the NOSpro-*nptII* target gene. Consistent with its suggested role in RdDM the NOSpro dsRNA was detected only in these silenced plants. Transcription through an IR produces a NOSpro RNA hairpin. To test whether open dsRNA would act as a *trans*-silencer, constructs designed to synthesize separate NOSpro sense and antisense RNAs were introduced into the same plant line homozygous for the NOSpro-*nptII* target locus (Mette et al. 2000). Three sense and four antisense lines were chosen and the presence of the respective NOSpro RNA in each line was confirmed by the RNase protection assay. Individually, these sense and antisense NOSpro RNAs did not trigger substantial silencing or methylation of the target NOSpro. No silencing was observed also in their intercrossed lines. This might have been due to inability of these sense and antisense NOSpro RNAs to "find" each other in the nucleus and form dsRNA. To overcome this limitation, the constructs were designed to synthesize overlapping NOSpro sense and antisense RNAs from the same locus. Again no *trans*-silencing was detected upon transformation of target tobacco lines with these constructs. Since dsRNA involved in PTGS in plants (and other species) is degraded to small (21-25 nt) RNAs, RNA samples from silenced and non-silenced plants were tested for the presence of such small RNAs. Indeed, both sense and antisense NOSpro 23-25 nt RNAs were detected in all silenced plant lines but not in the original target line or non-silenced lines. These small RNAs were no longer detectable after crossing to the 35S suppressive 271 line, which also repressed synthesis of the NOSpro dsRNA. NOSpro small RNAs were also detected in all silenced plants containing a transcribed NOSpro IR that had been created *in planta* by Cre recombinase but not in non-silenced plants harboring a transcribed NOSpro DR before Cre-mediated conversion. These results suggest that silencing and methylation of the NOSpro target promoter depend on synthesis of a NOSpro dsRNA that can be degraded to small RNAs in a manner similar to dsRNAs that induce PTGS. The size of these small RNAs approaches the lower limit of the DNA target length for RdDM (~30 bp) (Pélissier and Wassenegger, 2000). Small RNAs produced during PTGS appear to guide a dsRNA endonuclease to the homologous RNA and target it for degradation. It is conceivable that small RNAs also guide DNA-methyltransferase to homologous DNA sequences in the genome.

The identity of DNA-methyltransferases that are required for RdDM remained unknown. A second unanswered question was whether some specific alterations in chromatin structure are required to initiate and/or maintain RdDM-produced methylation. Since all relevant mutants are readily available in *Arabidopsis*, and the model of NOSpro-*nptII* silencing by

NOSpro IR dsRNA was successfully used in two *Arabidopsis* lines containing different NOSpro-*nptII* target genes (Mette et al. 2000), both questions could be evaluated by means of classical genetic analysis (Aufsatz et al. 2002a). A homozygous line that stably expresses a NOSpro-*NPTII* target gene was transformed with a 35Spro-NOSpro IR silencing construct. As it was anticipated, the silencing locus produces NOSpro dsRNA that is processed into the short ~21-24 nt RNAs similar to those observed in tobacco plant. In the presence of the silencing locus the target NOSpro-*NPTII* gene was efficiently inactivated at the transcriptional level. Transcriptional silencing of the NOSpro-*NPTII* target gene was accompanied by *de novo* methylation of the target NOSpro. When active, the target gene is normally unmethylated in the NOSpro region, as indicated by nearly complete digestion with the methylation-sensitive restriction enzymes *Sac*II, *Bst*UI, and *Nhe*I. In the presence of the silencing locus the NOSpro region becomes methylated at both symmetrical (CG and CNG) and nonsymmetrical (CNN) cytosines (Cs) as demonstrated both by resistance to the same restriction enzymes and by bisulfite DNA sequencing. Methylation did not spread significantly onto the *NPTII* coding sequences. Methylation was essentially eliminated, when the target and silencing loci segregate in progeny. Some residual methylation was probably caused by maintenance methylation at CG and/or CNG sites. The removal of the 35Spro with Cre recombinase fully eliminated the silencing potential of NOSpro IR. NOSpro short RNAs were no longer detectable, the target NOSpro-*NPTII* gene remained active in the presence of such "disarmed" NOSpro IR, and methylation of the NOSpro-*NPTII* target gene was reduced by ~30% at symmetrical cytosines in the *Sac*II and *Bst*UI sites and almost completely at nonsymmetrical C residues in the *Nhe*I site. The NOSpro dsRNA not only triggers methylation and silencing of the target NOSpro *in trans* but also methylation in *cis* of the NOSpro copies in the IR at the silencing locus itself. This was demonstrated by examining the methylation of the NOSpro IR before and after removal of the active 35Spro by Cre recombinase. The transcribed NOSpro IR in the unaltered silencing locus is heavily methylated at both symmetrical and nonsymmetrical Cs within the repeated region. On the contrary, the nontranscribed NOSpro IR with 35Spro removed is largely demethylated at nonsymmetrical C residues, whereas the methylation at symmetrical C residues is almost completely retained. To ascertain the effects of several mutations, known to affect gene silencing, on the NOSpro dsRNA-mediated TGS system, the double homozygous target/silencer line was crossed with lines homozygous for *ddm1*, *met1* and *mom1* recessive mutations. The *mom1* mutation was the only one that affects neither NOSpro silencing nor methylation of the target NOSpro. The *met1* mutation partially released silencing of the NOSpro-*NPTII* gene in F_2 progeny. In the first generation of crosses with the *ddm1* mutant the sporadic weak reactivation of the NOSpro-*NPTII* gene expression was observed. In both *met1* and *ddm1* mutants the strength of the NOSpro-*NPTII* expression improved in advanced generations, although continued to be nonuniform in the genotypically identical seedlings. The strongest expressing plants sustained significant losses of methylation from the target locus. These plants were homozygous for the respective mutations that caused global DNA demethylation. In contrast to the substantial reduction in methylation of the NOSpro target locus in such plants, the NOSpro IR at the silencing locus retains a considerable methylation. This was particularly evident in the *ddm1* mutant plants, where, similarly to wild type plants, virtually no digestion of the NOSpro IR by *Sac*II, *Bst*UI and *Nhe*I was observed. In *met1* plants, the methylation was reduced by ~20-30% at both symmetrical (*Sac*II, *Bst*UI) and nonsymmetrical (*Nhe*I) sites. NOSpro dsRNA continues to be synthesized at wild-type levels in the *met1* and *ddm1* mutant plants. Thus, NOSpro dsRNA induces *de novo* methylation of the target NOSpro at cytosines in any sequence context within the region of the RNA-DNA sequence identity. Removing the source of the dsRNA by either segregating away the silencing locus or its inactivation by removing 35Spro via Cre/*lox*-mediated recombination results in nearly complete loss of methylation at cytosines in nonsymmetrical sites, indicating that continuous *de novo* methylation at such sites is required. On the contrary, methylation at symmetrical CG and CNG sites can be maintained probably by the DNA-methyltransferases MET1 and CMT3, respectively. The *met1* and *ddm1* mutations, which reduce global methylation, partially alleviate silencing and reduce methylation of the NOSpro-*NPTII* target gene. In both mutants, losses of methylation in the target NOSpro can be substantial in F_3 and F_4 progeny that show the strongest expression of NOSpro-*NPTII.* Any slight methylation that persists is presumably caused by continued *de novo* methylation induced by NOSpro dsRNA. Two copies of the NOSpro in the IR at the silencing locus are methylated substantially at symmetrical and nonsymmetrical C residues. When dsRNA synthesis terminates following Cre-mediated removal of the 35Spro, methylations at CG and CNG are mainly retained, whereas methylation at nonsymmetrical sites is substantially reduced. After withdrawal of the silencer dsRNA the nontranscribed NOSpro IR maintains methylations at CG and CNG sites better than singlet copies of NOSpro at the target locus; it suggests that some intrinsic feature of the IR helps to maintain methylation at symmetrical sites. One possibility is that pairing of the IR in *cis* generates an unusual structure that is recognized by the maintenance DNA-methyltransferase activities. A general conclusion from this study is that both MET1 and DDM1 are required for an efficient maintenance of the RdDM-induced methylation and silencing. Despite the continued presence of the NOSpro dsRNA, significant losses of target NOSpro methylation occur after several generations in *met1* and *ddm1* mutants. The evident question unanswered in this study remained which DNA-methyltransferase catalyzes the *de novo* methylation step of RdDM. MET1 appears not to be the one, since the effects of *met1* mutations are somewhat delayed and partial. In an independent study, *Nicotiana benthamiana* lines carrying a single copy of a 35S-GFP transgene were infected with tobacco rattle virus (TRV) modified to carry the 3′-359 nucleotides of GFP (TRV-P), 347 nucleotides of the 35S promoter sequence (TRV-35S), or TRV with no additional insert (TRV-00) (Jones et al. 2001). Systemic infection with TRV-P and TRV-35S led to silencing of GFP (loss of green fluorescence), whereas TRV-00 did not affect GFP expression. Northern blot analysis of the GFP mRNA levels confirmed the visible silencing phenotypes. Although silencing of GFP could be achieved by targeting either transcribed or nontranscribed portions of the 35S-GFP transgene, the runoff transcription analyses showed that the reduced GFP mRNA accumulation in TRV-P-infected plants is due to posttranscriptional gene silencing (PTGS), whereas in TRV-35S-infected plants the reduction is at the level of the 35S-GFP transgene transcription. Furthermore, all progeny of selfed TRV-P-silenced plants were green fluorescent to the same extent as progeny of TRV-00-infected plants, indicating that the PTGS induced by TRV-P is not inherited. In contrast, the progeny of TRV-35S-silenced plants were red fluorescent, indicating that RNA-induced transcriptional silencing can be inherited. The inheritance of TRV-35S-induced silencing was not due to seed transmission of the virus, since viral RNA in progeny plants was not detectable. The young progeny seedlings from the F_1 generation of TRV-35S-infected plants were red fluorescent. Approximately 30% of these F_1 plants remained fully silenced during development, whereas the others reverted to producing nonsilenced green fluorescent leaves. The transition from silencing to nonsilencing was not associated with developmental sectors or sharp boundaries. The F_1 plants, that maintained a fully silenced red fluorescent phenotype throughout development, produced silenced F_2 progeny plants that were similar to their parents, namely, some of them remained fully silenced, whereas others reverted to the nonsilenced state. The F_2 progeny of F_1 plants that had reverted to green fluorescence were likewise all green fluorescent. To test the ability of a silenced 35S-GFP allele to *trans*-silence nonsilenced alleles, a series of crosses with silenced (S) and nonsilenced (NS) plants were carried out. TRV-35S-infected and -silenced plants or silenced F_1 progeny (F_1-S) were crossed in a reciprocal manner with nonsilenced plants. The progeny of five F_1-S × NS crosses were all nonsilenced, whereas eight crosses using a primary infected plant as a parent all produced both silenced and revertant nonsilenced progeny. Thus, *trans*-silencing requires a factor that is present in the TRV-35S-infected plants but absent in the F_1 silenced progeny. As it was inferred from digestion with methylation-sensitive restriction enzymes that cut within the 35S promoter and the GFP sequence, methylation in the 35S promoter of TRV-35S-infected plants is by ~40–50 times more than that in DNA from nonsilenced TRV-00-infected plants or PTGS-silenced TRV-P-infected plants, at both symmetrical (*Mae*II) and nonsymmetrical sites (Sau96I). *Vice versa*, GFP methylation was 23–70 times higher in samples from TRV-P-infected plants than in samples from TRV-00- or TRV-35S-infected plants. Thus, sequence-specific RNA-directed methylation of the 35S promoter can be detected in plants infected with TRV-35S, whereas methylation of the GFP sequence can be detected in plants infected with TRV-P. The methylation status of the 35S promoter and GFP sequences in the progeny of TRV-35S- and TRV-P-infected plants was also studied. For tissue prepared from F_1 plants that remained fully silenced the identical methylation patterns to those of the primary infected plants were observed for cytosine residues at symmetrical sites (*Mae*II and *Hga*I) in 35S promoter, indicating that methylation patterns at these sites are inherited. In contrast, the methylation patterns of cytosines at nonsymmetrical sites (*Sau96*I and *Xmn*I) in these plants were identical to those of nonsilenced plants. Thus, nonsymmetrical type of methylation in the primary infected plants is not maintained in the next generation. The F_1 progeny that reverted to a nonsilenced state showed the nonsilenced 35S methylation patterns. For TRV-P-infected plants the GFP-specific DNA methylation was only observed in the primary infected plants, and neither symmetrical nor nonsymmetrical methylation was passed to the progeny. Thus, for TGS the DNA methylation at symmetrical sites is inherited and correlates with silencing, whereas for PTGS, although GFP-specific DNA methylation is detected in the primary infected plants, it is not passed to the next generation. To examine the role of the maintenance DNA-methyltransferase MET1 in inheritance of RNA-triggered TGS, the TRV-induced silencing of *MET1* gene was used. A 180 bp fragment of the *N. benthamiana MET1* gene was cloned into the TRV vector, and the construct was used to infect silenced F_1 progeny. TGS reversal was clearly observed in these TRV-*MET1*-infected plants. This correlated with apparent demethylation of 35S promoter as well as other sequences of genomic DNA. The role of MET1 in the initiation of RdDM was addressed by coinoculating nonsilenced 35S-GFP transgenic plants with PVX-35S or PVX-P in combination with TRV-00 or TRV-*MET1*. Silencing of GFP initiated by PVX-35S or PVX-P was clearly visible in the newly emerging leaves, though general hypomethylation effects of TRV-*MET1* infection on genomic DNA sequences was quite clear-cut. The 35S and GFP sequences, in contrast, were found to be equally methylated in the presence of TRV-*MET1* or TRV-00, indicating that MET1 does not affect initiation of RNA-directed methylation. Thus, MET1 seems to be essential for the maintenance of gene silencing caused by RdDM but not for the initiation of RdDM.

To study the role of other DNA-methyltransferases in the maintenance of RdDM, a triple-mutant *drm1 drm2 cmt3* plant was crossed to a line homozygous for both silencer 35Spro-NOSpro IR and target NOSpro-*nptII* transgenes (Cao et al. 2003). F_1 plants were allowed to self pollinate, and F_2 progeny plants were screened using PCR-based molecular markers to identify 35Spro-NOSpro IR /35Spro-NOSpro IR NOSpro-*nptII*/NOSpro-*nptII* lines with no methyltransferase mutations, with *drm1 drm2*, with *cmt3* and with *drm1 drm2 cmt3*. These F_2 plants were allowed to self pollinate, and DNA was extracted from the F_3 plants for methylation analysis. The methylation patterns of the *NOSpro:NTPII* target locus were studied in each of these genotypes by bisulfite genomic sequencing. Since silencer 35Spro-NOSpro IR and target NOSpro-*nptII* transgenes had been together before crossing to the methyltransferase mutants, this experiment measured the effect of the methyltransferase mutations on the maintenance of preexisting RdDM. Consistent with MET1 function as the primary maintenance CpG-methyltransferase, the CpG methylation of *NOSpro:NTPII* was not reduced in the *drm1 drm2*, *cmt3*, or *drm1 drm2 cmt3* triple-mutant plants. The *drm1 drm2* double mutants and *cmt3* single mutants showed little reduction in DNA methylation at CpNpG sites, whereas in the *drm1 drm2 cmt3* triple mutant CpNpG methylation was completely lost. Thus, *DRM* and *CMT3* act redundantly to maintain RNA-directed CpNpG methylation. For cytosines in asymmetric sequence contexts, the *drm1 drm2* plants showed a major loss of methylation, while *cmt3* single mutant plants did not show a reduction. However, the residual 3% asymmetric methylation remaining in *drm1 drm2* double mutants was completely eliminated in the *drm1 drm2 cmt3* triple mutant plants. Thus, *DRM* and *CMT3* also act redundantly to maintain RNA-directed asymmetric methylation. The levels of *NPTII* mRNA produced in the *drm1 drm2*, *cmt3* and *drm1 drm2 cmt3* lines were measured by RT-PCR. A small level of *NOSpro:NPTII* reactivation was found. However, it was much lower than that in *NOSpro:NPTII* plants that did not contain the silencing 35Spro-NOSpro IR transgene. The *cmt3* single mutant causes reactivation at even lower level than *drm1 drm2* double mutant and *drm1 drm2 cmt3* triple mutant. These data suggest that *drm*, and to a lesser extent *cmt3*, can weakly reactivate NOS promoter. Probably, the remaining CpG methylation in these mutants can largely maintain gene silencing.

To investigate the relationship between siRNAs and the function of DNA-methyltransferase genes, the steady-state levels of siRNAs were measured in homozygous 35Spro-NOSpro IR/NOSpro-*nptII* plants, wild-type for DNA-methyltransferase genes and their *drm1 drm2 cmt3* siblings (Cao et al. 2003). NOSpro siRNA was readily detectable in both genotypes. Interestingly, it was increased in abundance in the *drm1 drm2 cmt3* triple mutant background. Thus, mutations affecting transcriptional gene silencing can cause feedback upregulation of siRNA accumulation. It was tested whether *drm1 drm2* or *cmt3* mutants would block the initiation of DNA methylation of the target transgene that normally occurs when the silencer 35Spro-NOSpro IR and target NOSpro-*nptII* transgenes are first brought together in a cross. RdDM can occur within one generation (Aufsatz et al. 2002a). Thus, lines homozygous for either *drm1 drm2* or *cmt3* in the 35Spro-NOSpro IR or NOSpro-*nptII* backgrounds were constructed. Such 35Spro-NOSpro IR *drm1 drm2* plants were then crossed with NOSpro-*nptII drm1 drm2* plants, 35Spro-NOSpro IR *cmt3* plants with NOSpro-*nptII cmt3* plants, and as a control 35Spro-NOSpro IR plants with the NOSpro-*nptII* plants. In the F_1 generation of these crosses the methylation patterns of the NOSpro*:nptII* target gene were examined by bisulfite genomic sequencing. Since the target NOSpro had no methylation before exposure to the silencer transgene, all methylation observed in the F_1 generation represented *de novo* RdDM. In the control plants the NOSpro region became methylated at both symmetrical (CpG and CpNpG) and asymmetrical sites. In the *cmt3* plants the methylation was also observed in all sequence contexts but it was significantly lower than in the control plants. In the *drm1 drm2* plants, no methylation was detected in any sequence context. This suggests that the *DRM* genes are required for the establishment of RdDM. The F_2 progeny resulting from self pollination of the F_1 plants was also studied. In the F_2 control plants, higher levels of CpG methylation were observed than in F_1 plants showing that full establishment of CpG methylation is progressive. In the *cmt3* homozygous plants the DNA methylation levels were similar to but slightly lower than in the control. This suggests that while full levels of RdDM are delayed to appear in the *cmt3* mutant, *CMT3* is not strictly required for establishment of RdDM. In the *drm1 drm2* plants no methylation in any sequence context was detected again showing that DRM activity is absolutely required for the initiation of RdDM.

The role of different DNA-methyltransferases in establishement and maintenance of RNA-directed DNA methylation, therefore, may be summarized as follows. The dsRNA-dependent *de novo* DNA methylation activity of DRM methyltransferases is absolutely required for initial establishement of RdDM in all sequence contexts. Both MET1 and CMT3 methyltransferases seem to be non-essential at this step. Maintenance of CpG methylation can occur in the absence of the triggering RNA signals and is dependent on the activity of MET1 exclusively. For the maintenance of CpNpG and asymmetric methylations, both DRMs and CMT3 are required. DRMs act redundantly with CMT3 in their maintenance capacity, since CpNpG and asymmetric methylations are only totally lost, when both gene types are mutated. The maintenance phase for these non-CpG methylation types seems to consist of persistent dsRNA-dependent *de novo* activity of DRMs and CMT3. Nevertheless, it is clearly distinct from the initiation phase, since DRMs alone are strictly required for the latter one.

The *DRM* and *CMT3* genes are required for non-CpG methylation at all loci that have been tested including endogenous genes such as *SUPERMAN*, *FWA*, and *MEDEA* (Cao, Jacobsen, 2002a), endogenous transposon sequences such as *AtSN1*, *AtMu1*, and *Ta3* (ibid; Zilberman et al. 2003) and at the NOSpro sequences just reviewed. In the last case there is a clear source of double-stranded RNA, which is required for the non-CpG methylation. There are some indications that such dsRNAs may play a role in non-CpG methylation at endogenous loci as well. For instance, the *AtSN1* retrotransposable elements are associated with 25 nt siRNAs, and the loss of these siRNAs correlates with a reduction of non-CpG

AtSN1 methylation (Hamilton et al. 2002; Zilberman et al. 2003). Full levels of non-CpG methylation at *SUPERMAN*, *MEDEA*, *AtSN1* and *AtMu1* depend on the activity of ARGONAUTE4, a protein normally associated with RNA interference and microRNA pathways (Zilberman et al. 2003). In the *PAI* gene silencing system, non-CpG methylation of the *PAI2* locus depends on transcription of the inverted repeat containing *PAI1-4* locus (Melquist, Bender, 2003). Thus, non-CpG methylation may largely, if not totally, be directed by dsRNA.

RNA-directed DNA methylation of a tissue-specific promoter was studied in a two-component transgene system consisting of a α'-GFP target construct (α′ is seed-specific promoter from the gene encoding the α′ subunit of a soybean seed storage protein β-conglycinin) and a 35S-α′proIR silencer construct containing α′ promoter fragment in the sense and antisense orientation with respect to 35S promoter. The two α′ promoter fragments are separated by a 298 bp spacer containing the NOSpro promoter fragment in the sense orientation with respect to 35S promoter (Kanno et al. 2004). Thus, silencing and methylation of an α′*GFP* reporter gene were triggered by the α′ promoter hairpin RNA that was transcribed from the inverted DNA repeat. To identify the proteins involved, the seeds of a homozygous silenced α′GFP line (DT7-3) were mutagenized by EMS, germinated, and the resulting F_1 plants were selfed to produce F_2 seeds. Silencing-defective (green fluorescent) mutants were selected and proved to belong to three complementation groups defective in RNA-directed DNA methylation (*drd* mutants). All these mutants were recessive, since resilencing of the α′*GFP* target gene readily occurred upon backcrossing to the wild-type DT7-3 plants. First group (*drd1*) appeared not to be defective in the synthesis of α′ promoter double-stranded RNA or its processing to α′ promoter short RNAs. Both methylation-sensitive restriction enzymes and bisulfite sequencing revealed a dramatic decrease in CpNpG and CpNpN methylations of the target α′ promoter in the *drd1* mutant, whereas CpG methylation was unaffected. Thus, non-CpG methylation induced by RNA in the α′-promoter silencing system requires DRD1. By contrast, neither CpG nor non-CpG methylation was detectably reduced in centromeric and rDNA repeats in the *drd1* mutant. Thus, DRD1 acts locally to regulate levels of non-CpG methylation. The *DRD1* mutation was found to code for a putative chromatin remodeling protein CHR35, a member of a plant-specific SNF2-like protein subfamily. All *drd1* mutations identified were found to affect a strongly conservative region of the SWI/SNF ATPase domain. Thus, similar to DDM1, DRD1 is another SNF2-like protein important for DNA methylation in plants.

To find out whether DRD1 is needed for RNA-directed *de novo* methylation of target sequences or for the maintenance of this methylation in the absence of the trigger RNA, F_1 plants were produced by crossing respective lines containing the target α′ promoter complex to lines containing the silencer complex encoding the α′ promoter dsRNA, in either wild-type (*D/D*) or homozygous *drd1* (*d/d*) background (Kanno et al. 2005a). In wild-type F_1 plants, the target α′ promoter had increased methylation in CGs and in non-CGs after introducing the silencer complex. The level of methylation observed in wild-type F_1 plants was similar to that seen in plants, in which the target complex and silencer complex had been together in the same genome for several generations. Thus, the maximum attainable level of RNA-directed methylation of the target α′ promoter is essentially reached in the first generation containing both transgene complexes. By contrast, the target α′ promoter did not acquire detectable methylation after being combined with the silencer complex in homozygous *drd1* plants, though the production of α′ promoter short RNAs in these plants was quite normal. Similar results were obtained with a systems of RNA-mediated silencing and methylation of the constitutive nopaline synthase (NOS) promoter (Aufsatz et al. 2002b). The only significant difference was that, in contrast to the α′ promoter, the level of NOS promoter methylation was less in F_1 progeny than in plants that have possessed the target and silencer for several generations. The efficiency of maintenance methylation was examined in wild-type and *drd1* plants after crossing out the respective silencer *H* complexes to remove the source of the RNA signals. In wild-type F_2 progeny descended from DRD1 parents the target α′ promoter lost both CG and non-CG methylations after segregating away the silencer complex. An identical pattern of methylation was observed in the wild-type F_2 progeny descended from the *drd1* mutant, except for some residual methylation at two CG dinucleotides. Thus, in the α′ promoter silencing system almost all methylation is lost in wild-type progeny, when the source of the RNA signal is withdrawn. Unexpectedly, however, in *drd1* progeny lacking the silencer complex the substantial CG methylation was detected even though non-CG methylation was lost. Similar but not identical results were obtained with the NOS promoter system. In contrast to α′ promoter, the target NOS promoter retained significant methylation in the absence of the silencer complex in wild-type plants: although non-CG methylation was lost, considerable CG methylation remained. Similarly to the α′ promoter, however, the NOS promoter showed increased CG methylation in *drd1* progeny. These findings show a previously unsuspected role of DRD1 in the complete erasure of CG methylation after segregating away the silencer complex that encodes the RNA trigger. Thus, DRD1 seems to have a dual role. First, it is required for RNA-directed *de novo* methylation of Cs in all sequence contexts including CG dinucleotides. Second, DRD1 is also necessary for efficient loss of methylation, particularly CG-type, once the source of the RNA signal is removed. Since all the *drd1* alleles identified contain mutations in functionally implicated regions of the SWI2/SNF2 ATPase domain, DRD1 seems to function as a chromatin-remodelling protein to disrupt histone–DNA contacts and/or displace nucleosomes. One possibility is that DRD1 is specialized to allow RNA signals to access homologous target DNA in a chromatin context. Depending on the availability of RNA signals and various DNA-modifying enzymes in different cell types, the DRD1 activity could facilitate RNA-guided *de novo* methylation catalyzed by DNA-methyltransferases or demethylation of CG dinucleotides catalyzed by DNA glycosylases.

DNA methylation at asymmetric sites, as it was already noted, is mostly controlled by the DNA-methyltransferase DRM2, which is targeted by short 24-nucleotide-long interfering RNAs (siRNAs) produced through RNA interference pathways. As a matter of fact, both siRNAs and DRM2 are indispensable for the initial *de novo* DNA methylation in all sequence contexts. Another proteins needed are two forms of the plant specific nuclear DNA-dependent RNA polymerase IV, whose subunits are encoded by *NRPD2a* (*DRD2*) and *NRPD1a* (*SDE4*), *NRPD1b* (*DRD3*) genes, the RNA-dependent RNA polymerase 2 *(RDR2)*, DICER-LIKE3 *(DCL3)* and ARGONAUTE4 *(AGO4)* (Chan et al. 2004, 2006a,b; Herr, 2005; Kanno et al. 2005b). RNA polymerase IVa seems to be necessary for initial transcription of endogenous loci to be silenced such as transposones or repeated sequences. The transcripts produced are further converted to dsRNAs by the RDR2 and processed to siRNAs by DCL3

(Herr et al. 2005). Both RNA polymerase IVb and DRD1 are required for the RNA-directed DNA methylation downstream of siRNA. Argonaute proteins associate with the RNA silencing effector complexes.

Whether DRD1 acts through the DRM2 and/or CMT3 methyltransferases in its control of non-CG methylation was tested by the effect of *drd1* mutation on the maintenance DNA methylation at different endogenous loci (Chan et al. 2006b). At the endogenous direct repeats present at *FWA* and *MEA-ISR*, the *drd1* lacked all non-CG methylation but did not affect CG methylation. At the SINE element *AtSN1,* the *drd1* mutant also lacked all non-CG methylation. This suggests that DRD1 can act through DRM2 and CMT3 that latter both have a redundant action at endogenous genes.

The DRD1-dependent non-CG DNA methylation at *AtSN1, FWA* and *MEA-ISR* is associated with the presence of endogenous siRNAs corresponding to these loci. In contrast, the pericentromeric retrotransposon *Ta3* lacks siRNAs (Lu et al. 2005), its CpNpG methylation does not depend on *AGO4* (Zilberman et al. 2003) and depends solely on the CMT3 DNA-methyltransferase (Cao, Jacobsen, 2002). Importantly, *drd1* mutants showed no defect in CNG methylation at *Ta3*. Thus, *Ta3* is a locus, where CMT3 maintains CNG DNA methylation independent of siRNAs and DRD1.

As it was already noted, the *drm1 drm2 cmt3* triple mutants display a pleiotropic set of developmental abnormalities that, unlike those observed in *met1* mutants, are mainly homogeneous and not intensified during successive generations of inbreeding (Cao, Jacobsen, 2002). Three major defects in such plants are twisted leaf shape, shorter stature and partial sterility. Further difference from *met1* phenotypes is that all these defects are entirely recessive and disappear upon crossing to wild type plants. Both *DRM2* and *CMT3,* when introduced to the *drm1 drm2 cmt3* triple mutants by *Agrobacterium*-mediated plant transformation, completely restored normal phenotype. Thus, the active signals that target non-CG DNA methylation are still present in the *drm1 drm2 cmt3* triple mutant. This restoration is consistent with a model, in which *drm1 drm2 cmt3* developmental phenotypes result mostly from genes that are overexpressed when silencing-associated non-CG methylation is lost. The plant defective for both RNA polymerase IV (*nrpd2a nrpd2b*) and CMT3 (*cmt3*) showed a developmental phenotype identical to that of *drm1 drm2 cmt3* triple mutants (Chan et al. 2006a,b). The same was found for the *drd1 cmt3* double mutants. Thus, the mutations in NRPD2 and DRD1 show the same effect as DRM2 mutation, when combined with mutation of CMT3. On the contrary, both the *nrpd2a nrpd2b drm1 drm2* quadruple mutants and the *drd1 drm1 drm2* triple mutants have a wild-type morphological phenotype. Thus, the developmental gene regulation by DRM2 requires RNAi and the RNA-directed DNA methylation factor DRD1. DRD1 does not control all developmental regulation by DRM2 and CMT3, however, because the single *drd1* mutant has a wild-type morphological phenotype. This contrasts to *AtSN1* non-CG methylation, where *drd1* has the same effect as *drm1 drm2 cmt3*. Since DRM2 requires DRD1 to establish and maintain DNA methylation at all loci tested, one should assume that CMT3 has a DRD1-independent targeting pathway as exemplified by CNG methylation at the *Ta3* retrotransposon. To say it in a more simple way, the control of normal gene expression by CMT3 is not solely directed by RNAi. Unlike the *drm1 drm2 nrpd2a nrpd2b* plants, the *drm1 drm2 kyp* plants displayed developmental abnormalities very similar to *drm1 drm2 cmt3*. This suggests that the loss of

KYP-mediated H3Lys9 methylation has the same effect as the loss of CMT3, when combined with mutations in *DRM* genes. Thus, both RNAi pathways and the histone H3Lys9 methylation can target non-CG DNA methylation to developmentally important genes (Figure 2).

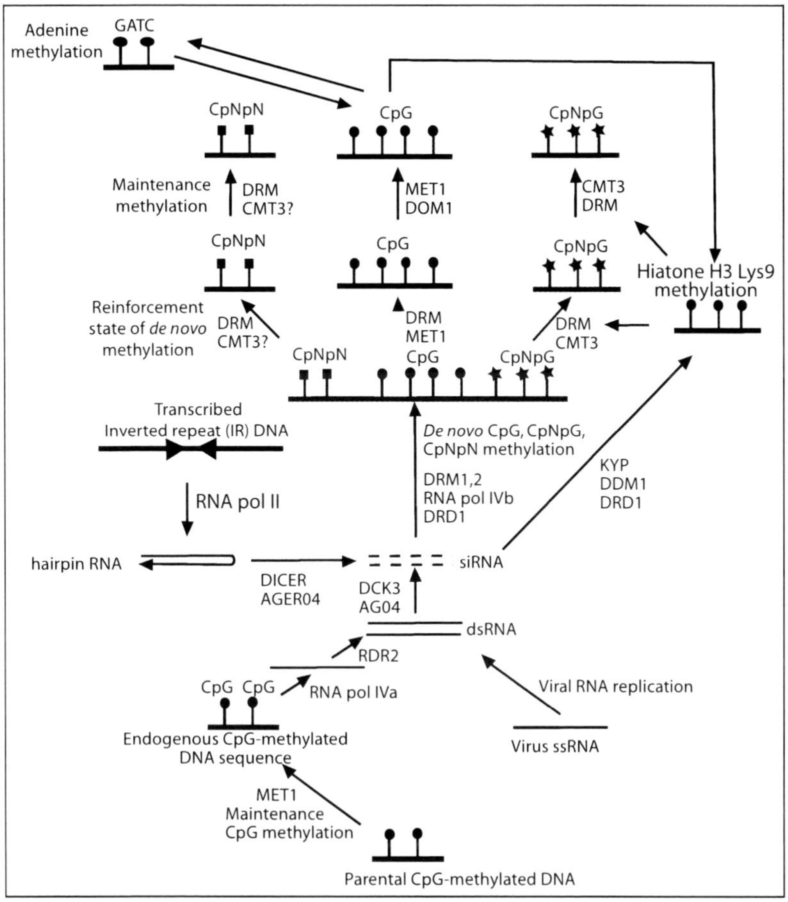

Figure 2. DNA and histone methylation events involved in plant epigenetics. The *de novo* DNA methylations at different sequence contexts seem to be mainly targeted by siRNA, though RNA independent pathway targeted by histone H3Lys9 methylation seems to exist at some DNA loci. Histone H3Lys9 methylation is targeted both by CpG specific DNA methylation and siRNA.

Several targeting pathways seem to exist that control the locus-specific propagation of the non-CG DNA methylation patterns. In one the 24-nucleotide siRNA pathway acts together with DRD1 to target the DRM2 DNA-methyltransferase. Certain loci, like *FWA* and *MEA-ISR*, appear to only use this pathway, since all non-CG methylation is lost at these loci in the RNAi mutants and in the *drd1* and *drm2* mutants. Other loci, such as *AtSN1,* appear to use a combination of the RNAi/DRD1/DRM2 pathway and a second pathway, in which CMT3 is guided by histone methylation through KYP. DRD1 seems to act in both pathways, which would explain why DRD1 can facilitate non-CG DNA methylation by both DRM2 and CMT3, even though these enzymes have locus-specific effects. Its chromatin remodeling activity may be necessary for both DRM2 and CMT3 to methylate nucleosomal DNA *in vivo.*

In yet a third pathway exemplified by the *Ta3* locus, CMT3 propagates CNG DNA methylation without siRNAs or DRD1.

As it was already noted, *cmt3* and *kyp* mutations were found among suppressors of the *clk-st* allele, a stably silenced form of *SUP* gene in *Arabidopsis thaliana*. Thus, both *CMT3* and *KYP* are required for the maintenance of silencing. *CMT3* encodes a DNA-methyltransferase, and *KYP* encodes a histone H3Lys9-specific methyltransferase (Lindroth et al. 2001; Jackson et al. 2002). Both *kyp* and *cmt3* mutants cause a loss of CpNpG methylation at *SUP* and all other loci tested. A third *clk-st* suppressor mutation appeared to be an allelic form of the *AGO4* gene coding for a member of the ARGONAUTE (AGO) protein family (Zilberman et al. 2003). The proteins of this family are known to be important in RNA-mediated silencing systems such as posttranscriptional gene silencing (PTGS) in plants, RNA interference in animals and quelling in fungi (Fagart et al. 2000; Carmell, 2002). A recessive allele of a *clk-st* suppressor gene was mapped to chromosome 2. By sequencing candidate genes, a mutation in the *AGO4* gene previously named on the basis of its sequence similarity to *AGO1* (Fagart et al. 2000) was identified. To confirm that the suppressor mutation is within *AGO4*, the mutant plants were transformed with *AGO4* gene and the original *clk-st* phenotype was restored. The effect of the *ago4* mutation on *SUP* DNA methylation was analyzed by bisulfite genomic sequencing. The mutant showed a 2.8-fold reduction in CpNpG and a 4.5-fold reduction in asymmetric cytosine methylation, whereas CpG methylation levels were unchanged. This methylation phenotype is strikingly reminiscent of that seen in *cmt3* and *kyp* mutants, except that *cmt3* and *kyp* showed a stronger reduction of CpNpG methylation than did *ago4*. It was previously found that *cmt3* and *kyp* showed a reduction in CpNpG but not CpG methylation at all loci tested (Lindroth et al. 2001; Jackson et al. 2002). Therefore, the effect of *ago4* on both CpG and CpNpG methylations was tested at three additional loci: the 180 bp centromeric repeat (*CEN*) sequence, the *Ta3* retrotransposon and the *FWA* gene. The *ago4* mutation did not affect either CpNpG or CpG methylation levels at these loci. The *FWA* locus also has a substantial degree of asymmetric methylation, and the bisulfite sequencing of *FWA* showed that the *ago4* mutation did not reduce this methylation. Thus, the methylation phenotype of *ago4* is locus-specific and different than that of the *cmt3* and *kyp* mutants. Three other loci, where *ago4* did influence DNA methylation, are *MEA-ISR*, *AtSN1* and *AtMu1*. *MEA-ISR* is an approximately 183 bp sequence present in seven direct repeats in an intergenic region adjacent to the imprinted *MEDEA* gene (Cao, Jacobsen, 2002). In the wild type the MEA-ISR locus contains cytosines methylated at 95% CpG, 58% CpNpG and 26% asymmetric sites. The *ago4* mutation essentially eliminated the CpNpG and asymmetric methylations but did not affect the CpG methylation at this locus. *AtSN1* is a retrotransposon sequence previously shown to be methylated (Hamilton et al. 2002). The wild-type *AtSN1* locus contains cytosines methylated at 75% CpG, 70% CpNpG, and 24% asymmetric sites. The *ago4* mutation greatly reduced the non-CpG methylation to 14% CpNpG and 0.8% asymmetric sites. The *AtMu1* sequence is the 3'-terminal inverted repeat of the *Arabidopsis* DNA transposon *Mu1* (Singer et al. 2001). The wild-type *AtMu1* shows 58% CpG, 35% CpNpG and 11% asymmetric methylations. The *ago4* mutation did not affect the CpG methylation but reduced the CpNpG methylation to 19% and the asymmetric methylation to 4.8%.

The locus-specific effects of *ago4* show that both *AGO4* dependent and independent mechanisms control the non-CpG methylation. *CEN*, *Ta3*, and *FWA* rely on an *AGO4* independent mechanism, *MEA-ISR* on an *AGO4* dependent mechanism, and *SUP*, *AtSN1* and *AtMu1* on both mechanisms. One explanation for *AGO4* independent non-CpG methylation is that another *AGO* gene (nine are present in the *Arabidopsis* genome) could act redundantly with *AGO4*. Alternatively, pathways that do not involve *AGO* genes could function at some loci. A comparison of the methylation phenotype of *ago4* with those of mutants of *CMT3* and *DRM* did not reveal any simple relationship: at *MEA-ISR ago4* mimicked the *drm1 drm2* double mutant, at both *SUP* and *AtSN1* it showed a reduction in the CpNpG methylation intermediate between the effects of the *cmt3* and *drm1 drm2* mutants and a reduction of asymmetric methylation that was stronger than the effect of either. These results suggest that both *CMT3* and *DRM* are involved in *AGO4* dependent methylation. It is known that *kyp* but not *cmt3* reduces H3Lys9 methylation at *SUP* (Johnson et al. 2002). Apparently CMT3 is targeted to H3Lys9-methylated chromatin regions, thus acting downstream of KYP (Jackson et al. 2002). The *ago4* mutation reduces H3Lys9 methylation at *SUP* relative to the wild-type strain *clk-st*. The simplest interpretation of these results is that AGO4 acts upstream of KYP to target H3Lys9 methylation. The *ago4* reduces H3Lys9 methylation at *AtSN1*, a locus where *ago4* also reduces DNA methylation. However, *ago4* does not reduce H3Lys9 methylation at *Ta3* or at the *CEN* repeats, where it shows no DNA methylation effects. Thus, the effects of *ago4* on H3Lys9 methylation are locus-specific and correlate with effects on DNA methylation.

Whether AGO4 function is associated with siRNAs was tested by probing Northern blots of RNA preparations that had been enriched with small RNAs. *AtSN1* is associated with a newly discovered class of long (approximately 25 nt) siRNAs (Hamilton et al. 2002). Such RNA could be easily detected in the wild-type Ler or *clk-st* strains and in the *cmt3* or *kyp* mutant strains. However, these siRNAs were reduced to below the level of detection in *ago4* mutant.

In order to learn, which genetic loci are required for transgene-induced PTGS in plants, a mutation analysis was carried out in *Arabidopsis* carrying two *GFP* transgenes (Dalmay et al. 2000a). Two parent lines of transgenic *Arabidopsis* plants were crossed to produce a *GFP*-silenced hybrid line. The first one served as a reporter line; it contained an actively expressed transgene construction 35S-*GFP* (plants were green fluorescent under UV-light). The second line served as a silencer one; it contained a potato virus X:*GFP* transgene, 35S-*PVX:GFP*. The hybrid line homozygous for both transgenes exhibited strong PTGS manifested as almost complete absence of *GFP* RNA (plants were red fluorescent under UV-light due to chlorophyll). At least, four loci are required for this transgene-induced PTGS. Mutations of two of them (designed *sde1* and *sde2* for <u>s</u>ilencing <u>de</u>fective) lead to a full green phenotype: plants are green fluorescent in all tissues and throughout development. Mutations of another one, *sde3*, lead to a delayed green phenotype: the hypocotyl and cotyledons are red fluorescent as in the wild-type plants, whereas the true leaves are green fluorescent though slightly less intense as compared to plants with *sde1* and *sde2* mutations. Mutations of the forth locus, *sde4*, lead to a transient green phenotype: the young newly emerging leaves are green fluorescent, whereas mature leaves are red fluorescent. In nuclear runoff analysis the rates of the 35S-*GFP* transgene transcription in the *sde1*, *sde2* and *sde3* mutant lines were the same as in their parental hybrid line, whereas the steady state levels of the *GFP* and *PVX:GFP* RNAs were much higher (Nothern blot hybridization). The *GFP* RNA from the 35S-*GFP* transgene was as abundant in the mutant lines as in the nonsilenced 35S-*GFP* line. The level of *PVX:GFP* RNA from the 35S-*PVX:GFP* transgene was higher than in the parental 35S-*PVX:GFP* line. Thus, *SDE* loci seem to encode factors required for PTGS. In silenced hybrid line the PTGS is associated with partial methylation of both 35S-*GFP* and 35S-*PVX:GFP* transgenes and appearance of a discrete 25 nt RNA hybridizing to *GFP*-specific probes (Dalmay et al. 2000b). In the parental 35S-*GFP* and 35S-*PVX:GFP* lines the *GFP* DNA is not methylated and 25 nt *GFP*-specific RNA is non-detectable (35S-*GFP* line) or present at a very low level (35S-*PVX:GFP* line). Since, neither the 25 nt *PVX*-specific RNA nor the full-length viral *PVX:GFP* RNAs were detected in wild-type parental hybrid line, the 25 nt *GFP* RNA is probably derived from the 35S-*GFP* mRNA rather than from replicating *PVX:GFP* RNA or 35S-*PVX:GFP* mRNA. In the *sde1* mutant the *GFP* DNA is not methylated, the 25 nt *GFP* RNA is 6-fold less abundant than in the wild-type (*SDE1*) plants and 25 nt *PVX* RNA, undetectable in the wild type, is more abundant. Most likely, this 25 nt *PVX* RNA is derived from the replicating *PVX:GFP* RNA that is present at elevated levels in the *sde1* mutant. Thus, *sde1* mutation seems to be specific for PTGS induced by a transgene but not a virus. The reduced level of the 25 nt *GFP* RNA and low level of 25 nt *PVX* RNA appeared in this mutant should be due to the replicating *PVX:GFP* RNA. Indeed, further analyses showed that *sde1* mutation does not affect accumulation of other viral RNAs (crucifer strain of tobacco mosaic virus, tobacco rattle virus and turnip crinkle virus) after inoculation into *Arabidopsis*: the viral genomic and subgenomic RNAs were equally abundant in the mutant and wild-type plants. Infection of wild-type *Arabidopsis* with a TRV vector harboring an insert of the phytoene desaturase sequence causes a striking photobleached phenotype. This is a direct consequence of suppressed photoprotective carotenoid production due to virus-induced PTGS of the endogenous phytoene desaturase gene. The same phenotype developed in the *sde1* plants at the same rate as in the wild-type plants. Thus, the *SDE1* locus is not required for virus-induced PTGS.

By cosegregation analysis of the *sde1* mutant phenotype with markers from each *Arabidopsis* chromosome the *SDE1* locus was mapped to a region on the bottom arm of chromosome 3 at position 61.4–68.2 cM (Dalmay et al. 2000a). Since this region contains a homolog of a gene *QDE-1* required for quelling in *N. crassa*, the possibility that *SDE1* is the *Arabidopsis* homolog of *QDE-1* was further tested and confirmed by cosegregation analysis of close flanking markers and direct DNA sequence analysis of mutant alleles. All four *sde-1* mutants tested had nucleotide deletions that disrupted the *SDE1* open reading frame. The SDE1 is a 113.7 kDa (1196 aa) protein encoded in a 4182 nt mRNA. Three additional SDE1 homologs were found in genome sequence of *Arabidopsis*. In addition, the similarity of SDE1 with tomato RNA-dependent RNA polymerase (RdRP) was detected. The proposed role of SDE1 is to produce a dsRNA activator of transgene-induced PTGS. It may be not required for virus-induced PTGS, because the virus-encoded RdRP produces dsRNA as an intermediate in the replication cycle. A role of dsRNA as silencer could explain earlier findings that inverted repeat transgenes and coexpressed sense and antisense RNA can induce PTGS (Hamilton et al. 1998; Waterhouse et al. 1998; Stam et al. 2000). Similarly, the dsRNA intermediate in virus replication could explain why RNA viruses induce PTGS (Ratcliffe et al. 1997, 1999; Ruiz et al. 1998). The SDE1 RdRP activity seems to be responsible for synthesis of dsRNA in transgene-induced PTGS, whereas in virus-induced PTGS such dsRNA is synthesized by a viral RdRP and is independent on SDE1.

The 35S-*GFP* transgene methylation is dependent on the combined presence of the 35S-*PVX:GFP* and the 35S-*GFP* transgenes. Thus, replicating *PVX:GFP* RNA initiates PTGS and leads (directly or indirectly) to transgene methylation. The finding that the transgenes in the *sde* mutants are not methylated despite the presence of viral *PVX:GFP* RNA shows that the transgene methylation is not directly due to presence of replicating viral *PVX:GFP* RNA. A more likely possibility is that transgene methylation is mediated by the 25 nt *GFP* RNA. The role of transgene methylation in PTGS is still not quite clear. In principle, there could be an SDE1-dependent cycle of PTGS operating purely at the RNA level without any epigenetic changes at the DNA or chromatin level. If methylation does play a causal role in the gene silencing, it could be in a secondary process that reinforces the proposed SDE1-dependent PTGS cycle.

SDE1 is not required for virus-induced PTGS and unlikely to be involved in the antiviral defense action of PTGS. None of the *SDE* loci are likely to play any significant role in plant development or basic cellular function, because all *sde* mutant plants grow and develop normally. Most likely, *SDE* proteins function in protection against transposable DNA.

The long siRNAs of tobacco TS SINE retroelements were shown not to mediate resistance to a virus carrying TS SINE sequences, suggesting that, unlike the 21- to 22 nt siRNAs, long 25 nt siRNAs do not participate in PTGS (Hamilton et al. 2002). In addition, mutants that affect RNA silencing were used to show a correlation of long siRNAs content with DNA methylation. In particular, mutants in *SDE1* (an RNA-dependent RNA polymerase), *SDE3* (an RNA helicase) and *SGS3* (a novel gene) did not suppress the accumulation of long siRNAs or affect DNA methylation of *AtSN1* but the *sde4* mutant suppressed both long siRNAs formation and DNA methylation. The *ago4* and *sde4* map to different chromosomes and are, therefore, not allelic.

Thus, *AGO4* and *SDE4* seem to encode components of a silencing system that generates long siRNAs specialized for gene silencing at the chromatin level. Presumably, a Dicer-like enzyme and an RNA-dependent RNA polymerase are involved in siRNA production. Once generated, the long siRNAs guide the KYP-dependent histone methylation, the CMT3- and DRM-dependent DNA methylations to specific regions of chromatin. The targeting of this system to transposable elements likely contributes to genome stability ant suppression of the transposon proliferation.

ADENINE DNA METHYLATION

N^6-Methyladenine in DNA of Eukaryotes

N^6-Methyladenine (m^6A) occurs as a minor base in DNA of various organisms. It was first detected in *E. coli* DNA (Dunn and Smith, 1955). Then it was shown to be obvious in most bacterial DNA (Vanyushin et al. 1968; Barras and Marinus 1989). It has also been found in DNA of algae (Pakhomova et al. 1968; Hattman et al. 1978; Babinger et al. 2001) and their viruses (Que et al. 1997; Nelson et al. 1998), fungi (Buryanov et al. 1970; Rogers et al. 1986), and protozoa (Gutierrez et al. 2000) including *Tetrahymena* (Gorovsky et al. 1973; Kirnos et al. 1980; Pratt and Hattman 1981), *Crithidia* (Zaitseva et al. 1974), *Paramecium* (Cummings et al. 1974), *Oxytricha* (Rae and Spear 1978), *Trypanosoma cruzi* (Rojas and Galanti 1990), and *Stylonychia* (Ammermann et al. 1981). In DNA of various algae, N^6-dimethyadenine was detected (Pakhomova 1974). About 0.8% of adenine residues are found as m^6A in DNA of the transcriptionally active macronuclei of *Tetrahymena* (Gorovsky et al. 1973; Kirnos et al. 1980). A methylation site is 5'-NAT-3' (Bromberg et al. 1982), and about 3% methylation sites are GATC (Harrison et al. 1986; Karrer and Van Nuland 1998). The adenine methylated GATC sites are preferentially located in linker DNA, unmethylated sites are generally in DNA of nucleosome cores, and histone H1 is not required for the maintenance of normal methylation patterns (Karrer and Van Nuland 2002). DNA of the slime mould *Physarum flavicomum* becomes sensitive to the *Dpn*I restriction endonuclease during encystment. This may be due to the appearance of m^6A residues in GATC sequences in this DNA (Zhu and Henney 1990). Early data on the presence of m^6A in mammalian sperm DNA were ambiguous (Unger and Venner 1966), and attempts to detect and isolate this minor base as well as adenine DNA-methyltransferase activity from DNA of many invertebrates and vertebrates were unsuccessful (Vanyushin et al. 1970; Lawley et al. 1972; Fantappie et al. 2001; Ratel et al. 2006; Wion and Casadesus, 2006). Nevertheless, it was judged from the different resistance of animal DNA to restriction endonucleases sensitive to methylation of adenine residues (*Taq*I, *Mbo*I and *Sau*3AI) that some genes (*Myo*-D1) (Kay et al. 1994) - steroid-5-α-reductase genes 1 and 2 (Reyes et al. 1997) - of mammals (mouse, rat) might contain m^6A residues. This indirectly suggests that animals may have adenine DNA-methyltransferases. It is interesting that addition of N^6-methyldeoxyadenosine (MedAdo) to C6.9 glioma cells triggers a differentiation process and the expression of the oligodendroglial marker 2',3'-cyclic nucleotide 3'-phosphorylase. The differentiation induced by N^6-methyldeoxyadenosine was also observed on pheochromocytoma and teratocarcinoma cell lines and on dysembryoplastic neuroepithelial tumour cells (Ratel et al. 2001). The precise mechanism by which modified nucleoside induces cell differentiation is still unclear, but it is considered to be related to cell cycle modifications. The incubation of C2C12 myoblasts in the presence of MedAdo induces myogenesis (Charles et al. 2004). It is remarkable that m^6A was detected by a method based on HPLC coupled to electrospray ionization tandem mass spectrometry in the DNA from MedAdo-treated cells (it remains undetectable in DNA from control cells). Furthermore, MedAdo regulates the expression of p21, myogenin, mTOR and MHC. Interestingly, in the pluripotent C2C12 cell line, MedAdo drives the differentiation towards myogenesis only (Charles et al. 2004). These results point to N^6-methyldeoxyadenosine as a novel inducer of myogenesis and further extend the differentiation potentialities of this methylated nucleoside. m^6A has been found in total DNA of higher plants (Vanyushin et al. 1971; Buryanov et al. 1972). It may be present in plastid (amyloplast) DNA (Ngernprasirtsiri et al. 1988). In wheat seedlings it is present in heavy (ρ = 1.718 g/cm^3) mitochondrial DNA (Vanyushin et al. 1988; Aleksandrushkina et al.1990; Kirnos et al. 1992a, b). Similar mtDNA containing m^6A were also found in many other higher plants including various archegoniates (mosses, ferns, and others) and angiosperms (monocots, dicots; Kirnos et al. 1992a). The synthesis of this unusual DNA takes place mainly in specific vacuolar vesicles containing mitochondria, and it is a sort of aging index in wheat and other plants (Kirnos et al. 1992b; Bakeeva et al. 1999; Vanyushin et al. 2004). There is some indirect evidence (based on the comparison of products of DNA hydrolysis with restriction endonucleases *Mbo*I and *Sau*3A) that some adenine residues in zein genes of corn can be methylated (Pintor-Toro 1987). The *DRM2* gene in *Arabidopsis* was found to be methylated at both adenine residues in some GATC sequences and at the internal cytosine residues in CCGG sites (Ashapkin et al. 2002). Thus, two different systems of the genome modification exist in higher plants. It is absolutely unknown how these systems may interact and to what degree they are interdependent. It appears that adenine methylation may influence the cytosine modification and *vice versa*. Interestingly, the adenine methylation of the *DRM2* gene observed is most prominent in wild-type plants and appears to be diminished by the presence of antisense *METI* transgenes. Anyway, a new sophisticated type of interdependent regulation of gene functioning in plants may exist, based on the combinatory hierarchy of certain chemically and biologically different methylations of the genome.

ADENINE DNA-METHYLTRANSFERASES

m^6A is formed in DNA due to the recognition and methylation of respective adenine residues in certain sequences by specific adenine DNA-methyltransferases. Adenine DNA-methyltransferases of bacterial origin can also methylate cytosine residues in DNA with the formation of m^4C (Jeltsch 2001).

Adenine DNA-methyltransferases of eukaryotes could be inherited from some prokaryotic ancestor. They may be homologous to known prokaryotic DNA-(amino)-methyltransferases due to the very conservative nature of DNA-methyltransferases in general. ORFs for putative adenine DNA-methyltransferases were found in nuclear but not mitochondrial DNA of protozoa (*Leishmania major*), fungi (*Saccharomyces cerevisiae*, *Schizosaccharomyces pombe*), higher plants (*A. thaliana*), and animals (*Drosophila melanogaster*, *Caenorhabditis elegans*, *Homo sapiens* (Shorning and Vanyushin 2001).

There is nothing currently known about the ORF expression detected or activity of respective eukaryotic proteins encoded in these organisms. The enzymatic activity of these DNA-methyltransferases may be very limited as is true, for example, with the transcription of the *Drosophila melanogaster* C^5-cytosine-DNA-methyltransferase gene [this insect DNA contains an extremely low amount of 5-methylcytosine (Gowher et al. 2000), and the DNA-methyltransferase gene is a component of a transposon-similar element expressed only in the early stages of embryonic development (Lyko et al. 2000)]. The amino acid sequences of putative eukaryotic DNA-(amino)-methyltransferases (Shorning and Vanyushin 2001) are very homologous to each other, as well as to real DNA-(amino)-methyltransferases of eubacteria, hypothetical methyltransferases of archaebacteria and putative HemK-proteins of eukaryotes (Bujnicki and Radlinska 1999). These putative eukaryotic adenine DNA-methyltransferases (ORF) share conservative motifs (I, IV) specific for DNA-(amino)methyltransferases and motifs II, III, V, VI and X. Motif I (it takes part in binding of the methionine part of the *S*-adenosylmethionine molecule and is specific for all AdoMet-dependent methyltransferases) was detected in all eukaryotic ORFs found. The amino acid composition of the catalytic centre in all putative DNA-(amino)-methyltransferases is practically the same; it is extremely conservative and does not have any mutations. In fully sequenced mitochondrial genomes of eukaryotes (the liverwort *Marchantia polymorpha*, *Arabidopsis thaliana*, sugar beet, the alga *Chrysodidymus synuroideus*) the nucleotide sequences with significant homology to genes of prokaryotic DNA-(amino)-methyltransferases were not observed (Shorning and Vanyushin 2001). It is most probable that an enzyme encoded in the nucleus is transported somehow into mitochondria.

The first eukaryotic (plant) N^6-adenine DNA-methyltransferase (*wadmtase*) was isolated from the vacuolar vesicle fraction of aging wheat coleoptiles (Fedoreyeva and Vanyushin 2002). The vesicles appear in plant apoptotic cells, they are enriched with Ca^{2+} and contain actively replicating mitochondria (Bakeeva et al. 1999; Vanyushin 2004). In the presence of *S*-adenosyl-L-methionine, the enzyme *de novo* methylates the first adenine residue in the TGATCA sequence in the single-stranded (ss) DNA or dsDNA substrates but it prefers single-stranded structures. Wheat adenine DNA-methyltransferase is a Mg^{2+}- or Ca^{2+}-dependent enzyme with a maximum activity at pH 7.5-8.0. About 2-3 mM $CaCl_2$ or $MgCl_2$ in the reaction mixture is needed for the maximal DNA methylation activity. The enzyme is strongly inhibited by ethylenediaminotetraacetate (EDTA). The optimal concentration of AdoMet in DNA methylation with *wadmtase* is about 10 μM. *Wadmtase* encoded in the wheat nuclear DNA may be homologous to the *A. thaliana* ORF (GenBank, BAB02202.1), which might be ascribed to putative adenine DNA-methyltransferases (Shorning and Vanyushin 2001). The methylated adenine residues found in Gm^{6}ATC sites of a *DRM2* gene in a nuclear DNA of *A. thaliana* (Ashapkin et al. 2002) could be a constituent part of a sequence TGATCA recognized and methylated by wheat adenine DNA-methyltransferase (*wadmtase*). Unfortunately, we do not know whether adenine DNA-methyltransferase in *Arabidopsis* cells has the same site specificity as it has in wheat plants. Since *wadmtase* is found in vesicles with mitochondrial actively-replicating DNA, its maximal activity is associated with mtDNA replication and it prefers to methylate ssDNA, the enzyme seems to operate mainly with replicating mtDNA. Similar to the known dam enzyme controlling plasmid replication in bacteria, *wadmtase* seems to control replication of mtDNA that are represented mainly by circular molecules in wheat seedlings (Kirnos et al. 1992a, b). As mitochondria could be evolutionarily of bacterial origin, the bacterial control for plasmid replication by adenine DNA methylation seems to be acquired by plant cells, and it is probably used for the control of mitochondria replication.

Putative Role of Adenine DNA Methylation in Plants

Unfortunately, the functional role of adenine DNA methylation in plants and most other higher eukaryotes is unknown. There is some data available showing that the character of transcription of many plant genes and the morphology and development of transformed plant cells and the plants are drastically changed after introduction into them of genetic constructs with expressed genes of prokaryotic adenine DNA-methyltransferases. For example, introduction and expression of the bacterial adenine DNA-methyltransferase (*dam*) gene is accompanied by GATC sequence methylation in DNA of transgenic tobacco plants and changes in the leaf and inflorescence morphology. The efficiency of adenine DNA methylation was directly proportional to expression levels of the *dam* construct, and methylation of all GATC sites was observed in a highly expressing line. This correlates with abnormal phenotypes affecting leaf pigmentation, apical dominance and leaf and floral structure (van Blokland et al. 1998). Moreover, *dam*-methylation of promoter regions in constructs with plant genes for alcohol dehydrogenase, ubiquitin and actin results in an increase in the transcription of these genes in tobacco and wheat tissues (Graham and Larkin 1995). This preliminary methylation of promoters is also important for transcription of *PR1* and *PR2* genes in constructs introduced into tobacco protoplasts by electroporation (Brodzik and Hennig 1998). Adenine methylation of the AG-motif sequence AGATCCAA in the promoter of NtMyb2 (a regulator of the tobacco retrotransposon Tto1) by bacterial dam methylase enhances activity of the AG-motif-binding protein (AGP1) in tobacco cells (Sugimoto et al. 2003). The presence of methylated adenine residues in the GATC sequence scattered in the reporter plasmid introduced into intact barley aleurone layers by a particle bombardment increased transcription from hormone-regulated α-amylase promoters two- to fivefold, regardless of the promoter strength, and proper hormonal regulation of transcription was maintained (Rogers and Rogers 1995). The methylated adenine effect was similar when the amount of reporter construct DNA used was varied over a 20-fold range, beginning with an amount that gave only a small increment of expression. Similar transcription-enhancing effects for methylated adenine residues in DNA were seen with the CaMV 35S, maize Adh1 and maize ubiquitin promoters (Rogers and Rogers 1995). It was shown that some proteins present in wheat germ nuclear extracts bound preferentially to adenine-methylated DNA rather than cytosine-methylated DNA. It seems that enhanced transcription of nuclear genes in barley due to the presence of m^6A residues in the vicinity of active promoters may be mediated by m^6A DNA-binding protein (Rogers and Rogers 1995). Hence, methylation of adenine residues in DNA may control gene expression in plants. It was hypothesized that modulation of methylation of adenine residues by incorporation of cytokinins (N^6-derivatives of adenine) into DNA may serve as a mechanism of phytohormonal regulation of gene expression and cellular differentiation in plants (Vanyushin 1984). Cytokinins (6-benzylaminopurine) can incorporate into the DNA of plants (Kudryashova and Vanyushin 1986) and *Tetrahymena pyriformis* (Mazin and Vanyushin 1986). In fact, 6-benzylaminopurine inhibits plastid DNA methylation in sycamore cell culture and induces in these cells the expression of enzymes involved in photosynthesis (Ngernprasirtsiri and Akazawa 1990). It cannot be ruled out that in this particular case, cytokinin may be involved in regulation of adenine DNA methylation in a plastid.

The data showing that adenine DNA methylation may be involved in a control for persistence of foreign DNA in a plant cell is of special interest. Unlike cytosine methylation, the adenine methylation alone is associated with marked foreign DNA instability (Rogers and Rogers 1992). Plant cells seem to have a system discriminating between adenine and cytosine DNA modifications, and the specific enzymes resembling to some extent bacterial restriction endonucleases could be responsible for selective elimination of impropriate adenine methylated DNA.

Recently we have isolated from wheat seedlings a few specific AdoMet-, Ca^{2+}, Mg^{2+}-dependent endonucleases discriminating between methylated and unmethylated DNAs

(Fedoreeva, Sobolev and Vanyushin, 2007). One of these wheat endonucleases, WEN1, is Ca^{2+}-, Mg^{2+}- dependent enzyme with molecular mass of about 27 kDa hydrolyzed methylated DNA of λ phage grown on *E. coli* dam^+, dcm^+ cells more effectively compared with DNA of the same phage grown on dam^-, dcm^- cells. Two pH activity maxima (pH 6.5-7.5 and 9.0-10.5) were observed when double-stranded DNA was hydrolyzed. WEN1 is stable at elevated temperatures (65°C) and in wide range of pH values. WEN1 is activated by S-adenosyl-L-methionine, S-adenosyl-L-homocysteine and S-isobutyladenosine. It is a first case to show that higher eukaryote endonuclease discriminates between DNA of various methylation states and is modulated by S-AdoMet and its analogs. WEN1 prefers single-stranded DNA; one of the target sequences cleaved by WEN1 in these DNA seems to be TGATCA. The functional role of WEN1 in plants is unknown. It seems that its action may be somehow co-ordinated with adenine DNA-methyltransferase *wadmtase* that like WEN1 is located in the same vesicles, recognizes TGATCA sequence and prefers to act on ssDNA. Both enzymes are AdoMet-, Ca^{2+}, Mg^{2+}-dependent, they discriminate the substrate DNA by methylation status and seem to act on the same target DNA structures. Thus, plants may have a system of DNA modification and of control for replication that in many features corresponds to R-M system in bacteria. It is very possible that this system operates in plant mitochondria replicating mtDNA very actively in vesicles of plant apoptotic cells. On the other hand, we cannot rule out that WEN1 may take part also in the nuclear DNA degradation and it might be a candidate for endonuclease that is one of the executors of the terminal apoptosis stages in plants.

CONCLUSION

DNA methylation controls plant development and is involved in tissue-specific gene silencing and parental imprinting. It also seems to be a regulatory mechanism that keeps expression of repeated sequences within an acceptable range for the plant well-being. Another important role is inactivation of potentially dangerous elements in the plant genome including both transposable sequences (methylation seems to be a major mechanism inactivating their transcription and transposition) and foreign DNA. As a matter of fact, DNA methylation seems to be only a part of a complicated multi-step process of gene silencing, though a very important one. Silenced state of genes is usually correlated with methylation of their promoters, whereas hypomethylation usually leads to reactivation of transcription. Nevertheless, the other steps of gene silencing also contribute to the maintanance of the silent state. Their breakage, for example, by *mom1* mutation, may lead to partial or full reactivation of transcription even when the affected gene remains heavily methylated. In contrast to the animals, where the "resetting" of epigenetic status occurs in each generation by extensive demethylation and subsequent *de novo* DNA methylation during gametogenesis and early development, the epigenetic states of plant genes are often stably inherited through generations.

One more striking difference between animals and plants is that in animals methylation affects mostly symmetric CpG sequences, whereas plant DNA is extensively methylated at two types of symmetric sequences, namely CpG and CpNpG, as well as at asymmetric ones.

Severe distortions in DNA methylation, whether by DNA hypomethylation mutations or chemical agents, are accompanied by essential changes in plant growth and morphology. The homozygous DNA hypomethylation mutants (*ddm1, met1*) show progressive accumulation of morphological abnormalities in successive generations of self-pollinated plant lines, probably caused by ectopic expression of undermethylated genes with tissue specific expression. But unlike animals, where *dmt1* knockout results in a block of development and is mostly lethal, plants lacking analogous enzyme MET1 survive. It may be associated with existence of a multicomponent partially redundant DNA methylation system in plants as well as reinforcing action of other epigenetic systems such as histone modifications and siRNA gene silencing. Plants have, at least, three gene families that code for cytosine DNA-methyltransferases, more than any other eukaryotes. The MET class consists of genes related to the mammalian *Dnmt1* (Finnegan, Kovac, 2000). These are the major CpG-specific maintenance DNA-methyltransferases that seem to be an essential component in determining processes of the developmental phase transitions and meristem determinacy. A second type of methyltransferase, the CMT "chromomethylase" class, is unique to plants. They have a novel chromo-domain amino acid motif inserted between two canonical methyltransferase motifs, I and IV. In addition to chromodomain and a conservative methyltransferase domain all chromomethylases contain a bromo adjacent homology (BAH) domain, a feature common with Dnmt1/MET methyltransferases that is believed to be implicated in linking DNA methylation, replication and transcriptional regulation. These two classes of methyltransferases both may be involved in the hereditary maintaining symmetric methylation patterns, with the Dnmt1/Met1 class acting on hemimethylated CpG sites and the chromomethylases methylating hemimethylated CpNpG sites soon after replication. This agrees with a nearly complete loss of genomic CpNpG methylation in null *CMT3* mutants. Unlike *met1* mutants that exhibit severe developmental abnormalities, the *cmt3* mutants are morphologically normal even after multiple generations of inbreeding. One explanation is that CpNpG and CpG methylations may act in a partially redundant fashion to silence most genes. Chromodomains are found in several proteins involved in the chromatin-level repression of transcription. Some recent findings suggest that chromomethylases may be targeted by their chromodomain to heterochromatic regions marked by H3 Lys-9 methylation. A third class of *Arabidopsis* methyltransferases is the "domain rearranged methyltransferases" or DRM class, which is most related to Dnmt3, except that the canonical methyltransferase motifs are organized in a novel order, namely VI, IX, X, I, II, III, IV, V, as if a rearrangement has taken place at a region of several amino acids between motifs X and I. A series of ubiquitin-association (UBA) domains is a unique feature of plant DRM methyltransferases. Thus, these enzymes may be controlled in a cell cycle by ubiquitin-mediated protein degradation or (and) the ubiquitinization may alter the cellular localization of these enzymes due to respective external signals. The *drm1 drm2* null mutations of *DRM* genes eliminate all asymmetric DNA methylation and cause some locus-dependent loss of the CpNpG methylation. Both *DRM* and *CMT3* are required for proper maintenance of the CpNpG methylation patterns. The combined *drm cmt* mutants lack all traces of asymmetric and CpNpG methylation, while CpG methylation levels are similar to the wild type. Thus, the *DRM* and *CMT3* genes encode DNA-methyltransferase enzymes that show overlapping roles in the control of asymmetric and CpNpG methylations. However, the activities of these methyltransferases are highly dependent on the locus under study, giving a surprising number of different patterns of dependence on either *DRM*, *CMT3*, or both. It seems quite likely that several factors could be involved in targeting particular sequences to different DNA-methyltransferases, including the DNA sequences themselves, chromatin modifications present at specific loci, and, last but not least, the cross talk between different kinds of DNA methylations. Nevertheless, all data available today are consistent with a general view that plant DRM members are *de novo* DNA-methyltransferases.

Developmental defects seen in *met1* mutants, though clearly result from loss of the CpG DNA methylation at the particular genes, nevertheless, segregate independently of their "parental" *met1* mutations. This is because the loss of CpG DNA methylation is inheritable and not readily regained, when these mutants are crossed to wild type plants. Loss of non-CpG methylation in plants with combined mutations in the *DRM* and *CMT3* genes also causes a suite of the developmental defects but these are not inherited independently of the *drm* and *cmt3* mutations. This disparity seems to be a consequence of a basic difference in the mode of maintenance of two types of DNA methylations in plant cell. The maintenance activity of MET1 can replicate CpG DNA methylation even when the initial trigger for DNA methylation is genetically removed. This may be explained in part by the fact that Dnmt1-type DNA-methyltransferases have a strong preference for hemimethylated substrates such as those left by DNA replication of a CpG sequence that was initially methylated in both strands. Non-CpG DNA methylation appears to require for its maintenance the active signals to continually target the DNA methylation regions. One of such signals seems to come from histones associated with DNA. The mutations of a gene coding for histone H3 Lys9-methyltransferase, KRYPTONITE, cause nearly complete elimination of both types of non-CpG methylations in plants, whereas mutations of DNA-methyltransferase genes have little effects on H3Lys9 methylation. LHP1, a plant homologue of HP1 protein, probably involved in heterochromatin-specific gene silencing, specifically binds to both Lys9-methylated histone H3 and CMT3. The Lys9-methylated N-tail of H3, therefore, may well serve as a signal targeting specific CpNpG sites for methylation by LHP1-mediated binding to CMT3. In any case, the cross-talk between the histone code and cytosine DNA methylation provides, at least, a tentative answer to the long-standing question of how DNA methylation patterns are established and maintained. Generally, the combination of two methylation systems (histones and DNA) could be the means to stabilize silent state of respective chromatin regions. How exactly DNA methylation is coupled to histone methylation is far from clear. One can imagine that methylated DNA attracts respective methyl-CpG-binding proteins, which in their turn recruit histone deacetylase complexes to deacetylate histone tails so that the tails become suitable as substrates for H3 Lys9-methylation. Alternatively, it is also possible that chromodomain-containing proteins bind to methylated histone tails and recruit cytosine DNA-methyltransferases to methylate adjacent DNA sequences. This latter possibility seems to be realized in the case of H3 Lys9-methylation dependent on DNA methylation by plant CMT3 methyltransferase.

During the last several years regulatory RNAs have been linked to various gene silencing phenomena in plants, animals and fungi. Different types of regulatory RNA were shown to act in distinct ways to induce gene silencing. The short RNAs (21-26 nt), which are derived via cleavage of double-stranded RNA (dsRNA) precursors, serve as specificity determinants for enzyme complexes that degrade, modify or inhibit the function of homologous nucleic acids. Gene silencing phenomena that are induced by nucleotide sequence-specific interactions mediated by RNA are termed collectively 'RNA silencing'. In plants, heavy *de novo* methylation and silencing of multiple transgene copies occurs frequently. The *de novo* methylation is directed by unusual structures that could arise by pairing of RNA molecules with their genomic counterparts. This process, RNA-directed DNA methylation (RdDM) is specifically targeted to DNA sequences complementary to the directing RNA. Most, if not all, cytosines within the putative RNA-DNA triplex region are methylated irrespective of their sequence context. The recognition of specific DNA structures that are formed during the RdDM process may strongly stimulate activity of *de novo* DNA-methyltransferase. As soon as the DNA-methyltransferase slips to flanking RNA-free region, it either leaves the template or its methylation activity ceases. Both strands of the target DNA sequence are heavily methylated at symmetric and asymmetric sites suggesting a mechanism operating on both strands more or less simultaneously. Different DNA-methyltransferases are involved in an establishement or maintenance of the RNA-directed DNA methylation. Namely, the dsRNA-dependent *de novo* DNA methylation activity of DRM methyltransferases is absolutely required for initial establishement of RdDM in all sequence contexts. Both MET1 and CMT3 methyltransferases seem to be non-essential at this step. Maintenance of CpG methylation can occur in the absence of the triggering RNA signals and is dependent on the activity of MET1 exclusively. Both DRMs and CMT3 are required for efficient maintenance of CpNpG and asymmetric methylations. The maintenance phase for these non-CpG methylation types seems to involve persistent dsRNA-dependent *de novo* activity of DRM. Nevertheless, it is clearly distinct from the initiation phase, where DRM alone is absolutely required.

Along with cytosine methylation the methylation of adenine in plant DNA was observed and specific adenine DNA-methyltransferase was described. The same plant gene may be methylated at both the adenine and cytosine residues. The functional role of adenine DNA methylation is still unknown. Anyway, two different systems of the genome modification based on methylation of adenines and cytosines exist in higher plants. It is yet unknown how these systems may interact and to what degree they are interdependent. It appears that adenine methylation may influence cytosine modification and *vice versa*, and mutual control for these genome modifications may be a part of the epigenetic control of gene activity in plants. The specific endonucleases discriminating between DNA methylated and unmethylated at adenine and cytosine residues seem to be present in plants. It means that plants may have a restriction-modification system.

Thus, DNA methylations desribed are very delicate and efficient natural means for regulation of gene activities and genome state and reproduction in the cell. These DNA modifications are very closely associated with many other known sophisticated epigenetic signals, genome and cell perturbances that all together, in fact, determine the life, its essence and quality, or in other words "to be or not to be". The discrtete chains and the total web of this interdependent signals are yet far from complete understanding but there is no doubt that DNA methylations play there very essential role and, hence, the further comprehensive investigations in this fascinating field are the most important and profitable goal of our age that is called by right as era of epigenetics.

ACKNOWLEDGEMENTS

This work was supported in part by the Russian Foundation for Basic Research (grant No. 05-04-48071)

REFERENCES

Akhtar, A., Zink, D. & Becker, P.B. (2000) Chromodomains are protein-RNA interaction modules. *Nature 407*: 405-409

Aleksandrushkina, N.I., Kudryashova, I.B., Kirnos, M.D. & Vanyushin, B.F. (1990) Synthesis of "heavy" miniplasmid-like mitochondrial DNA and its adenine methylation in coleoptile and initial leaf of wheat seedlings: influence of phytohormones. *Biokhimiya (Moscow) 55*: 2038-2045

Alvarez-Venegas, R., Pien, S., Sadder, M., Witmer, X., Grossniklaus, U. & Avramova, Z. (2003) ATX-1, an Arabidopsis Homolog of Trithorax, Activates Flower Homeotic Genes. *Curr Biol 13*: 627-637

Amedeo, P., Habu, Y., Afsar, K., Scheid, O.M. & Paszkowski, J. (2000) Disruption of plant gene *MOM* releases transcriptional silencing of methylated genes. *Nature 405*: 203-206

Ammermann, D., Steinbruck, G., Baur, R. & Wohlert, H. (1981) Methylated bases in the DNA of the ciliate *Stylonychia mytilus*. *Eur J Cell Biol 24*:154-156

Arnaud, P., Goubely, C., Pelissier, T. & Deragon, J.M. (2000) SINE retroposons can be used *in vivo* as nucleation centers for de novo methylation. *Mol Cell Biol 20*: 3434–3441

Ashapkin, V.V., Kutueva, L.I. & Vanyushin, B.F. (2002) The gene for domains rearranged methyltransferase (DRM2) in *Arabidopsis thaliana* plants is methylated at both cytosine and adenine residues. *FEBS Lett 532*: 367-372

Aufsatz, W., Mette, M.F., van der Winden, J., Matzke, A.J.M. & Matzke, M. (2002a) RNA-directed DNA methylation in *Arabidopsis*. *Proc Natl Acad Sci USA 94*: 11721-11725

Aufsatz, W., Mette, M.F., van der Winden, J., Matzke, M. & Matzke, A.J. (2002b) HDA6, a putative histone deacetylase needed to enhance DNA methylation induced by double-stranded RNA. *EMBO J 21*: 6832-6841

Aufsatz, W., Mette, M.F., Matzke, A.J.M. & Matzke, M. (2004) The role of MET1 in RNA-directed *de novo* and maintenance methylation of CG dinucleotides. *Plant Mol Biol 54*: 793–804

Babinger, P., Kobl, I., Mages, W. & Schmitt, R. (2001) A link between DNA methylation and epigenetic silencing in transgenic *Volvox carteri*. *Nucleic Acids Res 29*: 1261-1271

Bakeeva, L.E., Kirnos, M.D., Aleksandrushkina, N.I., Kazimirchyuk, S.B., Shorning, B.Yu., Zamyatnina, V.A., Yaguzhinsky, L.S. & Vanyushin, B.F. (1999) Subcellular reorganization of mitochondria producing heavy DNA in aging wheat coleoptiles. *FEBS Lett 457*: 122-125

Barras, F. & Marinus, M.G. (1989) The great GATC: DNA methylation in *E. coli*. *Trends Genet 5*: 139-145

Bartee, L. & Bender, J. (2001) Two *Arabidopsis* methylation-deficiency mutations confer only partial effects on a methylated endogenous gene family. *Nucleic Acids Res 29*: 2127-2134

Bartee, L., Malagnac, F. & Bender, J. (2001) Arabidopsis cmt3 chromomethylase mutations block non-CG methylation and silencing of an endogenous gene. *Genes Dev 15*: 1753-1758

Baumbusch, L.O., Thorstensen, T., Krauss, V., Fischer, A., Naumann, K., Assalkhou, R., Schulz, I., Reuter, G. & Aalen, R.B. (2001) The Arabidopsis thaliana genome contains at least 29 active genes encoding SET domain proteins that can be assigned to four evolutionarily conserved classes. *Nucl Acids Res 29*: 4319-4333

Bender, J. (2001) A vicious cycle: RNA silencing and DNA methylation in plants. *Cell 106*: 129-32

Bender, J. (2004) DNA methylation and epigenetics. *Annu Rev Plant Biol 55*: 41-68

Bender, J. & Fink, G.R. (1995) Epigenetic control of an endogenous gene family is revealed by a novel blue fluorescent mutant of Arabidopsis. *Cell 83*: 725-734

Bernstein, B.E., Kamal, M., Lindblad-Toh, K., Bekiranov, S., Bailey, D.K., Huebert, D.J., McMahon, S., Karlsson, E.K., Kulbokas, E.J., III, Gingeras, T.R., Schreiber, S.L. & Lander, E.S. (2005) Genomic Maps and Comparative Analysis of Histone Modifications in Human and Mouse. *Cell 120*: 169-181

Bestor, T., Laudano, A., Mattaliano, R. & Ingram, V. (1988) Cloning and sequencing of a cDNA encoding DNA methyltransferase of mouse cells. The carboxyl-terminal domain of the mammalian enzymes is related to bacterial restriction methyltransferases. *J Mol Biol 203*: 971-983

Bianchi, M.W. & Viotti, A. (1988) DNA methylation and tissue specific transcription of storage protein genes of maize. *Plant Mol Biol 11*: 203–214

Bird, A. (1992) The essentials of DNA methylation. *Cell 70*: 5–8

Bourc'his, D. & Bestor, T.H. (2002) Helicase homologues maintain cytosine methylation in plants and mammals. *Bioessays 24*: 297-299

Brodzik, R. & Hennig, J. (1998) Adenine methylation of GATC sequences regulates activity of tobacco PR-1 and PR-2 promoters in electroporated protoplasts. *Plant Physiol Biochem 36*: 401-406

Bromberg, S., Pratt, K. & Hattman, S. (1982) Sequence specificity of the DNA adenine methylase in the protozoan *Tetrahymena thermoiphila. J Bacteriol 150*: 993-996

Brutnell, T.P. & Dellaporta, S.L. (1994) Somatic inactivation and reactivation of *Ac* associated with changes in cytosine methylation and transposase expression. *Genetics 138*: 213-225

Bujnicki, J.M. & Radlinska, M. (1999) Is the HemK family of putative S-adenosylmethionine-dependent methyltransferases a "missing" ζ subfamily of adenine methyltransferases? A hypothesis. *IUBMB Life, 48*: 247-249

Burn, J.E., Smyth, D.R., Peacock, W.J. & Dennis, E.S. (1993) Genes conferring late flowering in *Arabidopsis thaliana. Genetica 90*: 147-155

Buryanov, Y.I., Eroshina, N.V., Vagabova, L.M. & Iliin, A.V. (1972) On the detection of 6-methylaminopurine in DNA of higher plant pollen. *Dokl Akad Nauk SSSR 206*: 992-994

Bushell, C., Spielman, M. & Scott, R.J. (2003) The basis of natural and artificial postzygotic hybridization barriers in *Arabidopsis* species. *Plant Cell 15*: 1430–1442

Callebaut, I., Courvalin. J.C. & Mornon, J.P. (1999) The BAH (bromo-adjacent homology) domain: A link between DNA methylation, replication and transcriptional regulation. *FEBS Lett 446*: 189–193

Cao, X., Springer, N.M., Muszynskil, M.G., Phillips, R.L., Kaeppler, S. & Jacobsen, S.E. (2000) Conserved plant genes with similarity to mammalian *de novo* DNA methyltransferases. *Proc Natl Acad Sci USA 97*: 4979-4984

Cao, X. & Jacobsen, S.E. (2002a) Locus-specific control of asymmetric and CpNpG methylation by the *DRM* and *CMT3* methyltransferase genes. *Proc Natl Acad Sci USA 99, Suppl 4*: 16491-16498

Cao, X. & Jacobsen, S.E. (2002b) Role of the *Arabidopsis DRM* methyltransferases in *de novo* DNA methylation and gene silencing. *Curr Biol 12*: 1138–1144

Cao, X., Aufsatz, W., Zilberman, D., Mette, M.F., Huang, M.S., Matzke, M. & Jacobsen, S.E. (2003) Role of the DRM and CMT3 methyltransferases in RNA-directed DNA methylation. *Curr Biol 13*: 2212-2217

Carmell, M.A., Xuan, Z., Zhang, M.Q. & Hannon, G.J. (2002) The Argonaute family: tentacles that reach into RNAi, developmental control, stem cell maintenance, and tumorigenesis. *Genes Develop 16*: 2733-2742

Cavalli, G. & Paro, R. (1998) Chromo-domain proteins: linking chromatin structure to epigenetic regulation. *Curr Opin Cell Biol 10*:354-360

Chan, S.W., Zilberman, D., Xie, Z., Johansen, L.K., Carrington, J.C. & Jacobsen, S.E. (2004) RNA silencing genes control *de novo* DNA methylation. *Science 303*: 1336

Chan, S.W., Henderson, I.R. & Jacobsen, S.E. (2005) Gardening the genome: DNA methylation in *Arabidopsis thaliana. Nat Rev Genet 6*: 351-360

Chan, S.W.L., Zhang, X., Bernatavichute, Y.V. & Jacobsen, S.E. (2006a) Two-step recruitment of RNA-directed DNA methylation to tandem repeats. *PLoS Biol 4*: e363.

Chan, S.W.L., Henderson, I.R., Zhang, X., Shah, G., Chien, J.S. & Jacobsen, S.E. (2006b) RNAi, DRD1, and histone methylation actively target developmentally important non-CG DNA methylation in *Arabidopsis. PLoS Genet 2*: e83

Charles, M.P., Ravanat, J.L., Adamski, D., D'Orazi, G., Cadet, J., Favier, A., Berger, F. & Wion, D. (2004) N(6)-Methyldeoxyadenosine, a nucleoside commonly found in prokaryotes, induces C2C12 myogenic differentiation. *Biochem Biophys Res Commun 314*: 476-482

Chaudhury, A.M., Koltunow, A., Payne, T., Luo, M., Tucker, M.R., Dennis, E.S. & Peacock, W.J. (2001) Control of early seed development. *Annu Rev Cell Dev Biol 17*: 677-699

Cheng, X., Kumar, S., Posfai, J., Pflugrath, J.W. & Roberts, R.J. (1993) Crystal structure of the HhaI DNA methyltransferase complexed with S-adenosyl-L-methionine. *Cell 74*: 299-307

Cheung, P. & Lau, P. (2005) Epigenetic Regulation by Histone Methylation and Histone Variants. *Mol Endocrinol 19*: 563-573

Chopra, S., Cocciolone, S.M., Bushman, S., Sangar, V., McMullen, M.D. & Peterson, T. (2003) The maize unstable factor for orange1 is a dominant epigenetic modifier of a tissue specifically silent allele of pericarp color1. *Genetics 163*: 1135-1146

Chvojka, L., Sulimova, G.E., Bulgakov, R., Bashkite, E.A. & Vanyushin, B.F. (1978) Changes in the 5-methylcytosine content in plant DNA associated with flowering gradient. *Biokhimiya (Moscow) 43*: 996-1000

Colot, V. & Rossignol, J.L. (1999) Eukaryotic DNA methylation as an evolutionary device. *Bioessays 21*: 402-11

Cummings, D.J., Tait, A. & Godard, J.M. (1974) Methylated bases in DNA from *Paramecium aurelia. Biochim Biophys Acta 374*: 1-11

Dalmay, T., Hamilton, A., Rudd, S., Angell, S. & Baulcombe, D.C. (2000a) An RNA-Dependent RNA Polymerase Gene in Arabidopsis Is Required for Posttranscriptional Gene Silencing Mediated by a Transgene but Not by a Virus. *Cell 101*: 543-553

Dalmay, T., Hamilton, A., Mueller, E. & Baulcombe, D.C. (2000b) Potato virus X amplicons in *Arabidopsis* mediate genetic and epigenetic gene silencing. *Plant Cell 12*: 369-379

Dennis, E.S., Bilodeau, P., Burn, J., Finnegan, E.J., Genger, R., Helliwell, C., Kang, B.J., Sheldon, C.C. & Peacock, W.J. (1998) Methylation controls the low temperature induction of flowering in *Arabidopsis. Symp Soc Exp Biol 51*: 97-103

Dieguez, M.J., Vaucheret, H., Paszkowski, J. & Mittelsten Scheid, O. (1998) Cytosine methylation at CG and CNG sites is not a prerequisite for the initiation of transcriptional gene silencing in plants, but it is required for its maintenance. *Mol Gen Genet 259*: 207-215

Dodge, J.E., Ramsahoye, B.H., Wo, Z.G., Okano, M. & Li, E. (2002) De novo methylation of MMLV provirus in embryonic stem cells: CpG versus non-CpG methylation. *Gene 289*: 41-48

Dunn, D.B. & Smith, J.D. (1955) Occurrence of a new base in the deoxyribonucleic acid of a strain of *Bacterium coli. Nature 175*: 336-337

Fagard, M., Boutet, S., Morel, J.-B., Bellini, C. & Vaucheret, H. (2000) AGO1, QDE-2, and RDE-1 are related proteins required for post-transcriptional gene silencing in plants, quelling in fungi, and RNA interference in animals. *Proc Nat Acad Sci USA 97*: 11650-11654

Fantappie, M.R., Gimba, E.R. & Rumjanek, F.D. (2001) Lack of DNA methylation in *Schistosoma mansoni. Exp Parasitol 98*: 162-166

Fedoreyeva, L.I. & Vanyushin, B.F. (2002) N^6-adenine DNA-methyltransferase in wheat seedlings. *FEBS Lett 514*: 305-308

Fedoreyeva, L.I., Sobolev, D.E. & Vanyushin, B.F. (2007) Wheat endonuclease WEN1 dependent on S-adenosyl-L-methionine and sensitive to DNA methylation status. *Epigenetics 2*: 50-53

Fedoroff, N.V. (1995) DNA methylation and activity of the maize Spm transposable element. *Curr Top Microbiol Immunol 197*: 143-164

Finnegan, E.J. & Dennis, E.S. (1993) Isolation and identification by sequence homology of a putative cytosine methyltransferase from *Arabidopsis thaliana. Nucleic Acids Res 21*: 2383-2388

Finnegan, E.J. & Kovac, K.A. (2000) Plant DNA methyltransferases. *Plant Mol Biol 43*: 189-210

Finnegan, E.J., Peacock, W.J. & Dennis, E.S. (1996) Reduced DNA methylation in *Arabidopsis thaliana* results in abnormal plant development. *Proc Natl Acad Sci USA 93*: 8449-8454

Finnegan, E.J., Genger, R.K., Kovac, K., Peacock, W.J. & Dennis, E.S. (1998a) DNA methylation and the promotion of flowering by vernalization. *Proc Natl Acad Sci USA 95*: 5824-5829

Finnegan, E.J., Genger, R.K., Peacock, W.J. & Dennis, E.S. (1998b) DNA methylation in plants. *Ann Rev Plant Physiol Plant Mol Biol 49*: 223-247

Finnegan, E.J., Peacock, W.J. & Dennis, E.S. (2000) DNA methylation, a key regulator of plant development and other processes. *Curr Opin Genetics Developm 10*: 217–223

Finnegan, E.J., Sheldon, C.C., Jardinaud, F., Peacock, W.J. & Dennis, E.S. (2004) A cluster of *Arabidopsis* genes with a coordinate response to an environmental stimulus. *Curr Biol 14*: 911-916

Fischle, W., Wang, Y., Jacobs, S.A., Kim, Y., Allis, C.D. & Khorasanizadeh, S. (2003) Molecular basis for the discrimination of repressive methyl-lysine marks in histone H3 by Polycomb and HP1 chromodomains. *Genes Develop 17*:1870-1881

Fransz, P.F., Armstrong, S., de Jong, J.H., Parnell, L.D., van Drunen, C., Dean, C., Zabel, P., Bisseling, T. & Jones, G.H. (2000) Integrated cytogenetic map of chromosome arm 4S of *A. thaliana*: structural organization of heterochromatic knob and centromere region. *Cell 100*: 367-376

Fulnecek, J., Matyasek, R. & Kovarik, A. (2002) Distribution of 5-methylcytosine residues in 5S rRNA genes in *Arabidopsis thaliana* and *Secale cereale*. *Mol Genet Genomics 268*: 510-517

Gendrel, A.-V., Lippman, Z., Yordan, C., Colot, V. & Martienssen, R.A. (2002) Dependence of heterochromatic histone H3 methylation patterns on the *Arabidopsis* gene *DDM1*. *Science 297*: 1871-1873

Gendrel, A.-V. & Colot, V. (2005) *Arabidopsis* epigenetics: when RNA meets chromatin. *Curr Opin Plant Biol 8*: 142-147

Genger, R.K., Kovac, K.A., Dennis, E.S., Peacock, W.J. & Finnegan, E.J. (1999) Multiple DNA methyltransferase genes in *Arabidopsis thaliana*. *Plant Mol Biol 41*: 269-278

Goll, M.G., Kirpekar, F., Maggert, K.A., Yoder, J.A., Hsieh, C.L., Zhang, X., Golic, K.G., Jacobsen, S.E. & Bestor, T.H. (2006) Methylation of tRNAAsp by the DNA methyltransferase homolog Dnmt2. *Science 311*: 395-398

Goodrich, J., Puangsomlee, P., Martin, M., Long, D., Meyerowitz, E.M. & Coupland, G. (1997) Polycomb-group gene regulates homeotic gene expression in Arabidopsis. *Nature 386*: 44-51

Gorovsky, M.A., Hattman, S. & Pleger, G.L. (1973) [N^6]methyladenine in the nuclear DNA of eucaryote, *Tetrahymena pyrifirmis*. *J Cell Biol 56*: 697-701

Goubely, C., Arnaud, P., Tatout, C., Heslop-Harrison, J.S. & Deragon, J.M. (1999) S1 SINE retroposons are methylated at symmetrical and non-symmetrical positions in *Brassica napus*: identification of a preferred target site for asymmetrical methylation. *Plant Mol Biol 39*: 243-255

Gowher, H., Leismann, O. & Jeltsch, A. (2000) DNA of *Drosophila melanogaster* contains 5-methylcytosine. *EMBO J 19*: 6918-6923

Graham, M.W. & Larkin, P.J. (1995) Adenine methylation at dam sites increases transient gene expression in plant cells. *Transgenic Res 4*: 324-331

Grossniklaus, U., Vielle-Calzada, J.-P., Hoeppner, M.A. & Gagliano, W.B. (1998) Maternal Control of Embryogenesis by MEDEA, a Polycomb Group Gene in Arabidopsis. *Science* 280: 446 - 450

Gruenbaum, Y., Naveh-Many, T. & Cedar, H. (1981) Sequence specificity of methylation in higher plant DNA. *Nature 292*: 860-862

Guseinov, V.A. & Vanyushin, B.F. (1975) Content and localization of 5-methylcytosine in DNA of healthy and wilt-infected cotton plants. *Biochim Biophys Acta 395*: 229-238

Gutierrez, J.C., Callejas, S., Borniquel, S. & Martin-Gonzalez, A. (2000) DNA methylation in ciliates: implications in differentiation processes. *Int Microbiol 3*: 139-46

Hamilton, A.J., Brown, S., Yuanhai, H., Ishizuka, M., Lowe, A., Solis, A.-G.A. & Grierson, D. (1998) A transgene with repeated DNA causes high frequency, post-transcriptional suppression of ACC-oxidase gene expression in tomato. *Plant J 15*: 737-746

Hamilton, A.J. & Baulcombe, D.C. (1999) A species of small antisense RNA in posttranscriptional gene silencing in plants. *Science 286*: 950-952

Hamilton, A., Voinnet, O., Chappell, L. & Baulcombe, D. (2002) Two classes of short interfering RNA in RNA silencing. *EMBO J 21*: 4671-4679

Harrison, G.S., Findly, R.C. & Karrer, K.M. (1986) Site-specific methylation of adenine in the nuclear genome of a eukaryote, *Tetrahymena thermophila. Mol. Cell Biol 6*: 2364-2370

Hashida, S.N., Kitamura, K., Mikami, T. & Kishima, Y. (2003) Temperature shift coordinately changes the activity and the methylation state of transposon Tam3 in *Antirrhinum majus. Plant Physiol 132*: 1207-1216

Hattman, S., Kenny, C., Berger, L. & Pratt, K. (1978) Comparative study of DNA methylation in three unicellular eucaryotes. *J Bacteriol 135*: 1156-1157

Henikoff, S. & Comai, L. (1998) A DNA methyltransferase homologous with a chromodomain exists in multiple polymorphic forms in *Arabidopsis. Genetics 149*: 307-318

Herr, A.J. (2005) Pathways through the small RNA world of plants. *FEBS Lett 579*: 5879–5888

Herr. A.J., Jensen, M.B., Dalmay, T. & Baulcombe, D.C. (2005) RNA polymerase IV directs silencing of endogenous DNA. *Science* 308: 118-120

Hirochika, H., Okamoto, H. & Kakutani, T. (2000) Silencing of retrotransposons in *Arabidopsis* and reactivation by the *ddm1* mutation. *Plant Cell 12*: 357-369

Holliday, R. & Pugh, J.E. (1975) DNA modification mechanisms and gene activity during development. *Science 187*: 226-232

Honma, T. & Goto, K. (2001) Complexes of MADS-box proteins are sufficient to convert leaves into floral organs. *Nature 409*: 525-529

Jackson, J.P., Lindroth, A.M., Cao, X. & Jacobsen, S.E. (2002) Control of CpNpG DNA methylation by the KRYPTONITE histone H3 methyltransferase. *Nature 416*: 556-560

Jackson, J.P., Johnson, L., Jasencakova, Z., Zhang, X., PerezBurgos, L., Singh, P.B., Cheng, X., Schubert, I., Jenuwein, T. & Jacobsen, S.E. (2004) Dimethylation of histone H3 lysine 9 is a critical mark for DNA methylation and gene silencing in *Arabidopsis thaliana*. *Chromosoma 112*: 308-315

Jacobsen, S.E. & Meyerowitz, E.M. (1997) Hypermethylated *SUPERMAN* epigenetic alleles in *Arabidopsis*. *Science 277*: 1100-1103

Jacobsen, S.E., Sakai, H., Finnegan, E.J., Cao, X. & Meyerowitz, E.M. (2000) Ectopic hypermethylation of flower-specific genes in *Arabidopsis*. *Curr Biol 10*: 179-186

Jeddeloh, J.A., Bender, J. & Richards, E.J. (1998) The DNA methylation locus DDM1 is required for maintenance of gene silencing in *Arabidopsis*. *Genes Devel 12*: 1714-1725

Jeddeloh, J.A., Stokes, T.L. & Richards, E.J. (1999) Maintenance of genomic methylation requires a SWI2/SNF2-like protein. *Nat Genet 22*: 94-97

Jeltsch, A. (2001) The cytosine N4-methyltransferase M.PvuII also modifies adenine residues. *J Biol Chem 382*: 707-710

Johnson, L.M., Cao, X. & Jacobsen, S.E. (2002) Interplay between two epigenetic marks. DNA methylation and histone H3 lysine 9 methylation. *Curr Biol 12*: 1360–1367

Jones, L., Ratcliff, F. & Baulcombe, D.C. (2001) RNA-directed transcriptional gene silencing in plants can be inherited independently of the RNA trigger and requires Met1 for maintenance. *Curr Biol 11*: 747-757

Kakutani, T. (2002) Epi-alleles in plants: inheritance of epigenetic information over generations. *Plant Cell Physiol 43*: 1106-1111

Kakutani, T., Jeddeloh, J.A. & Richards, E.J. (1995) Characterization of an *Arabidopsis thaliana* DNA hypomethylation mutant. *Nucl Acids Res 23*: 130-137

Kakutani, T., Jeddeloh, J.A., Flowers, S.K., Munakata, K. & Richards, E.J. (1996) Developmental abnormalities and epimutations associated with DNA hypomethylation mutations. *Proc Natl Acad Sci USA 93*: 12406-12411

Kakutani, T., Munakata, K., Richards, E.J. & Hirochika, H. (1999) Meiotically and mitotically stable inheritance of DNA hypomethylation induced by *ddm1* mutation of *Arabidopsis thaliana*. *Genetics 151*: 831-838

Kankel, M.W., Ramsey, D.E., Stokes, T.L., Flowers, S.K., Haag, J.R., Jeddeloh, J.A., Riddle, N.C., Verbsky, M.L. & Richards, E.J. (2003) *Arabidopsis MET1* cytosine methyltransferase mutants. *Genetics 163*: 1109-1122

Kanno, T., Mette, M.F., Kreil, D.P., Aufsatz, W., Matzke, M. & Matzke, A.J. (2004) Involvement of putative SNF2 chromatin remodeling protein DRD1 in RNA-directed DNA methylation. *Curr Biol 14*: 801-805

Kanno, T., Aufsatz, W., Jaligot, E., Mette, M.F., Matzke, M. & Matzke, A.J.M. (2005a) A SNF2-like protein facilitates dynamic control of DNA methylation. *EMBO Rep 6*: 649–655

Kanno, T., Huettel, B., Mette, M.F., Aufsatz, W., Jaligot, E., Daxinger, L., Kreil, D.P., Matzke, M. & Matzke, A.J.M. (2005b) Atypical RNA polymerase subunits required for RNA-directed DNA methylation. *Nature Genet 37*: 761-765

Karrer, K.M. & Van Nuland, T.A. (1998) Position effect takes precedence over target sequence in determination of adenine methylation patterns in the nuclear genome of a eukayote, *Tetrahymena thermophila*. *Nucleic Acids Res 26*: 4566-4573

Karrer, K.M. & Van Nuland, T.A. (2002) Metthylation of adenine in the nuclear DNA of *Tetrahymena* is internucleosomal and independent of histone H1. *Nucleic Acids Res 30*: 1364-1370

Kato, M., Miura, A., Bender, J., Jacobsen, S.E. & Kakutani, T. (2003) Role of CG and non-CG methylation in immobilization of transposons in *Arabidopsis*. *Curr Biol 13*: 421-426

Kay, P.H., Pereira, E., Marlow, S.A., Turbett, G., Mitchell, C.A., Jacobsen, P.F., Holliday, R. & Papadimitriou, J.M. (1994) Evidence for adenine methylation within the mouse myogenic gene Myo-D1. *Gene 151*: 89-95

Kinoshita, T., Miura, A., Choi, Y., Kinoshita, Y., Cao, X., Jacobsen, S.E., Fischer, R.L. & Kakutani, T. (2004) One-way control of FWA imprinting in *Arabidopsis* endosperm by DNA methylation. *Science 303*: 521-523

Kirnos, M.D., Merkulova, N.A., Borkhsenius, S.N. & Vanyushin, B.F. (1980) Character of macronucleus DNA methylation in protozoan *Tetrahymena pyriformis*. *Dokl Akad Nauk SSSR 255*: 225-227

Kirnos, M.D., Aleksandrushkina, N.I. & Vanyushin, B.F. (1981) 5-Methylcytosine in pyrimidine sequences of plant and animal DNA: specificity of DNA methylation. *Biokhimiya (Moscow) 46*: 1458-1474

Kirnos, M.D., Alexandrushkina, N.I., Goremykin, V.V., Kudryashova, I.B. & Vanyushin, B.F. (1992a) "Heavy" mitochondrial DNA in higher plants. *Biokhimiya (Moscow) 57*: 1566-1573

Kirnos, M.D., Alexandrushkina, N.I., Zagorskaya, G.Ya., Kireev, I.I. & Vanyushin, B.F. (1992b) Superproduction of heavy minicircular mitochondrial DNA in aging wheat coleoptiles. *FEBS Lett 298*: 109-112

Kishimoto, N., Sakai, H., Jackson, J., Jacobsen, S.E., Meyerowitz, E.M., Dennis, E.S. & Finnegan, E.J. (2001) Site specificity of the *Arabidopsis* METI DNA methyltransferase demonstrated through hypermethylation of the superman locus. *Plant Mol Biol 46*: 171-183

Kitamura, K., Hashida, S.N., Mikami, T. & Kishima, Y. (2001) Position effect of the excision frequency of the *Antirrhinum* transposon Tam3: implications for the degree of position-dependent methylation in the ends of the element. *Plant Mol Biol 47*: 475-490

Klimasauskas, S., Kumar, S., Roberts, R.J. & Cheng, X. (1994) HhaI methyltransferase flips its target base out of the DNA helix. *Cell 76*: 357-369

Klotti, A., He, X., Potrykus, I., Hohn, T. & Futterer, J. (2002) Tissue-specific silencing of a transgene in rice. *Proc Natl Acad Sci USA 99*: 10881-10886

Kooter, J.M., Matzke, M.A. & Meyer, P. (1999) Listening to the silent genes: transgene silencing, gene regulation and pathogen control. *Trends Plant Sci 4*: 340-347

Kouzminova, E. & Selker, E.U. (2001) *dim-2* encodes a DNA methyltransferase responsible for all known cytosine methylation in *Neurospora. EMBO J 20*: 4309–4323

Kovalchuk, O., Burke, P., Arkhipov, A., Kuchma, N., James, S.J., Kovalchuk, I. & Pogribny, I. (2003) Genome hypermethylation in Pinus silvestris of Chernobyl - a mechanism for radiation adaptation? *Mutat Res 529*: 13-20

Kovarik, A., Matyasek, R., Leitch, A., Gazdova, B., Fulnecek, J. & Bezdek, M. (1997) Variability in CpNpG methylation in higher plant genomes. *Gene 204*: 25-33

Kovarik, A., Van Houdt, H., Holy, A. & Depicker, A. (2000b) Drug-induced hypomethylation of a posttranscriptionally silenced transgene locus of tobacco leads to partial release of silencing. *FEBS Lett 467*: 47-51

Kudryashova, I.B. & Vanyushin, B.F. (1986) Incorporation of cytokinins in DNA of wheat seedlings. *Biokhimiya (Moscow) 51*: 321-327

Kumpatla, S.P. & Hall, T.C. (1998) Longevity of 5-azacytidine-mediated gene expression and reestablishment of silencing in transgenic rice. *Plant Mol Biol 38*: 1113-1122

Kutueva, L.I., Ashapkin, V.V. & Vanyushin, B.F. (1996) The methylation pattern of a cytosine DNA-methyltransferase gene in *Arabidopsis thaliana* plants. *Biochem Mol Biol Intern 40*:347-353

Lachner, M., O'Sullivan, R.J. & Jenuwein, T. (2003) An epigenetic road map for histone lysine methylation. *J Cell Sci 116*: 2117-2124

Lauria, M., Rupe, M., Guo, M., Kranz, E., Pirona, R., Viotti, A. & Lund, G. (2004) Extensive maternal DNA hypomethylation in the endosperm of *Zea mays*. *Plant Cell 16*: 510-522

Lawley, P.D., Crathorn, A.R., Shah, S.A. & Smith, B.A. (1972) Biomethylation of deoxyribonucleic acid in cultured human tumor cells (HeLa). Methylated bases other than 5-methylcytosine not detected. *Biochem J 128*: 133–138

Li, E., Bestor, T.H. & Jaenisch, R. (1992) Targeted mutation of the DNA methyltransferase gene results in embryonic lethality. *Cell 69*: 915–926

Li, G., Hall, T.C. & Holmes-Davis, R. (2002) Plant chromatin: development and gene control. *Bioessays 24*: 234-243

Lindroth, A.M., Cao, X., Jackson, J.P., Zilberman, D., McCallum, C.M., Henikoff, S. & Jacobsen, S.E. (2001) Requirement of chromomethylase3 for maintenance of CpXpG methylation. *Science 292*: 2077-2080

Liu, B. & Wendel, J.F. (2003) Epigenetic phenomena and the evolution of plant allopolyploids. *Mol Phylogenet Evol 29*: 365-379

Lo Schiavo, F., Pitto, L., Giuliano, G., Torti, G., Nutti Ronchi, V., Marazziti, D., Vergara, M.R., Orselli, S. & Terzi, M. (1989) DNA methylation of embryonic carrot cell culture and its variations as caused by mutation, differentiation, hormones and hypomethylating drugs. *Theor Appl Genet 77*: 325-331

Lu, C., Tej, S.S., Luo, S., Haudenschild, C.D., Meyers, B.C. & Green, P.J. (2005) Elucidation of the small RNA component of the transcriptome. *Science* 309: 1567-1569

Luff, B., Pawlowski, L. & Bender, J. (1999) An inverted repeat triggers cytosine methylation of identical sequences in *Arabidopsis. Mol Cell 3*: 505–511

Lund, G., Ciceri, P. & Viotti, A. (1995a) Maternal-specific demethylation and expression of specific alleles of zein genes in the endosperm of *Zea mays* L. *Plant J* 8: 571–581

Lund, G., Messing, J. & Viotti, A. (1995b) Endosperm-specific demethylation and activation of specific alleles of α-tubulin genes of *Zea mays* L. *Mol Gen Genet 246*: 716–722

Lund, G., Lauria, M., Guldberg, P. & Zaina, S. (2003) Duplication-dependent CG suppression of the seed storage protein genes of maize. *Genetics 165*: 835-848

Luo, M., Bilodeau, P., Dennis, E.S., Peacock, W.J. & Chaudhury, A. (2000) Expression and parent-of-origin effects for FIS2, MEA, and FIE in the endosperm and embryo of developing *Arabidopsis* seeds. *Proc Natl Acad Sci USA 97*: 10637-10642

Lyko, F., Ramsahoye, B.H., Kashevsky, H., Tudor, M., Mastrangelo, M.-A., Orr-Weaver, T.L. & Jaenisch, R. (1999) Mammalian (cytosine-5) methyltransferases cause genomic DNA methylation and lethality in Drosophila. *Nature Genet 23*: 363 - 366

Lyko, F., Ramsahoye, B.H. & Jaenisch, R. (2000) DNA methylation in *Drosophila melanogaster. Nature 408*: 538-540

Madlung, A., Masuelli, R.W., Watson, B., Reynolds, S.H., Davison, J. & Comai, L. (2002) Remodeling of DNA methylation and phenotypic and transcriptional changes in synthetic *Arabidopsis* allotetraploids. *Plant Physiol 129*: 733-746

Malagnac, F., Wendel, B., Goyon, C., Faugeron, G., Zickler, D., Rossignol, J.L., Noyer-Weidner, M., Vollmayr, P., Trautner, T.A. & Walter, J. (1997) A gene essential for de novo methylation and development in Ascobolus reveals a novel type of eukaryotic DNA methyltransferase structure. *Cell 91*: 281-290

Martienssen, R. & Baron, A. (1994) Coordinate Suppression of Mutations Caused by Robertson's Mutator Transposons in Maize. *Genetics 136*: 1157–1170

Martienssen, R.A. & Colot, V. (2001) DNA methylation and epigenetic inheritance in plants and filamentous fungi. *Science 293*: 1070-1074

Matzke, M.A. & Matzke, A.J.M. (1995) Homology-dependent gene silencing in transgenic plants: what does it really tell us? *Trends Genet 11*: 1-3

Matzke, M.A., Mette, M.F. & Matzke, A.J. (2000) Transgene silencing by the host genome defense: implications for the evolution of epigenetic control mechanisms in plants and vertebrates. *Plant Mol Biol 43*: 401-415

Matzke, M., Aufsatz, W., Kanno, T., Daxinger, L., Papp, I., Mette, M.F. & Matzke, A.J.M. (2004) Genetic analysis of RNA-mediated transcriptional gene silencing. *Biochim Biophys Acta 167*: 129-141

Mazin, A.L., Gimadutdinov, O.A., Turkin, S.I., Burtseva, N.N. & Vanyushin, B.F. (1985) Non-enzymatic DNA methylation by S-adenosylmethionine resulting in formation of minor 5-methylcytosine and thymine from cytosine residues. *Mol. Biologiya (Moscow) 19*: 903-914

Mazin, A.L. & Vanyushin, B.F. (1986) Incorporation of cytokinin (6-benzylaminopurine) in DNA of *Tetrahymena pyriformis. Izv Akad Nauk USSR, Ser Biol 1*: 122-124

McCallum, C.M., Comai, L., Greene, E.A. & Henikoff, S. (2000) Targeted screening for induced mutations. *Nature Biotechnol 18*: 455-457

McClintock, B. (1967) Genetic systems regulating gene expression during development. *Dev Biol Suppl 1*: 84-112

Melquist, S. & Bender, J. (2003) Transcription from an upstream promoter controls methylation signaling from an inverted repeat of endogenous genes in *Arabidopsis. Genes Dev 17*: 2036-2047

Mette, M.F., van der Winden, J., Matzke, M.A. & Matzke, A.J.M. (1999) Production of aberrant promoter transcripts contributes to methylation and silencing of unlinked homologous promoters *in trans*. *EMBO J 18*: 241–248

Mette, M.F., Aufsatz, W., van der Winden, J., Matzke, M.A. & Matzke, A.J. (2000) Transcriptional silencing and promoter methylation triggered by double-stranded RNA. *EMBO J 19*: 5194-5201

Meyer, P. (1995) DNA methylation and transgene silencing in *Petunia hybrida. Curr Top Microbiol Immunol 197*: 15-28

Miura, A., Yonebayashi, S., Watanabe, K., Toyama, T., Shimada, H. & Kakutani, T. (2001) Mobilization of transposons by a mutation abolishing full DNA methylation in *Arabidopsis. Nature 411*: 212-214

Mlotshwa, S., Voinnet, O., Mette, M.F., Matzke, M., Vaucheret, H., Ding, S.W., Pruss, G. & Vance, V.B. (2002) RNA silencing and the mobile silencing signal. *Plant Cell 14*: S289-S301

Moehrle, A. & Paro, R. (1994) Spreading the silence: epigenetic transcriptional regulation during *Drosophila* development. *Dev Genet 15*: 478-484

Monk, M., Boubelik, M. & Lehnert, S. (1987) Temporal and regional changes in DNA methylation in the embryonic, extraembryonic and germ cell lineages during mouse embryo development. *Development 99*: 371-382

Montgomery, M.K. (2004) RNA interference: historical overview and significance. *Methods Mol Biol 265*: 3-21

Nakano, Y., Steward, N., Sekine, M., Kusano, T. & Sano, H. (2000) A tobacco NtMET1 cDNA encoding a DNA methyltransferase: molecular characterization and abnormal phenotypes of transgenic tobacco plants. *Plant Cell Physiol 41*: 448-457

Nakayama, J.-i., Rice, J.C., Strahl, B.D., Allis, C.D. & Grewal, S.I.S. (2001) Role of histone H3 lysine 9 methylation in epigenetic control of heterochromatin assembly. *Science* 292: 110 - 113

Nelson, M., Burbank, D.E. & Van Etten, J.L. (1998) *Chlorella* viruses encode multiple DNA methyltransferases. *Biol Chem 379*: 423-428

Ngernprasirtsiri, J. & Akazawa, T. (1990) Modulation of DNA methylation and gene expression in cultured sycamore cells treated by hypomethylating base analog. *Eur J Biochem 194*: 513-520

Ngernprasirtsiri, J., Kobayashi, H. & Akazawa, T. (1988) DNA methylation as a mechanism of transcriptional regulation in nonphotosynthetic plastids in plant cells. *Proc Natl Acad Sci USA 85*: 4750-4754

Okano, M., Xie, S. & Li, E. (1998) Dnmt2 is not required for *de novo* and maintenance methylation of viral DNA in embryonic stem cells. *Nucl Acids Res 26*: 2536-2540

Okano, M., Bell, D.W., Haber, D.A. & Li, E. (1999) DNA methyltransferases Dnmt3a and Dnmt3b are essential for *de novo* methylation and mammalian development. *Cell 99*: 247–257

Pakhomova, M.V. (1974) N^6-dimethylaminopurine in DNA of algae species. *Dokl Akad Nauk SSSR 214*: 1202-1205

Pakhomova, M.V., Zaitseva, G.N. & Belozerskii, A.N. (1968) The presence of 5-methylcytosine and 6-methylaminopurine in the composition of DNA in some algae. *Dokl Akad Nauk SSSR 182*: 712-714

Palmgren, G., Mattson, O. & Okkels, F.T. (1991) Specific levels of DNA methylations in various tissues, cell lines, and cell types of *Daucus carota. Plant Physiol 95*: 174-178

Papa, C.M., Springer, N.M., Muszynski, M.G., Meeley, R. & Kaeppler, S.M. (2001) Maize chromomethylase Zea methyltransferase2 is required for CpNpG methylation. *Plant Cell 13*: 1919-1928

Park, Y.D., Papp, I., Moscone, E.A., Iglesias, V.A., Vaucheret, H., Matzke, A.J. & Matzke, M.A. (1996) Gene silencing mediated by promoter homology occurs at the level of transcription and results in meiotically heritable alterations in methylation and gene activity. *Plant J 9*: 183-194

Paszkowski, J. & Whitham, S.A. (2001) Gene silencing and DNA methylation processes. *Curr Opin Plant Biol 4*: 123-129

Pelaz, S., Ditta, G.S., Baumann, E., Wisman, E. & Yanofsky, M.F. (2000) B and C floral organ identity functions require SEPALLATA MADS-box genes. *Nature 405*: 200-203

Pelissier, T., Thalmeir, S., Kempe, D., Sanger, H.-L. & Wassenegger, M. (1999) Heavy *de novo* methylation at symmetrical and non-symmetrical sites is a hallmark of RNA-directed DNA methylation. *Nucleic Acids Res 27*: 1625-1634

Pelissier, T. & Wassenegger, M. (2000) A DNA target of 30 bp is sufficient for RNA-directed DNA methylation. *RNA 6*: 55-65

Pintor-Toro, J.A. (1987) Adenine methylation in zein genes. *Biochem Biophys Res Commun 147*: 1082-1087

Pradhan, S. & Adams, R.L. (1995) Distinct CG and CNG DNA methyltransferases in *Pisum sativum*. *Plant J 7*: 471-81

Pradhan, S., Cummings, M., Roberts, R.J. & Adams, R.L.P. (1998) Isolation, characterization and baculovirus-mediated expression of the cDNA encoding cytosine DNA methyltransferase from *Pisum sativum*. *Nucleic Acids Res 26*: 1214-1222

Pradhan, S., Bacolla, A., Wells, R.D. & Roberts, R.J. (1999) Recombinant human DNA (cytosine-5) methyltransferase I. Expression, purification, and comparison of *de novo* and maintenance methylation. *J Biol Chem 274*: 33002-33010

Prakash, A.P., Kush, A., Lakshmanan, P. & Kumar, P.P. (2003) Cytosine methylation occurs in a CDC48 homologue and a MADS-box gene during adventitious shoot induction in *Petunia* leaf explants. *J Exp Bot 54*: 1361-1371

Pratt, K. & Hattman, S. (1981) Deoxyribonucleic acid methylation and chromatin organization in *Tetrahymena thermophila*. *Mol Cell Biol 1*: 600-608

Pray-Grant, M.G., Daniel, J.A., Schieltz, D., Yates, J.R. 3rd & Grant, P.A. (2005) Chd1 chromodomain links histone H3 methylation with SAGA- and SLIK-dependent acetylation. *Nature 433*: 434-438

Que, Q., Zhang, Y., Nelson, M., Ropp, S., Burbank, D.E. & Van Etten, J.L. (1997) *Chlorella* virus SC-1A encodes at least five functional and one nonfunctional DNA methyltransferases. *Gene 190*: 237-244

Rae, P.M. & Spear, B.B. (1978) Macronuclear DNA of the hypotrichous ciliate *Oxytricha fallax*. *Proc Natl Acad Sci USA 75*: 4992-4996

Ratcliff, F., Harrison, B.D. & Baulcombe, D.C. (1997) A similarity between viral defense and gene silencing in plants. *Science 276*: 1558 – 1560

Ratcliff, F., MacFarlane, S.A. & Baulcombe, D.C. (1999) Gene silencing without DNA: RNA-mediated cross-protection between viruses. *Plant Cell 11*: 1207-1216

Ratel, D., Boisseau, S., Davidson, S.M., Ballester, B., Mathieu, J., Morange, M., Adamski, D., Berger, F., Benabid, A.L. & Wion, D. (2001) The bacterial nucleoside N(6)-methyldeoxyadenosine induces the differentiation of mammalian tumor cells. *Biochem Biophys Res Commun 285*: 800-805

Ratel, D., Ravanat, J.L., Berger, F. & Wion, D. (2006) N^6-methyladenine: the other methylated base of DNA. *Bioessays 28*: 309-315

Razin, A. (1998) CpG methylation, chromatin structure and gene silencing - a three-way connection. *EMBO J* 17: 4905–4908

Razin, A. & Riggs, A.D. (1980) DNA methylation and gene function. *Science 210*: 604-610

Rea, S., Eisenhaber, F., O'Carroll, D., Strahl, B.D., Sun, Z.-W., Schmid, M., Opravil, S., Mechtler, K., Ponting, C.P., Allis, C.D. & Jenuwein, T. (2000) Regulation of chromatin structure by site-specific histone H3 methyltransferases. *Nature 406*: 593-599

Reyes, E.M., Camacho-Arroyo, I., Nava, G. & Cerbon, M.A. (1997) Differential methylation in steroid 5 alpha-reductase isozyme genes in epididymis, testis, and liver of the adult rat. *J Androl 18*: 372-377

Richards, E.J. (1997) DNA methylation and plant development. *Trends Genet 319*: 319-323

Riggs, C.D. & Chrispeels, M.J. (1999) The expression of phytohemagglutinin genes in *Phaseolus vulgaris* is associated with organ-specific DNA methylation patterns. *Plant Mol Biol 14*: 629-532

Rogers, S.W. & Rogers, J.C. (1992) The importance of DNA methylation for stability of foreign DNA in barley. *Plant Mol Biol 18*: 945-961

Rogers, J. & Rogers, S.W. (1995) Comparison of the effects of N6-methyldeoxyadenosine and N5-methyldeoxycytosine on transcription from nuclear gene promoters in barley. *Plant J 7*: 221–233

Rogers, S.D., Rogers, M.E., Saunders, G. & Holt, G. (1986) Isolation of mutants sensitive to 2-aminopurine and alkylating agents and evidence for the role of DNA methylation in *Penicillium chrysogenum. Curr Genet 10*: 557-560

Rojas, M.V. & Galanti, N. (1990) DNA methylation in *Trypanosoma cruzi. FEBS Lett 263*: 113-116

Ronchi, A., Petroni, K. & Tonelli, C. (1995) The reduced expression of endogenous duplications (REED) in the maize R gene family is mediated by DNA methylation. *EMBO J 14*: 5318–5328

Ronemus, M.J., Galbiati, M., Ticknor, C., Chen, J. & Dellaporta, S.L. (1996) Demethylation-induced developmental pleiotropy in *Arabidopsis. Science 273*: 654-657

Ros, F. & Kunze, R. (2001) Regulation of activator/dissociation transposition by replication and DNA methylation. *Genetics 157*: 1723-1733

Rossi, V., Motto, M. & Pellegrini, L. (1997) Analysis of the methylation pattern of the maize Opaque-2(02) promoter and *in vitro* binding studies indicate that the O2 B-Zip protein and other endosperm factors can bind to methylated target sequences. *J Biol Chem 272*: 13758-13765

Ruiz, M.T., Voinnet, O. & Baulcombe, D.C. (1998) Initiation and maintenance of virus-induced gene silencing. *Plant Cell 10*: 937-946

Sano, H. (2002) DNA methylation and Lamarckian inheritance. *Proc Japan Acad 78 Ser B*: 293-298

Santos-Rosa, H., Schneider, R., Bannister, A.J., Sherriff, J., Bernstein, B.E., Emre, N.C., Schreiber, S.L., Mellor, J. & Kouzarides, T. (2002) Active genes are tri-methylated at K4 of histone H3. *Nature 419*: 407–411

Santos-Rosa, H., Schneider, R., Bernstein, B.E., Karabetsou, N., Morillon, A., Weise, C., Schreiber, S.L., Mellor, J. & Kouzarides, T. (2003) Methylation of histone H3 K4 mediates association of the Isw1p ATPase with chromatin. *Molecular Cell 12*: 1325-1332

Saze, H., Scheid, O.M. & Paszkowski, J. (2003) Maintenance of CpG methylation is essential for epigenetic inheritance during plant gametogenesis. *Nat Genet 34*: 65-69

Schlappi, M., Raina, R. & Fedoroff, N. (1994) Epigenetic regulation of the maize *Spm* transposable element: novel activation of a methylated promoter by TnpA. *Cell 77*: 427-437

Schübeler, D., MacAlpine, D.M., Scalzo, D., Wirbelauer, C., Kooperberg, C., van Leeuwen, F., Gottschling, D.E., O'Neill, L.P., Turner, B.M., Delrow, J., Bell, S.P. & Groudine, M. (2004) The histone modification pattern of active genes revealed through genome-wide chromatin analysis of a higher eukaryote. *Genes Develop 18*:1263-1271

Scott, R.J. & Spielman, M. (2004) Epigenetics: imprinting in plants and mammals - the same but different? *Curr Biol 14*: R201-R203

Sheldon, C.C., Finnegan, E.J., Rouse, D.T., Tadege, M., Bagnall, D.J., Helliwell, C.A., Peacock, W.J. & Dennis, E.S. (2000) The control of flowering by vernalization. *Curr Opin Plant Biol 3*: 418-422

Sherman, J.D. & Talbert, L.E. (2002) Vernalization-induced changes of the DNA methylation pattern in winter wheat. *Genome 45*: 253-260

Shorning, B.Yu. & Vanyushin, B.F. (2001) Putative DNA-(amino)methyltransferases in eukaryotes. *Biochemistry (Moscow) 66*: 753-762

Singer, T., Yordan, C. & Martienssen, R.A. (2001) Robertson's mutator transposons in *A. thaliana* are regulated by the chromatin-remodeling gene decrease in DNA methylation (*DDM1*). *Genes Dev 15*: 591-602

Smith, S.S., Laayoun, A., Lingeman, R.G., Baker, D.J. & Riley, J. (1994) Hypermethylation of telomere-like foldbacks at codon 12 of the human c-Ha-ras gene and the trinucleotide repeat of the FMR-1 gene of fragile X. *J Mol Biol 243*: 143–151

Soppe, W.J.J., Jacobsen, S.E., Alonso-Blanco, C., Jackson, J.P., Kakutani, T., Koornneef, M. & Peeters, A.J.M. (2000) The late flowering phenotype of *fwa* mutants is caused by gain-of-function epigenetic alleles of a homeodomain gene. *Mol Cell 6*: 791-802

Soppe, W.J.J., Jasencakova, Z., Houben, A., Kakutani, T., Meister, A., Huang, M.S., Jacobsen, S.E., Schubert, I. & Fransz, P.F. (2002) DNA methylation controls histone H3 lysine 9 methylation and heterochromatin assembly in *Arabidopsis*. *EMBO J 21*: 6549-6559

Spielman, M., Vinkenoog, R. & Scott, R.J. (2003) Genetic mechanisms of apomixis. *Philos Trans R Soc Lond B Biol Sci 358*: 1095-1103

Stam, M., de Bruin, R., van Blokland, R., van der Hoorn, R.A.L., Mol, J.N.M. & Kooter, J.M. (2000) Distinct features of post-transcriptional gene silencing by antisense transgenes in single copy and inverted T-DNA repeat loci. *Plant J 21*: 27-42

Steimer, A., Schob, H. & Grossniklaus, U. (2004) Epigenetic control of plant development: new layers of complexity. *Curr Opin Plant Biol 7*: 11-19

Steward, N., Kusano, T. & Sano, H. (2000) Expression of *ZmMET1*, a gene encoding a DNA methyltransferase from maize, is associated not only with DNA replication in actively proliferating cells, but also with altered DNA methylation status in cold-stressed quiescent cells. *Nucleic Acids Res 28*: 3250-3259

Steward, N., Ito, M., Yamaguchi, Y., Koizumi, N. & Sano, H. (2002) Periodic DNA methylation in maize nucleosomes and demethylation by environmental stress. *J Biol Chem 277*: 37741-37746.

Stokes, T. (2003) DNA-RNA-protein gang together in silence. *Trends Plant Sci 8*: 53-55

Sugimoto, K., Takeda, S. & Hirochika, H. (2003) Transcriptional activation mediated by binding of a plant GATA-type zinc finger protein AGP1 to the AG-motif (AGATCCAA) of the wound-inducible Myb gene NtMyb2. *Plant J 36*: 550-564

Tamaru, H. & Selker, E.U. (2001) A histone H3 methyltransferase controls DNA methylation in *Neurospora crassa. Nature 414*: 277-283

Tamaru, H., Zhang, X., McMillen, D., Singh, P.B., Nakayama, J., Grewal, S.I., Allis, C.D., Cheng, X. & Selker, E.U. (2003) Trimethylated lysine 9 of histone H3 is a mark for DNA methylation in *Neurospora crassa. Nature Genet 34*: 75-79

Tang, W. & Leisner, S. (1998) Methylation of nonintegrated multiple copy DNA in plants. *Biochem Biophys Res Commun 245*: 403-406

Tariq, M. & Paszkowski, J. (2004) DNA and histone methylation in plants. *Trends Genet 20*: 244-251

Tariq, M., Saze, H., Probst, A.V., Lichota, J., Habu, Y. & Paszkowski, J. (2003) Erasure of CpG methylation in *Arabidopsis* alters patterns of histone H3 methylation in heterochromatin. *Proc Natl Acad Sci USA 100*: 8823-8827

Tompa, .R., McCallum, C.M., Delrow, J., Henikoff, J.G., van Steensel, B. & Henikoff, S. (2002) Genome-wide profiling of DNA methylation reveals transposon targets of chromomethylase3. *Curr Biol 12*: 65-68

Unger, G. & Venner, H. (1966) Nachweis von Minoritatsbasen in Sperma Desoxyribonucleinsaure. *Hoppe-Seyler's Z Physiol Chem 344*: 280-282

Van Blokland, R., Ross, S., Corrado, G., Scollan, C. & Meyer, P. (1998) Developmental abnormalities associated with deoxyadenosine methylation in transgenic tobacco. *Plant J 15*: 543-551

Vanyushin, B.F. (1984) Replicative DNA methylation in animals and higher plants. *Curr Top Microbiol Immunol 108*: 99-114

Vanyushin, B.F. (2005) Enzymatic DNA methylation is an epigenetic control for genetic functions of the cell. *Biochemistry (Moscow) 70*: 488-499

Vanyushin, B.F. (2006) DNA methylation in plants. *Curr Top Microbiol Immunol 301*: 67-122

Vanyushin, B.F. & Belozersky, A.N. (1959) Nucleotide composition of deoxyribonucleic acids in higher plants. *Dokl Akad Nauk SSSR 129*: 944-947

Vanyushin, B.F., Belozersky, A.N., Kokurina, N.A. & Kadirova, D.X. (1968) 5-methylcytosine and 6-methylaminopurine in bacterial DNA. *Nature 218*: 1066-1067

Vanyushin, B.F., Tkacheva, S.G. & Belozersky, A.N. (1970) Rare bases in animal DNA. *Nature 225*: 948-949

Vanyushin, B.F., Kadyrova, D.K., Karimov, K.K. & Belozersky, A.N. (1971) Minor bases in DNA of higher plants. *Biokhimiya (Moscow) 36*: 1251-1258

Vanyushin, B.F., Singkh, S.S. & Sonvol, G. (1979) Changes in DNA methylation in alfalfa plants infected with *Cuscuta* and tissue differences in DNA methylation of the parasite plants. *Biokhimiya (Moscow) 44*: 864-867

Vanyushin, B.F., Aleksandrushkina, N.I. & Kirnos, M.D. (1988) N^6-methyladenine in mitochondrial DNA of higher plants. *FEBS Lett 223*: 397-399

Vanyushin, B.F., Sevostyanova, S.S., Kirnos, M.D. & Saidova, N.S. (1990) An increase in the protein content in wheat seeds induced by inhibitor of DNA methylation, 5-azacytidine. *Izv Akad Nauk SSSR Ser. Biol 1*: 75-83

Vanyushin, B.F., Shorning, B.Yu., Seredina, A.V. & Aleksandrushkina, N.I. (2002) The effects of phytohormones and 5-azacytidine on apoptosis in etiolated wheat seedlings. *Russian J Plant Physiol 49*: 501-506

Vanyushin, B.F., Bakeeva, L.E., Zamyatnina, V.A. & Aleksandrushkina, N.I. (2004) Apoptosis in plants: specific features of plant apoptotic cells and effect of various factors and agents. *Intern Rev Cytology 233*: 135-179

Vaucheret, H. & Fagard, M. (2001) Transcriptional gene silencing in plants: targets, inducers and regulators. *Trends Genet 17*: 29-35

Vielle-Calzada, J.-P., Thomas, J., Spillane, C., Coluccio, A., Hoeppner, M.A. & Grossniklaus, U. (1999) Maintenance of genomic imprinting at the *Arabidopsis* medea locus requires zygotic DDM1 activity. *Genes Develop 13*: 2971-2982

Vinkenoog, R., Bushell. C., Spielman, M., Adams, S., Dickinson, H.G. & Scott, R.J. (2003) Genomic imprinting and endosperm development in flowering plants. *Mol Biotechnol 25*: 149-184

Voinnet, O. (2002) RNA silencing: small RNAs as ubiquitous regulators of gene expression. *Curr Opin Plant Biol 5*: 444–451

Vongs, A., Kakutani, T., Martienssen, R.A. & Richards, E.J. (1993) *Arabidopsis thaliana* DNA methylation mutants. *Science 260*: 1926-1928

Wada, Y., Ohya, H., Yamaguchi, Y., Koizumi, N. & Sano, H. (2003) Preferential *de novo* methylation of cytosine residues in non-CpG sequences by a domains rearranged DNA methyltransferase from tobacco plants. *J Biol Chem 278*: 42386-42393

Wassenegger, M. (2000) RNA-directed DNA methylation. *Plant Mol Biol 43*: 203-220

Wassenegger, M. (2002) Gene silencing. *Int Rev Cytol 219*: 61-113

Wassenegger, M., Heimes, S., Riedel, L. & Sanger, H.L. (1994) RNA directed *de novo* methylation of genomic sequences in plants. *Cell 76*: 567-576

Waterhouse, P.M., Graham, M.W., Wang, M.-B. (1998) Virus resistance and gene silencing in plants can be induced by simultaneous expression of sense and antisense RNA. *Proc Nat Acad Sci USA 95*: 13959-13964

Wilkinson, C.R., Bartlett, R., Nurse, P. & Bird, A.P. (1995) The fission yeast gene pmt1+ encodes a DNA methyltransferase homologue. *Nucl Acids Res 23*: 203-210

Wion, D. & Casadesus, J. (2006) N^6-methyladenine: an epigenetic signal for DNA–protein interactions. *Nature Rev Microbiol 4*: 183-192

Xiao, W., Gehring, M., Choi, Y., Margossian, L., Pu, H., Harada, J.J., Goldberg, R.B., Pennell, R.I. & Fischer, R.L. (2003) Imprinting of the MEA Polycomb gene is controlled by antagonism between MET1 methyltransferase and DME glycosylase. *Dev Cell 5*: 891-901

Xie, S., Wang, Z., Okano, M., Nogami, M., Li, Y., He, W.-W., Okumura, K. & Li, E. (1999) Cloning, expression and chromosome locations of the human *DNMT3* gene family. *Gene 236*: 87–95

Xu, G.L., Bestor, T.H., Bourc'his, D., Hsieh, C.L., Tommerup, N., Bugge, M., Hulten, M., Qu, X., Russo, J.J. & Viegas-Pequignot, E. (1999) Chromosome instability and immunodeficiency syndrome caused by mutations in a DNA methyltransferase gene. *Nature 402*: 187-291

Yi, H., Riddle, N.C., Stokes, T.L., Woo, H.R. & Richards, E.J. (2004) Induced and natural epigenetic variation. *Cold Spring Harb Symp Quant Biol 69*: 155-159

Yoder, J.A., Soman, N.S., Verdine, G.L. & Bestor, T.H. (1997) DNA (cytosine-5)-methyltransferases in mouse cells and tissues. Studies with a mechanism-based probe. *J Mol Biol 270*: 385–395.

Yokochi, T. & Robertson, K.D. (2002) Preferential methylation of unmethylated DNA by mammalian *de novo* DNA methyltransferase Dnmt3a. *J Biol Chem 277*: 11735-11745

Zaitseva, G.N., Kolesnikov, A.A., Iatsenko, I.A., Kirnos, M.D. & Vaniushin, B.F. (1974) Primary structure of DNA from *Crithidia oncopelti* kinetoplasts. *Dokl Akad Nauk SSSR 219*: 243-245

Zegerman, P., Canas, B., Pappin, D. & Kouzarides, T. (2002) Histone H3 lysine 4 methylation disrupts binding of nucleosome remodeling and deacetylase (NuRD) repressor complex. *J Biol Chem 277*: 11621-11624

Zhang, X., Yazaki, J., Sundaresan, A., Cokus, S., Chan, S.W., Chen, H., Henderson, I.R., Shinn, P., Pellegrini, M., Jacobsen, S.E. & Ecker, J.R. (2006) Genome-wide high-resolution mapping and functional analysis of DNA methylation in *Arabidopsis*. *Cell 126*:1189-1201

Zhang, Y. & Reinberg, D. (2001) Transcription regulation by histone methylation: interplay between different covalent modifications of the core histone tails. *Genes Develop 15*: 2343-2360

Zhu, C.M. & Henney, H.R. Jr. (1990) DNA methylation pattern during the encystment of *Physarum flavicomum. Biochem Cell Biol 68*: 944-948

Zilberman, D., Cao, X. & Jacobsen, S.E. (2003) ARGONAUTE4 control of locus-specific siRNA accumulation and DNA and histone methylation. *Science 299*: 716-719

In: Progress in DNA Methylation Research
Editor: Hans P. Neumann, pp. 91-116
ISBN: 978-1-60021-722-7

Chapter II

NEW AND IMPROVED METHODS FOR ANALYSIS OF METHYLATED DNA USING CAPILLARY ELECTROPHORESIS: FRAGMENT ANALYSIS, DIRECT SEQUENCING, AND TROUBLESHOOTING SEQUENCING OF BISULFITE-CONVERTED GENOMIC DNA

Victoria L. Boyd, Benjamin G. Schroeder, Linda G. Lee and Gerald Zon*
Applied Biosystems, 850 Lincoln Centre Drive, Foster City, California 94404 USA.

ABSTRACT

Treating gDNA with bisulfite selectively deaminates non-methylated cytosine (C) moieties thus converting C into uracil (U). Subsequent PCR results in amplicons wherein thymidine (T) replaces previously unmethylated C (as well as replacing pre-existing Ts), and C replaces 5-methylcytosine (5mC). Since methylation of C almost always occurs in the context of CpG motifs (5mCpG), it is possible to determine the positions and relative extent of methylation of gDNA by analyzing DNA following bisulfite conversion. Such methods include methylation specific priming, sequencing, cloning and sequencing, single-base extension, oligonucleotide ligation assays (OLA), fluorogenic (e.g., TaqMan®) assays, and hybridization. Bisulfite conversion and efficient, selective PCR amplification are fundamental to each of these sequence detection methods in order to obtain accurate methylation analysis. During our investigation of improving sequencing methodology, and development of a novel fragment analysis method by capillary electrophoresis (CE), we have compiled data that illustrate unique challenges of

* Correspondence concerning this article should be addressed to Victoria L. Boyd: tel. no. (650) 638-5506; fax (650) 638-6666

analyzing the 3-base bisulfite-converted genome using techniques that had been originally optimized for 4-base genomes. We will summarize our findings in a comprehensive and practical manner aimed at providing guidelines to readers for achieving successful methylation analysis by CE. We will provide protocols, as well as real-life examples, for readers to rationalize and troubleshoot on their own.

INTRODUCTION

Methylation of C as 5mC in CpG dinucleotides in the promoter region of a gene has been associated with transcriptional silencing and plays a central role in epigenetics [1-6]. Analysis of this type of methylation in gDNA can be achieved directly by the use of methylation-sensitive restriction enzymes, or after acid-catalyzed conversion of gDNA with bisulfite that is selective for C compared to 5mC [7], which was discovered by Hayatsu [8] in 1970. As shown in Figure 1, stepwise reaction of bisulfite with protonated C leads to deamination to form uracil (U) via reversible formation of an intermediate sulfonated adduct. This highly selective deamination of C to U, without significant conversion of 5mC to T, is presumably due, at least in part, to greater steric interference between bisulfite and CH_3 vs. H during cis-addition to the 5,6-double bond, as illustrated in Figure 2. Regardless of mechanistic details, the change in DNA sequence upon complete conversion by bisulfite allows methylation analysis by DNA sequencing (and other) methods. To obtain sufficient analyte after bisulfite conversion, PCR is generally performed, which leads to amplicons wherein Ts replace Us (i.e., former Cs) and Cs replace 5mC. On a statistical basis, only $1/4^{th}$ of all possible CpN dinucleotides are present as a CpG motif in gDNA, and are therefore eligible for the action of methyltransferases. Consequently, bisulfite treatment eliminates all Cs that are not present as 5mCs by replacement with Ts, and thus creates a nearly C-less sequence of mostly 3-base-DNA having predominantly A, G, and (~50%) T. This reduction in sequence complexity, relative to normal 4-base-DNA, can lead to lower specificity of hybridization with probes or primers [9,10]. It follows that PCR- and other probe-based methods for detection of bisulfite-converted, 3-base-DNA loci share a common constraint regarding specificity. Examples of problematic PCR and sequencing traces are presented herein to illustrate diagnosis of problematic bisulfite-sequencing data. Some aspects of the problems described are not unique to bisulfite-treated gDNA, and are encountered in other types of DNA analyses. The troubleshooting guide presented herein will focus on improvements to bisulfite-sequencing, which most methylation researchers depend on at some point during their investigations.

BISULFITE CONVERSION

Achieving essentially complete and highly selective bisulfite conversion of non-methylated Cs among 3 billion bases in the human genome, without significant interfering side-reactions or extensive cleavage of gDNA is a remarkable achievement. Treatment of denatured gDNA with relatively high concentrations (~3-9 M) [11] of bisulfite in freshly prepared solutions with added antioxidant and carefully adjusted pH at elevated temperature for extended times is required. An often cited (e.g., [12-14]) study has reported [15] that as much as 95% of gDNA is degraded during these types of forcing reaction conditions for bisulfite conversion. However, our recently published [16] finding of excellent recoveries of converted DNA using centrifugal filtration for post-bisulfite purification indicates that degradation of DNA during the conversion process may not be as problematic as is generally thought based on earlier findings [15]. We believe that, under appropriate reaction conditions, gDNA is not extensively degraded but may instead be lost during recovery using techniques originally optimized for duplex DNA. The bisulfite-converted gDNA intermediate is single-stranded, due to replacement of Cs with Us, and perhaps more importantly is more negatively charged due to an additional sulfonate group bonded to every U (cf. Figure 1). The sulfonated U moieties are desulfonated while still contained in the filtration device, thus eliminating the possibility of biased fractionation and/or other means of loss of material during isolation of converted gDNA. This centrifugal filtration protocol [16] thoroughly washes away all of the bisulfite, and allows complete dissolution of the intermediate sulfonated gDNA in 0.1 M NaOH for desulfonation. In our experience, the final bisulfite-converted gDNA obtained in this manner can be stored for long periods in a refrigerator, in contrast to other reported protocols that recommend immediate use or frozen storage. Bisulfite-converted gDNA purified by the centrifugal filtration protocol [16] and stored in a refrigerator has been reanalyzed multiple times over many months and with essentially identical results.

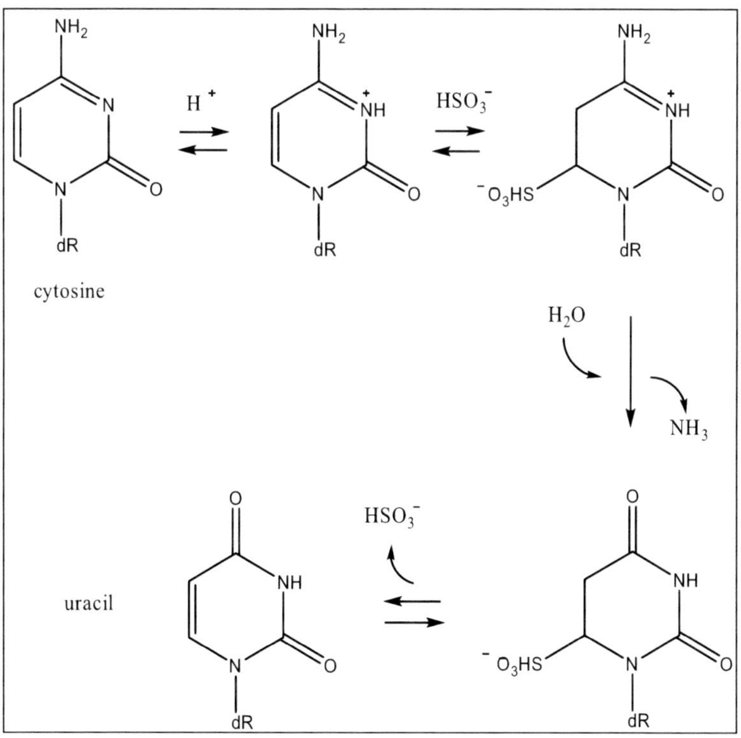

Figure 1. Stepwise acid-catalyzed conversion of cytosine (C) to uracil (U); dR = 2'-deoxyribosyl moiety.

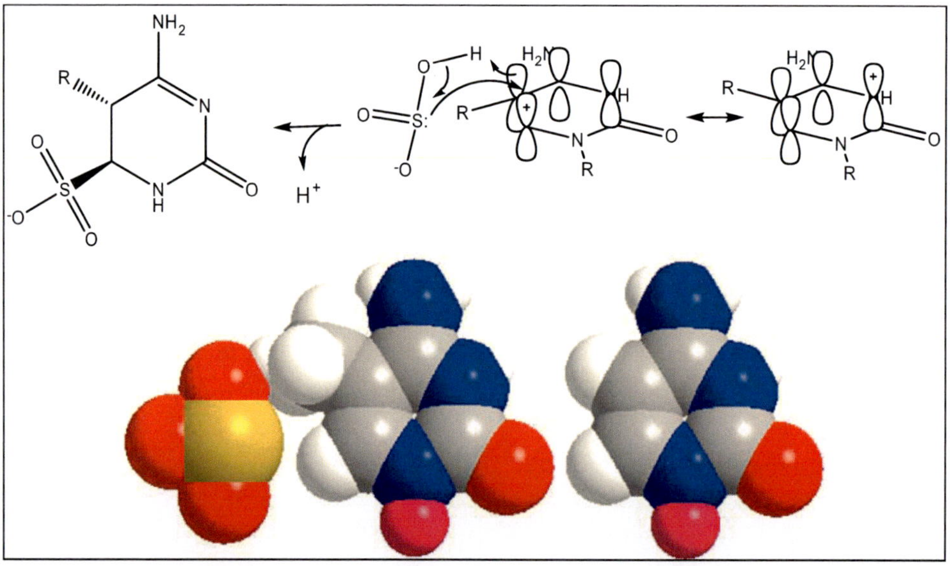

Figure 2. Acid-catalyzed cis-addition of bisulfite (HSO_3^-) to the 5,6-double bond of cytosine (C; R = H) is favored relative to 5-methylcytosine (5mC; R = CH_3) (top) due in part to greater steric repulsion between bisulfite and CH3 vs. H, as illustrated by space-filling 3-D structures (bottom).

Selecting a Region of Interest for Methylation Analysis

Methylation is a dynamic process that varies with health, age, diet, environment, and is probably heritable. Eukaryotes have both de novo methyltransferases and maintenace methyltransferases [17]. Demethylation may be passive, due to inefficient or lack of maintenace methylation during cell division, or actively mediated by enzymatic removal by processes that have been recently elucidated [18]. Methylation in cancer is believed to be a progressive event, whereby the percentage of a region that is methylated increases as the disease progresses [19]. One model proposes that methylation is initiated near the transcription start site, and spreads out over the entire region as loss of gene expression (and cancer) becomes more advanced. CpG motifs are underrepresented in the human genome; however, a significant percentage of genes have relatively high CpG-density ("islands") in promoter regions. Readers are encouraged to consult a recent publication by Berg and coworkers [20] for an excellent update on the sequence-based definition and distribution of CpG islands in the human genome that provides new and important perspectives on this fundamental topic.

PRIMER CONSIDERATIONS FOR BISULFITE-CONVERTED GENOMES

The sequences obtained from methylated gDNA and unmethylated gDNA are fundamentally different after bisulfite conversion. The sequence from methylated gDNA will still have Cs at CpGs, while that obtained from unmethylated gDNA will have no Cs. Primers for bisulfite-converted gDNA can be designed to anneal to a sequence specific for methylated gDNA, or unmethylated gDNA, or designed to a region without CpGs so that PCR amplification is not dependent on methylation status. Methylation-specific PCR (MSP), which was first reported in 1996 by Herman and coworkers [21], has become a widely used technique that is good for a "yes or no" indication concerning methylation of CpG sites. Two primer sets are designed to anneal to a region containing CpG motifs. The methylated primer set assumes the CpG's are fully methylated, and the primer will have all 4 bases in the sequence. The unmethylated primer set anneals to gDNA that is not methylated in the (same) primer binding site, and therefore will have T in place of C in the primer sequence. Design guidelines generally have a C or T near the 3' end of the two primer sets, respectively, where any mismatches will be discriminated against by the polymerase. It is important to test the primer sets with a control gDNA of known methylation status along with the gDNA of unknown methylation status. The properly designed methylated primer set will only amplify the control methylated gDNA, and not unmethylated gDNA, and the unmethylated primer set will be postive for only unmethylated gDNA.

Primers designed outside of a CpG region of interest will, in principle, amplify the target regardless of the methylation state of the internal sequence, absent other complications (see below). MSP excludes DNA that is partially or variably methylated, but primers for bisulfite sequencing permit full representation of the entire methylation state of the sample. The percentage of methylation, or methylation patterns are believed to be dynamic and changing during the life of a cell. Bisulfite sequencing provides an inherently more accurate assessment of the methylation state of a sample compared to PCR primers (or probes) that select for presupposed fully methylated or fully unmethylated complementary sequences. Investigative studies that are designed to determine a role of methylation in gene expression generally require bisulfite sequencing. However, once a biological condition or disease is established, methylation is generally observed throughout an entire region. Pilot genome-wide bisulfite sequencing studies [22] have demonstrated that methylation is approximately bimodal, i.e., while sequences were mostly either methylated or unmethylated throughout the region of interest there are a relatively low level of sequences with intermediate methylation states. Consequently, the nature of the methylation information being sought (i.e., exact at every CpG in a clonally pure allele vs. approximate average over all alleles) will determine whether primers should be designed to anneal in CpG-free regions or not.

FRAGMENT ANALYSIS

A newly developed and very simple workflow for methylation analysis after bisulfite conversion involves PCR using bisulfite-sequencing primers and CE analysis of the PCR amplicon (Figure 3) [23]. Formation of a correct-sized amplicon serves as proof of the presence of the intended target sequence. When there are a relatively large number of CpGs in an amplicon (i.e., ~1 CpG dinucleotide per ~10-12 bases), CE can lead to separation of amplicons derived from fully methylated and fully unmethylated gDNA regions (Figure 4). Moreover, the amplicon used for this easy fragment-anaylsis-based method of methylation analysis can also be directly sequenced. Amplicons from bisulfite-converted methylated and unmethylated gDNA are co-amplified with the same CpG-free primers using a fluorescent dye label (e.g., FAM) on one of the primers. A fluorescent label on the forward primer during PCR provides detection of a C or T at CpG sites, whereas the variable positions are G or A with a dye on the reverse strand. The amplicon with Cs (FAM dye on forward primer) will migrate faster than the corresponding amplicon wherein all the Cs are Ts. The cumulative effect of a lower mass/charge for a C vs. a T was discovered to be large enough in CpG-rich sequences to permit near baseline resolution in POP4 CE polymer at 60 °C. Separation of amplicons from fully methylated or unmethylated gDNA was greater in POP6 or POP7 CE polymer. A mathematical algorithm [23] that predicts the extent of separation can be applied to pre-select amplicons that are candidates for methylation ratio determination by this novel and easy method of fragment analysis by CE.

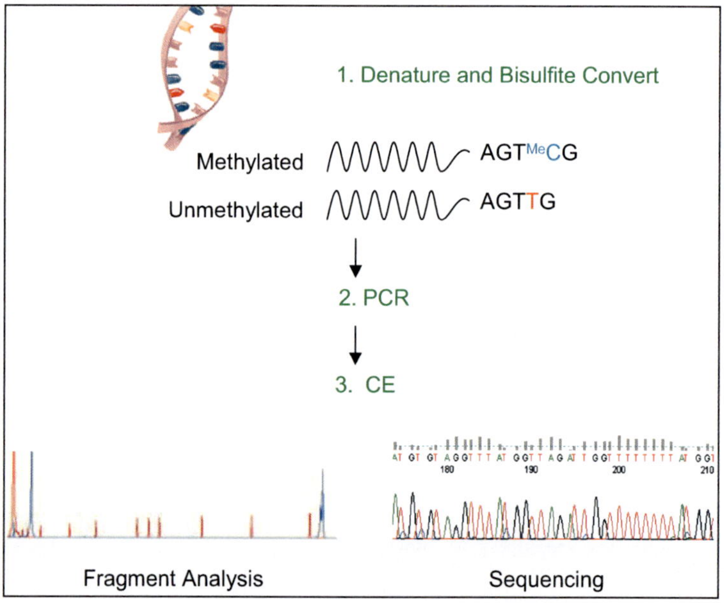

Figure 3. Workflow for methylation-dependent fragment separation (23). Step 1, bisulfite conversion of Cs to Us, except for methylated CpG's. Step 2, PCR amplification using a primer set designed in a non-CpG region to generate amplicons regardless of the methylation status. The forward primer has a fluorescent dye-label (FAM). Step 3, CE separation of the faster migrating C-rich strand from slower migrating T-rich strand derived from methylated (Me) and unmethylated (UnMe) DNA, respectively. The same amplicon is also analyzed by direct sequencing.

Limitations of this CE method for fragment analysis to determine methylation ratios include amplicons with >9 Ts. Homopolymeric runs of Ts lead to polymerase-related "slippage" phenomena that are seen as multiple sizes or a single broad signal, depending on CE resolution, analogous to what is known from PCR of short tandem repeat sequences [24]. Amplicons need to be in a CpG rich region, and greater than 200 bp. Amplicons from gDNA with highly variable methylation states would require sequencing. However, fragment analysis can serve as a simple screening tool in situations where a region is fully methylated or not methylated.

An important additional benefit of fragment analysis is that it will reveal the presence of primer dimer [25] and/or secondary amplicons that do not match the anticipated size. Real-time PCR methods using SYBR green detection or a method that selects for only a targeted sequence with a hybridization probe, as in MethyLight [26], do not detect such competing side reactions. Real-time PCR side reactions contribute to the measured SYBR green-derived Ct values. Post-PCR dissociation curves sometimes discriminate for the presence of secondary amplicons, but cannot serve to correct Ct values.

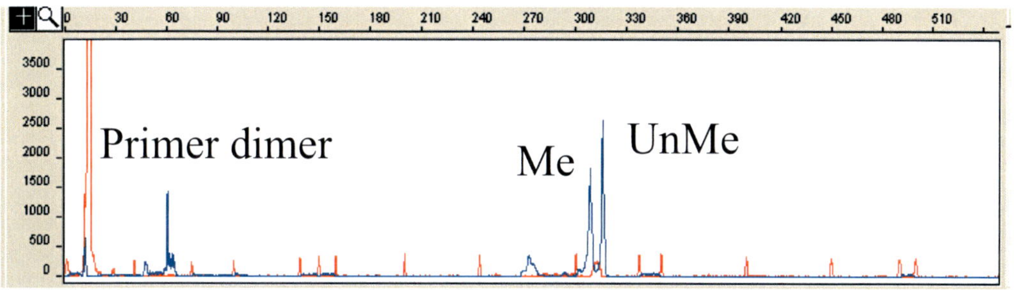

Figure 4. CE fragment-size trace of ~300-bp FAM-labeled amplicons (blue) derived from the p15 gene locus in human gDNA following bisulfite-conversion vs. size standards (red). Amplicons of the same length but different base composition were derived from methylated (Me) and unmethylated (UnMe) gDNA using the same primer set. Due to sequence composition differences (C vs. T) the p15 amplicon from methylated gDNA migrates faster than the amplicon from unmethylated gDNA (23). Primer dimer is seen at ~60 bp.

PCR CONSIDERATIONS

Primers directed at CpG flanking sequences in bisulfite-converted gDNA have only 3 bases and are T-rich. This oftentimes leads to primers >30 bases in length for obtaining typical PCR T_m-values of ~60 °C. These relatively long 3-base T-rich primers are apparently more prone to mismatch hybridization since multiple amplification products are frequently obtained. The reverse primer is A-rich and thus provides favorable conditions for primer dimer formation with the T-rich forward primer. Once the first cycle of amplification wrongly amplifies a mismatched sequence, subsequent PCR cycles are correctly matched, so PCR proceeds as efficiently as it would for the correctly matched sequence. Smaller amplicons from either primer dimer or mismatched secondary amplicons have greater amplification efficiency, and can out-compete amplification of the amplicon of interest. The presence of undesired amplicons usually diminishes the PCR efficiency of the intended amplicon.

Additional Considerations

After bisulfite conversion of gDNA, double-stranded nucleic acid is transformed into single stranded template, and is comprised of 5 different bases: A, G, T, U and 5mC. Methylated promoter regions will still be 5mCpG-rich, and likely have single-stranded secondary structure. 5mC reportedly enhances the T_m of 5mC-G base pairing relative to C-G by 1.2 °C [27]. C-G base-pair-rich regions are known to be difficult to PCR amplify. The polymerase must be capable of reading U and 5mC during the first round of synthesis of the reverse-complement strand, much like "first strand" synthesis in reverse transcription PCR. A hot start polymerase in conjunction with relatively high temperature should be used to avoid mismatch amplification. High-fidelity polymerases from archaebacteria such as Vent or pfu DNA polymerase do not accommodate U-containing template [28]. It should be noted that PCR master mixes containing uracil DNA glycosylase (UNG) should not be used. UNG cleaves U-containing DNA and will degrade U-containing template produced by bisulfite conversion of gDNA. Also, archaebacterial DNA polymerases are strongly inhibited by the presence of small amounts of uracil-containing DNA [28].

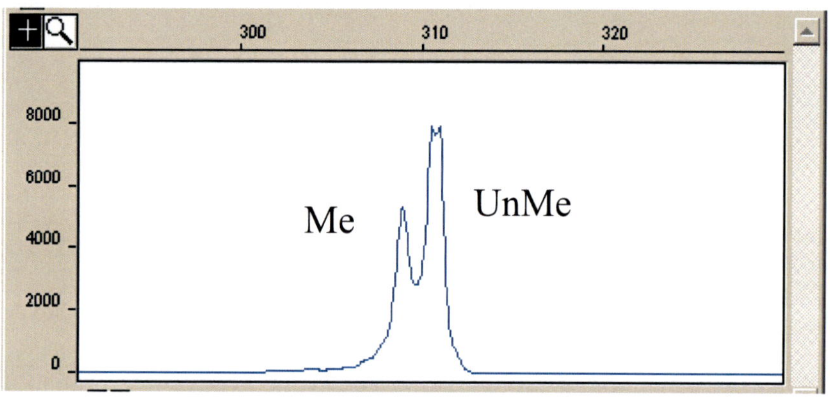

Figure 5. CE-based fragment analysis of the estrogen receptor (ER) gene in a 50/50 mixed methylated (Me)/unmethylated (UnMe) sample; size-standard not shown. Amplicons were generated using a dye-labeled (FAM) forward primer such that the methylated-derived amplicon, which has 26 CpGs, migrates faster than the amplicon from unmethylated gDNA, which instead has 26 TpGs. Note the bias in favor of the amplicon from unmethylated gDNA. The PCR conditions described in the protocols section significantly reduce the bias. Sequencing results for this mixture of amplicons are shown in Figure 6.

PCR Bias/Quantitation

Amplification efficiencies for bisulfite-converted templates obtained from methylated samples and unmethylated samples are known to be different [29]. Regions of interest are usually CpG islands so that a large number of C-G base pairs are present in an amplicon from a methylated sample, and correspond to T-G base pairs in a related unmethylated sample. Regions rich in C-G base pairs are widely recognized to be relatively difficult to PCR amplify. The degree of amplification bias relative to a corresponding region rich in T-G base pairs varies with the specific details of the amplicon. However, the template derived from the unmethylated strand is frequently amplified more efficiently than the template derived from the methylated strand, and will therefore dominate in a mixed sample, as shown in Figure 5.

PCR RECOMMENDATIONS

Data presented in Figires 4-6 were generated using bisulfite-sequencing primers with T_m ~55 °C and tailed with M13 forward and reverse sequences [30]. The well-known 18-base-21M13 sequence, or other suitable sequence, provides a universal sequencing primer binding site. These universally tailed regions of locus-specific primers have all 4 bases present, and after a first round of PCR amplification, the longer primer binding site provides for higher T_m (and higher specificity). During thermal cycling, the annealing temperature can be raised after the first (few) cycle(s). This higher temperature PCR provides enhanced selectivity as evidenced by less background in resulting sequencing traces and fragment analysis. Reduced formation of primer dimers and/or secondary amplicons was observed in preliminary experiments using "touchdown" PCR [31] in combination with raising the annealing temperature as described above.

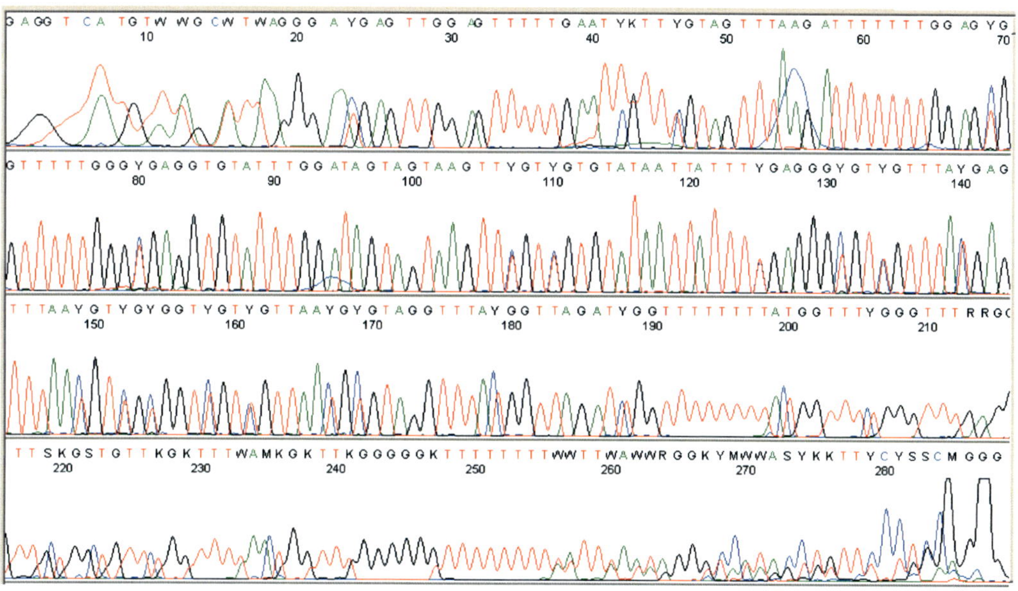

Figure 6. CE sequencing results for ER amplicons analyzed by fragment analysis as described in Figure 5. A mixed C/T signal is seen at all 5mCpG/CpG sites due to the PCR template being a 50/50 mixture derived from methylated and unmethylated gDNA. Cs not adjacent to Gs at the end of the sequence are present in the M13-primer tail. Full-length FAM-labeled amplicons from fully methylated and unmethylated gDNA are seen at the end of the sequence run as two relatively large G signals.

PCR reaction conditions generally include glycerol, a denaturant which reduces PCR bias during amplification of templates derived from bisulfite conversion of fully methylated and fully unmethylated sample. Betaine [32] also reduced such bias, but did so with less consistency, relative to glycerol, among several gene targets that were investigated. A 451-bp amplicon [23], which previously exhibited significant bias in favor of the amplicon from unmethylated gDNA template, resulted in PCR amplicons of equal intensity from a 50/50 mixed sample using the PCR conditions provided at the end of this chapter (cf. Figure 8). These results were obtained using M13 tailed primers, and increasing the annealing temperature after the first 5 cycles of PCR. A 3 minute extension time in the PCR thermal cycling may reduce PCR bias by permitting the polymerase to read through CG-rich regions. The 60 minutes at 72 degrees at the end of cycling permits the non-templated "A" addition-activity of the polymerase [33-35], which is important prior to fragment analysis.

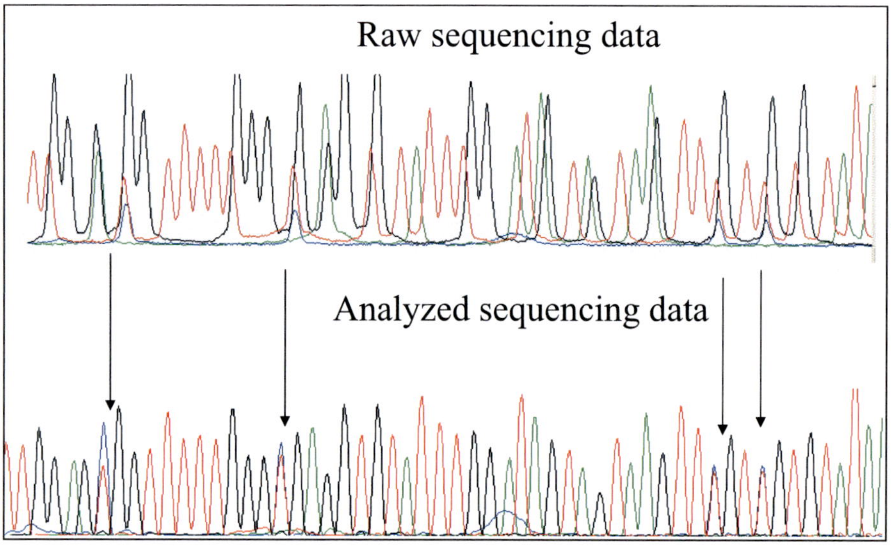

Figure 7. In bisulfite sequencing of amplicons derived from mixed methylated and unmethylated gDNA, unmethylated Cs are absent due to conversion to Ts via Us throughout the amplicon, while methylated Cs are detected as mixed-base signals of C (blue) and T (red). A section of the sequencing trace in Figure 6 is re-presented here comparing the raw signal (top) to the analyzed (i.e., software-processed) trace (bottom). The software normalizes the data so that each of the four bases has approximately equal signal strength, which therefore artificially distorts the actual C/T ratio that is better (but not exactly, due to fluorescence differences) represented by the raw data.

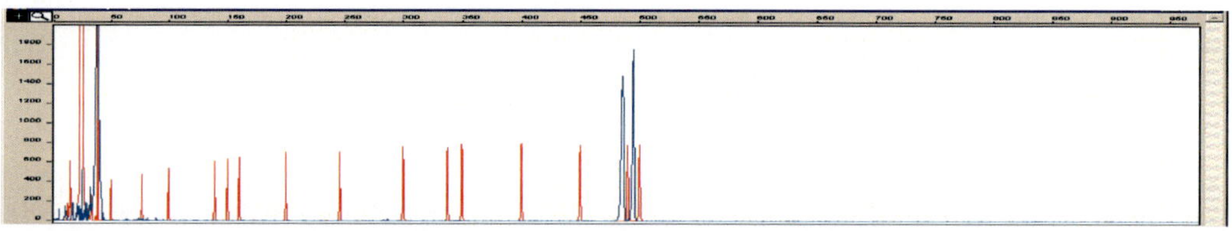

Figure 8. CE-based fragment analysis of amplicons (blue) [size standard (red)] of the RASSF gene using the presently reported protocol at the end of this chapter shows unbiased amplification (in contrast to our findings using an earlier protocol that does not include extra glycerol and BSA during the PCR (23)).

The PCR anomalies such as amplification bias due to variable CG-content, slippage, primer dimer, competing amplification of secondary amplicons due to lower primer specificity, and incomplete bisulfite conversion militate against quantitative accuracy. One region may be more difficult to analyze relative to a nearby region that has fewer CpG's, or avoids homoploymer sequences. Accurate determination of methylation status may require analyses of multiple sites within a given region.

A summary of recommendations to reduce PCR bias when amplifying templates of mixed methylated states is provided below:

- Hot start (AmpliTaq Gold®) polymerase
- Tailed primers with all 4 bases in design
- PCR denaturant, such as glycerol
- Annealing temperature approx 2-5 degreees above calculated Tm (gene specific portion)
- Increased annealing temperature after the first few PCR cycles
- Touchdown PCR
- Increase extension time/temperature during PCR
- Decrease primer concentration (to reduce primer-dimer)

Elimination of PCR bias is reportedly [36] dependent on the amount of genomic bisulfite-converted DNA used in the PCR. The recommended protocol at the end of this chapter using 3 ng of bisulfite converted gDNA per 5-uL reaction gives a concentration in excess of the minimum of 10 ng per 25-uL reaction reported [36] for reproducible amplification prior to pyrosequencing. In addition to further optimization of PCR conditions, we are investigating alternative protocols for bisulfite sequencing that will eliminate PCR bias concerns.

Bisulfite Sequencing

Bisulfite sequencing is the "gold standard" for the analysis of methylation and is used for both discovery and routine analysis. Essentially all DNA methylation researchers depend either directly or indirectly on bisulfite-sequencing data obtained for their region(s) of interest. Sequencing primers with or without universal tails are generally designed to anneal to non-CpG regions flanking each region of interest, and thus amplify bisulfite-converted gDNA regardless of methylation status. Consequently, if one of the primers is fluorescently labeled at the 5' end, the resultant amplicon can be analyzed by the CE-based fragment analysis method described above and also used for sequencing. The amplicon can be either directly sequenced, as in any other so-called "PCR-sequencing" protocol for DNA in general, or is cloned and then sequenced. Bisulfite sequencing of clonal regions of interest provides unambiguous data on methylation patterns. Bisulfite sequencing also provides information regarding possible incomplete bisulfite conversion as well as detection of mutations or single nucleotide polymorphisms (SNPs).

As noted above, bisulfite sequencing primers are generally designed to anneal to non-CpG-containing loci. Such primer annealing sites are relatively restricted due to limited availability of regions that span a CpG island yet exclude CpGs in the primer annealing site.

Selection of these primers can be further constrained by the relatively large number of closely spaced CpG motifs, which lead to primer sites too short to achieve a desirable T_m. An additional constraint is the presence of relatively long (>9 bases) poly(T) sequences in bisulfite-converted template that results in non-specific primer annealing, or poor amplification due to polymerase slippage (Figure 8). Primer selection is therefore often limited to a single locus with limited ability to lengthen or shorten the primer. If there are no suitable primers, the antisense strand derived from the reverse complement of the original input gDNA sequence can be used. The complementary nature of the sense and antisense strands is eliminated following bisulfite conversion. The same number of CpGs (and methylation state) will be present in both the sense and antisense strands due to the symmetry of the CpG motif, and action of methyltransferase, whereas the bisulfite-converted sequence of the sense and antisense strands will differ considerably.

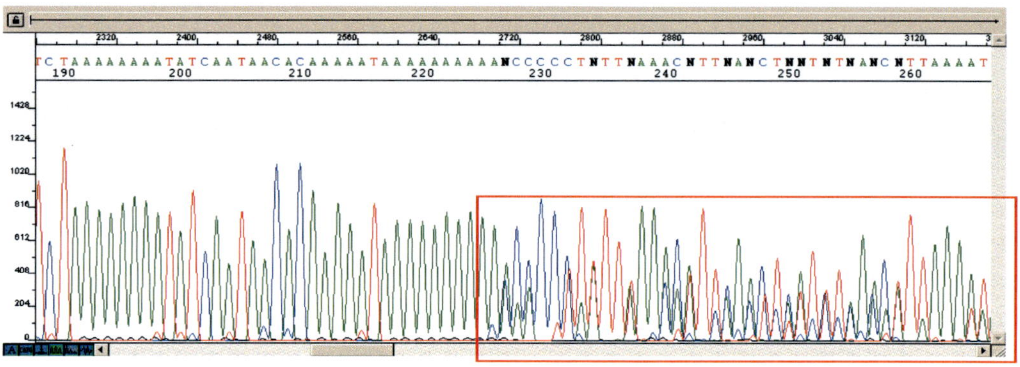

Figure 9. PCR slippage resulting in loss of callable antisense-strand sequencing (boxed sequence). The sequence from bases 188-216 is clearly resolved. However, following the homopolymer stretch of As from bases 217-226, there is both insertion and deletion of As due to polymerase slippage during PCR amplification that precedes sequencing. In CE-based fragment analysis, this PCR product mixture would be detected as a distribution of N+1 and N-1 peaks centered around the correct-size amplicon.

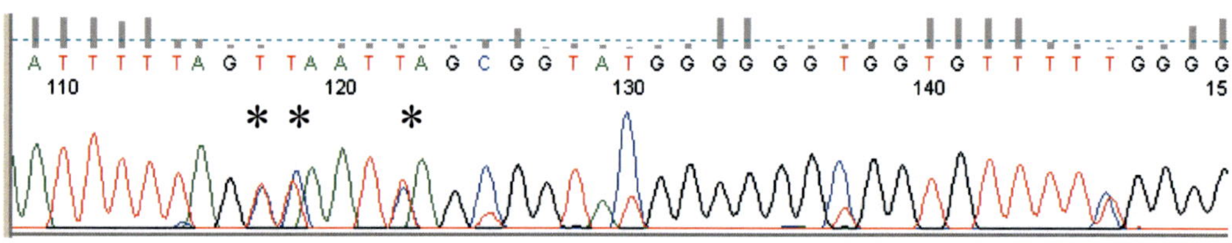

Figure 10. Positions indicated by an asterisk (*) indicate incomplete bisulfite conversion, and are readily identified as Cs (blue) not 5'-adjacent to Gs (black) (i.e., non-CpGs) and therefore are not due to methylation of C. All gDNA samples had incomplete bisulfite conversion in this region of the E-cadherin gene (CDH1). The purity of the gDNA sample used for this bisulfite conversion reaction may conceivably influence the completeness of the conversion. Primers designed to anneal to regions that are inherently difficult to bisulfite convert (e.g., due to strong secondary structure) can provide incomplete, inaccurate, or misleading data.

Sequencing reveals the extent of bisulfite conversion, which if incomplete will lead to the appearance of Cs that are not adjacent to the 5' side of Gs. For this reason, sequencing provides its own "internal reference standard" for completeness of bisulfite conversion.

Sequencing primers are designed to anneal to the bisulfite-converted sequence and thus select for fully bisulfite-converted gDNA template. Some regions, however, within an amplicon do not appear to bisulfite-convert to completion, as shown by the example in Figure 10.

Direct sequencing of a PCR amplicon derived from bisulfite-converted gDNA may lead to observation of superimposed signals due to contamination by secondary, co-amplified sequence(s) and/or primer dimer sequence(s). The presence of these types of shorter amplicons can contribute to off-scale signals at the beginning of the sequence trace, and depletion of the sequencing reactants, which causes rapid drop-off of signal intensity for the remaining sequence (Figure 11). Consequently, sequence-callable extension products may not be obtained even if the amplicon of interest is present. Such problems can be addressed by use of hot start PCR methods and/or application of "nested PCR" [37] protocols prior to sequencing to enrich for the amplicon of interest. Nested PCR, which requires a second set of internal PCR primers, is widely used in other applications to select for the desired amplicon.

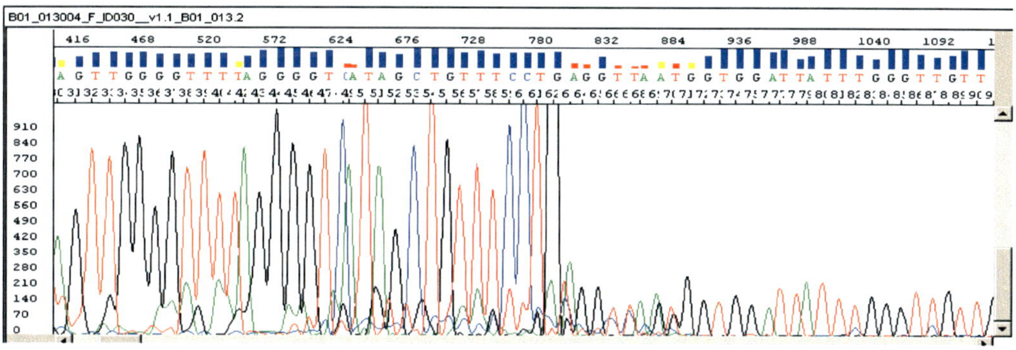

Figure 11. Primer-dimer signals will oftentimes dominate the initial sequence trace, as seen here for bases 30-62, followed by much lower intensity signals due to the amplicon of interest.

As for all sequencing reactions, an estimate of the amount of the PCR amplicon added to the sequencing reaction is needed to prevent off-scale signals. For subsequent clean-up of the sequencing reaction using the recently introduced BigDye® XTerminator™ Purification Kit, $1/5^{th}$ the amount of amplicon commonly used in a sequencing reaction is recommended to prevent off-scale signals during analysis.

M13-primer-tailed amplicons obtained from a bisulfite-PCR reaction are sequenced using the standard BigDye® Terminator v1.1 protocol. BigDye® Terminator v1.1 is recommended because it provides better signal resolution at the beginning of the sequence, although BigDye® Terminator v3.1 can be used. The sequence of the entire amplicon can be seen when using M13 tailed primers, and the universal tail simplifies the sequencing workflow when processing many samples. BigDye® Terminator v1.1 and a 2-temperature thermal cycling profile is used in place of a 3-temperature profile to maintain the level of signal strength. Figures 12-14 illustrate variations of sequencing signal intensity. When sequencing the forward strand (but not the reverse), signal intensity fades into the baseline more quickly for the unmethylated amplicon, which is T-rich relative to the methylated amplicon that has a more equal representation of all 4 bases (Figures 12a and 12b). A sample with mixed methylation status will appear to loose the T signal relative to the C signal over the length of the amplicon. The degree to which this occurs is dependent on the amplicon, its length, and the number of methylated sites (i.e., Cs). In a sample with a mixed population of bisulfite-converted methylated and unmethylated amplicons, more rapid loss of signal strength of the bisulfite-converted unmethylated sample relative to the bisulfite-converted methylated sample will cause loss of quantitative accuracy.

As seen in Figure 14, signal strength obtained by sequencing the forward strand of amplicons from bisulfite-converted methylated and unmethylated gDNA is more equivalent when cycle sequencing is changed from the standard 3-temperature program to a 2-temperature program using annealing and extension at 50 °C. The amount of template and the dilution-factor of the BigDye® Terminator ready-reaction mixture also influence the levelness of the signal strength. Sequencing of the reverse strand did not exhibit significant decline of signal strength compared to forward strand sequencing.

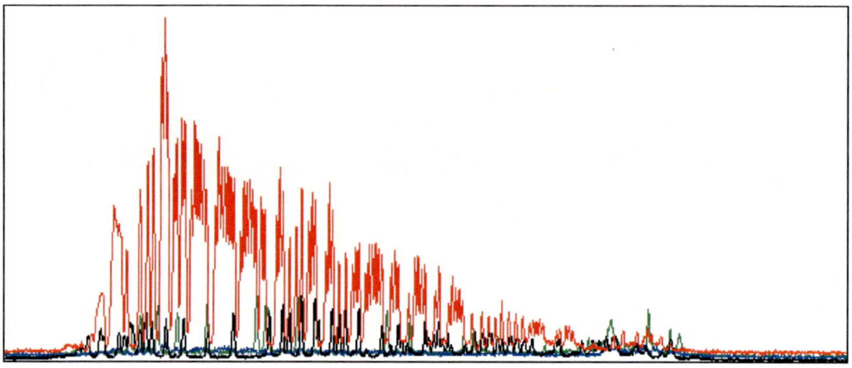

Figure 12a. Four-color "raw" (i.e., unprocessed) sequencing data for the T-rich (red) amplicon obtained from bisulfite-converted unmethylated gDNA.

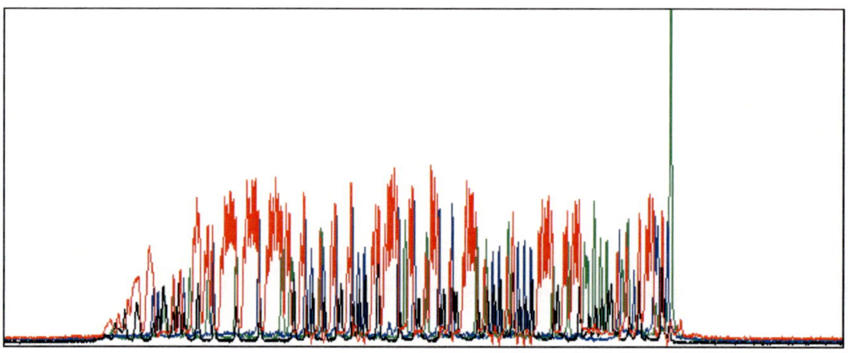

Figure 12b. Four-color "raw" (i.e., unprocessed) sequencing data for the amplicon obtained from bisulfite-converted methylated gDNA; T = red.

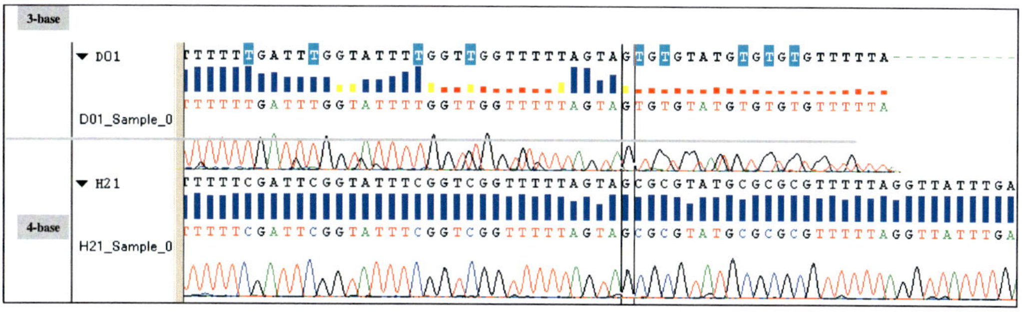

Figure 13. Sequencing traces for an amplicon obtained from bisulfite-converted unmethylated gDNA (top) and corresponding amplicon obtained from bisulfite-converted methylated gDNA (bottom). The signal strength decreases more rapidly in the trace from the relatively T-rich unmethylated vs. methylated amplicon. This type of rapid decline in signal strength is also seen in the raw data in Figure 12a.

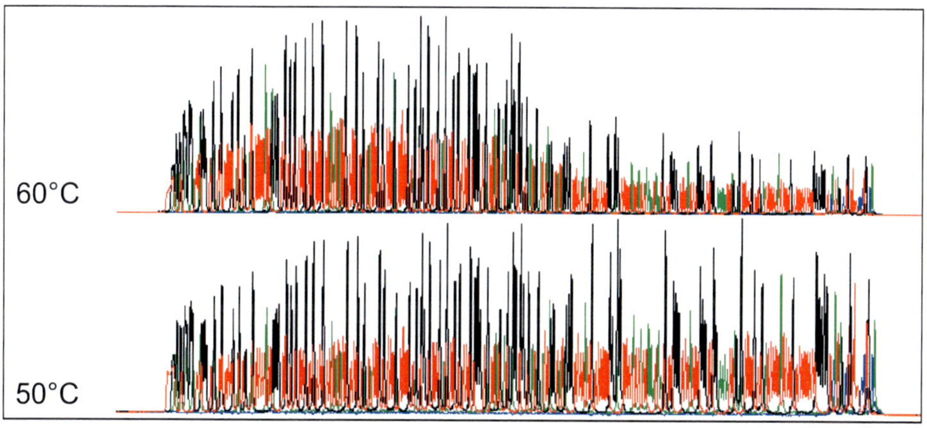

Figure 14. Four-color "raw" (i.e., unprocessed) sequencing data for a “standard” 3- temperature cycle sequencing (top), which has a 4-minute extension time at 60 °C, was replaced with 2-step cycle sequencing (bottom) that has a 4-minute extension time at 50 °C. The lower temperature provides better “signal balance” when applied to a mix of amplicons derived from bisulfite-converted methylated and unmethylated gDNA.

Characteristics of the Bisulfite-Sequencing Trace

Direct sequencing will detect mixed bases at CpG sites when obtained from samples with mixed methylation states. Control studies where the amplicon from bisulfite-converted methylated and unmethylated gDNA is mixed in known ratios indicated that accurate mixed-base signals were obtained if the minor component is ≥15%. Below 15% the minor-component signal may either be not detected or have exaggerated intensity in the analyzed data. However, this latter problem can be mitigated by use of Applied Biosystems KB basecaller software, which adjusts for sequences with skewed base composition in such a way that the underrepresented color is not artificially overly exaggerated. Software-related factors also account for the fact that amplicon sequences produced using universally tailed primers, or cloned sequences having sequence-content from the cloning vector, usually lead to more level sequence traces without background signal problems. The FAM dye on the full length amplicon-template from PCR is seen at the end of a sequence trace as a large, black colored "G" signal (Figure 6). The FAM dye adds the missing black colored "G" when sequencing the reverse strand, which fortuitously aids normalization of the signal strength for actual G bases.

Misalignment of signals is a problem encountered when sequencing longer bisulfite-converted amplicons (>300 bp). There are two sources for this misalignment, which is seen more frequently in bisulfite sequencing than in sequencing in general. Due to sequence-content that is T-rich (or A-rich), PCR slippage is common, and is typically seen when there are >9 sequential Ts. The onset of misalignment is acute, and N+1 (or N-1) shift occurs immediately following the homopolymer sequence. Sequencing is unambiguous up to the point of slippage, and is sometimes callable after the misalignment occurs. A second cause of misaligned signals that appears more gradually is related to the large number of C vs. T (or G vs. A) sites in the bisulfite-converted amplicon for samples that have a mixed methylation status. At each mixed-base site, the difference in CE migration increases for an amplicon derived from unmethylated gDNA vs. methylated gDNA; consequently, longer sequencing fragments no longer co-migrate. Migration differences in samples derived from mixed methylation states has also been observed in a pilot study of the Human Epigenome Project [22]. Signals gradually broaden, and then split, and for longer amplicons, will appear as an N-1 sequence superimposed on the signals of the more predominant amplicon (Figure 15). Sequencing in both the forward and reverse direction permits sequence reads up to the point of the misalignment(s). Shortened PCR amplicons that avoid homopolymer sequences are generally purer templates for sequencing. PCR bias is reduced for shorter amplicons, and the base composition, proportionately, is better represented by all 4 bases due to the tailed sequence representing a greater percentage of the amplicon length.

Based on the foregoing discussion, troubleshooting can be complicated due to multiple factors that may underlie poor sequencing results. Each step in the bisulfite sequencing workflow (bisulfite conversion, PCR and sequencing) may contribute to obtaining poor sequencing data. Troubleshooting typically requires optimization of more than just one parameter.

The following is a list of recommendations for direct sequencing of amplicons derived from bisulfite-converted gDNA:

- Quantification of PCR amplicons prior to sequencing
- Use of M13-tailed primers
- Use of full-strength BigDye® Terminator v1.1 Ready Reaction mix
- 2-temperature cycle-sequencing
- Use of Xterminator clean-up of sequencing reactions
- Analysis with KB basecaller software

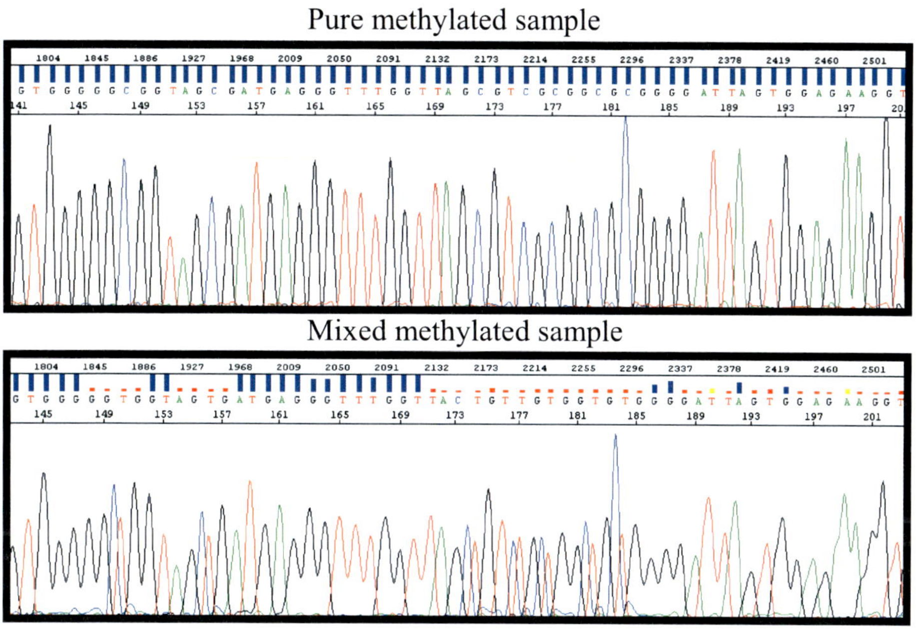

Figure 15. Bisulfite sequencing traces for an amplicon derived from a pure gDNA template (top) and a template that is a 50/50 mixture of methylated and unmethylated states (bottom). The sequencing products derived from the methylated and unmethylated sample do not co-migrate because of multiple C vs. T differences in the sequence (see text).

CLONING AND SEQUENCING

Regardless of the technique used, quantitative PCR applied to bisulfite-converted gDNA is potentially unreliable due to loss of specificity of primers and probes. Background signals are often unavoidable for oligonucleotides designed to anneal to bisulfite-converted non-CpG or unmethylated sequences. Primers and probes designed to anneal to bisulfite-converted CpG-rich regions, and which anneal over several of such positions, can have 4 bases, if methylated, and thus exhibit greater specificity, so that quantification of the methylated sample has a higher probability of being more accurate vs. unmethylated sample. There are several other limitations to accuracy of methods intended to provide quantification of methylation:

- Purity of the sample extracted from biological sources
- Completeness of bisulfite conversion
- PCR bias, which is a variable for each sample, and each analysis

Additionally, for cloning and sequencing:

- Cloning bias
- Limited number of clones sequenced

The accuracy of quantification by cloning and sequencing is limited by potential bias in the PCR amplification process as well as the cloning process. Relative to direct sequencing of amplicons from bisulfite-converted gDNA templates, sequencing of clones provides much "cleaner" sequencing due to a single amplicon insert per clone; there are no secondary sequences, no PCR slippage, no mixed bases, misaligned sequences due to mobility differences are eliminated, and all 4 bases are represented for signal normalization due to sequence content from the cloning vector. The improved appearance of the sequencing data permits semi-quantitation and determination of variable methylation patterns. However, the workflow is time-consuming, and often relatively few clones are sequenced—well below the number needed for statistical accuracy. PCR (and cloning bias) may still contribute to distortion of the data when attempting methylation quantification. Lack of variability seen in the clones indicates a low copy number of template going into the PCR, or PCR bias during amplification. Any sequence analysis method used, including pyrosequencing [36], will be subject to all the sample handling biases introduced during the PCR and cloning steps.

Future Directions for Sanger Sequencing Sample Preparation

In preliminary studies at Applied Biosystems, the cloning step in bisulfite sequencing has been replaced by PCR of single-molecules obtained by dilution. Suitable dilution of gDNA immediately after bisulfite conversion provides some wells with a single molecule of bisulfite-converted DNA as a template for PCR, thus precluding any biases during PCR amplification of different templates in a given well. Emulsion-based PCR [38,39] versions of single-molecule dilution offer the same advantages as cloning—a single template or sequence is present per aqueous compartment, and a large number of "clones" can be processed simultaneously. An automated selection step can then be used to isolate compartments that have PCR product from other compartments. PCR of single-molecule templates in emulsions has been recently shown to eliminate PCR bias [39], which is particularly advantageous when applied to different templates derived from bisulfite-converted methylated and unmethylated gDNA. Sequencing traces obtained from such amplicons provide data equivalent to clonal readouts. In addition, there are no mobility differences in the sequencing trace and methylation patterns can be readily obtained (Figures 15&16). Because of these various significant advantages, there are ongoing investigations at Applied Biosystems aimed at providing tools and workflows for easily implementing "PCR-cloning" and sequencing for bisulfite-converted samples.

The gDNA templates used in the single-molecule PCR experiments related to Figure 16 were obtained from a formalin-fixed paraffin-embedded (FFPE) clinical specimen, and did not afford an amplifiable intact template in every well (see below). However, for wells that produced amplicons, methylation patterns were apparently uniform and were likely derived from a single bisulfite-converted gDNA template. One of the three positive wells had a single methylated site. Acquired (somatic) changes in methylation patterns possibly occur in single individual cells. Any enzymatic event in a biological system, including the action of methyltransferases, may not be 100% consistent. Each cell is compartmentalized, so that variability due to somatic changes, as in tumor samples, can be more carefully assayed if each cell is independently assayed.

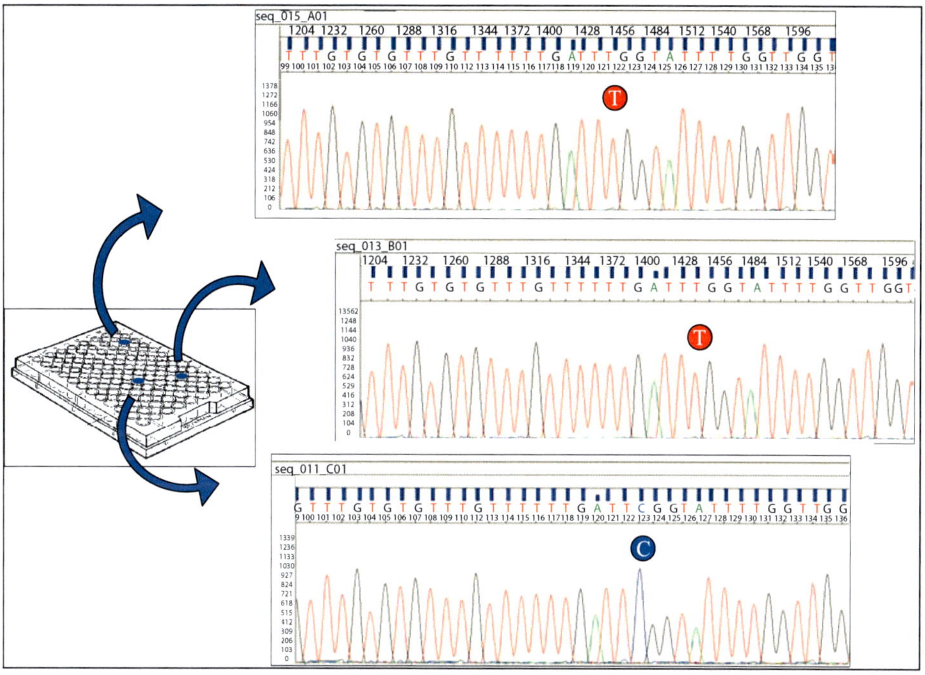

Figure 16. PCR and sequencing after partitioning the bisulfite-converted gDNA into a single-molecule template per well may provide more accurate quantification of methylation status by eliminating PCR and cloning biases. See text for discussion.

DNA analyses of FFPE specimens are particularly challenging in that the DNA is presumably highly crosslinked and is known to be much more degraded [40] compared to fresh or frozen samples. Only a fraction of the original gDNA in FFPE samples is recoverable for analysis, and may not be fully representative of the original methylation status. Sample-to-sample variability during formalin treatment and paraffin embedding, length of time the sample was stored, and gDNA extraction protocol differences may significantly influence the accuracy of methylation analysis of FFPE. Consequently, it is not surprising that there are only a few publications on DNA methylation analysis of FFPE samples [41,42]. We are therefore investigating new protocols aimed at providing improved results with FFPE samples.

Bisulfite Sequencing Clean-Up

Complete removal of fluorescently labeled dideoxy terminators, dNTPs and salts is needed prior to CE analysis of the sequencing extension products. Bisulfite sequencing reactions can be purified using Centri-Sep columns (96-well plate) or ethanol precipitation. Both these purification methods, which require careful attention to detail by laboratory personnel, encounter some loss of sample during processing. Gel purification (Centri-Sep) requires careful pipeting into the center of the gel-bed. A new product and protocol from Applied Biosystems, Big Dye® XTerminator™ Purification Kit, offer a simpler alternative that reduces sample loss. After the sequencing reaction is carried out, a slurry of absorptive, ionic beads is added directly to each sequencing well, which are then sealed and vortexed for 30 minutes. After centrifugation to collect the beads at the bottom of the wells, the plate is directly used with a Model 3730 or 3130 instrument for sequencing. A run module must first be downloaded to adjust the Z-axis of the autosampler to permit direct electrokinetic injection from the top of each well. The resulting sequence is free of so-called "dye-blobs" and provides much stronger signal than alternative purification protocols, due to desalting of the sample by the beads. Purification using Xterminator is both time-saving and very cost-effective.

Summary of Workflow for Methylation Analysis

- Bisulfite conversion with methylSEQr™ Bisulfite kit (P/N 4379580)
- User selected gene, sequence 500-bp +/- transcription start site
- Methyl Primer Express® software for selection of 55 °C bisulfite (PCR) sequencing primers
- PCR using M13-tailed primers, (and FAM-dye on the Forward primer), as per the conditions described below
- Optional: fragment analysis of the PCR amplicon to obtain the ratio of amplicons derived from bisulfite-converted methylated and unmethylated gDNA.
- ExoSap-IT® removal of PCR primers and dNTP
- Direct sequencing of amplicon using Big Dye® Terminator v1.1, conditions described below
- Sequencing clean-up with Big Dye® XTerminator™ Purification Kit
- Model 3730 or 3130 sequencing and analysis with KB basecaller software

Protocol for Bisulfite Sequencing

Bisulfite Conversion

Bisulfite conversions were performed using the methylSEQr™ bisulfite conversion kit (Applied Biosystems) according to manufacturers directions.

PCR

PCR conditions described below were optimized for bisulfite PCR using M13-tailed primers with a 55 °C T_m for the gene specific portion of the primer. When analyzing by fragment analysis, a dye-label (FAM) is included on the 5' end of one of the primers.

-21 M13

Forward Primer
5′TGTAAAACGACGGCCAGT 3′

M13 Reverse

Primer
5′CAGGAAACAGCTATGACC 3′
A denaturant is included in the PCR reaction to reduce PCR bias. Prepare a solution containing 5 mg/mL BSA and 5% glycerol:
250 uL of 20 mg/mL BSA solution (Sigma B8667)
700 uL of molecule biology grade water (Sigma W4502)
50 uL of molecular biology-certified glycerol (Shelton Scientific IB15760)
Required Materials for PCR:
AmpliTaq® Gold, 10X Gold buffer and 25 mM $MgCl_2$ (Applied Biosystems p/n 4311814)
dNTP's (P/N N808-0007)
For a 5 uL reaction:

AmpliTaq Gold 10X buffer	0.5
dNTP 2.5 mM each	0.4
$MgCl_2$ 25 mM	0.4
AmpliTaq Gold polymerase (5U/uL)	0.1
Fwd primer 5 uM	0.25
Rev Primer 5 uM	0.25
Bisulfite-gDNA template 6ng/uL	0.5
BSA-glycerol solution	0.5
Water	2.1

The following thermal cycling conditions are used with M13 tailed primers. The additional 60 minutes at 60 °C at the end of the thermal cycling is to allow full conversion to the non-templated A-addition amplicon-product. Complete A-addition (and not partial) is important when analyzing by fragment analysis, but not necessary if the amplicon is for sequencing only.

95 °C /5 min	(activate AmpliTaq® Gold)
5X	95 °C /30 sec
	60 °C /2:00 min
	72 °C / 3:00 min
35X	95 °C /30 sec
	65 °C /1:00 min
	72 °C / 3:00 min
	60 °C /60 min
	4 °C hold until storage

Estimate of PCR Amplicon Concentration

Run 1/10th of the PCR reactions on a 2% E-Gel to determine quality of PCR products. Use Invitrogen's low molecular weight standard as size/quantitation standard. If the amplicons are FAM labeled and are analyzed by fragment analysis by capillary electrophoresis using Genemapper, an estimate of the amount of amplicon can be obtained from the CE analysis. The fragment analysis protocol has been reported elsewhere [23].

ExoSAP-IT® Treatment

Required Materials:
ExoSAP-IT® (USB) (P/N 78201)
Protocol:

1. Add 2 uL of ExoSAP-IT® to each 5-uL well of PCR products. Make sure the enzymes are added and mixed well with the PCR products.
2. Cover plate and thermal cycle in a 9700 Thermal Cycler with the following cycling profile:

37°C	30 min
80°C	15 min
4°C	hold until storage

3. Remove plate from thermal cycler and spin down contents. Make sure contents are in the bottom of each well.
4. Adjust all amplicons to 1-5 ng/uL concentration based on the agarose gel results.
5. The ExoSAP-IT® treated samples can now be used for sequencing reactions.

Cycle Sequencing Protocol

Required Materials:
BigDye® Terminator v1.1 ((P/N 4337450) or 3.1 (P/N 4337455)
M13 Fwd or Rev primer (3.2 uM)
Protocol:

1. Add the following reagents in a well to set up your sequencing reaction. Use 1 ng or less when sequencing reactions are cleaned-up with Xterminator:

PCR amplicon (bisulfite treated, 1-5 ng/uL)	1 uL
BigDye® Terminator Ready Reaction Mix	8 uL
Primer (M13 Forward or Reverse, 3.2 uM))	1 uL
DH2O	qs
Total volume	20 uL

2. Seal the plate with Thermal seal and quickly spin down the contents in each well in a centrifuge. Place the plate in a 9700 Thermal Cycler. Run the following Sequencing Cycling Profile on 9700 Thermal Cycler:

96° C	1 min
96° C	10 sec \
25X	
50° C	4 min /
4° C	hold until storage

Post Cycle-Sequencing Clean-Up Protocol

The preferred protocol for removing unincorporated dye terminators and unused primer is with Applied Biosystems Big Dye® XTerminator™ Purification Kit (P/N 4376486, following manufacturer instructions). A run module to accommodate injection from a larger volume with the purification resin settled on the bottom of the well must first be installed on a Model 3730 or 3130.

Alternative Post-sequencing Clean-Up Protocol with Centri-Sep® Columns

A pre-treatment of samples with hot SDS to help with removal of unincorporated dye terminators is needed when using column clean-up methods.

Required Materials:

2.2% SDS

9700 Thermal Cycler

SDS Pre-treatment Protocol:

1. Prepare 2.2% SDS in deionized water from stock solution. This SDS solution is stable at room temperature.
2. Add an appropriate amount of the 2.2% SDS solution to your sample to bring the final SDS concentration to 0.2%.

For example: Add 2 uL of 2.2% SDS to each 20uL completed cycle sequencing reaction.

3. Seal the plate and mix thoroughly.
4. Heat the plate to 98°C for 5 min and then allow the plate to cool to ambient temperature before proceeding to the next step.

NOTE: A convenient way to perform this heating/cooling cycle is to place the plate in a thermal cycler and set it as follows:

98°C for 5 min

25°C for 10 min

5. Spin down the contents of the plate briefly.
6. Continue with spin column or 96-well plate purification.

Sequencing Cleanup Protocol

Required Materials: Princeton Separations Centri-Sep® Products (Columns, 8-Strip, or 96-well) Please follow manufacturer's recommended protocol when performing post cycle sequencing clean-up.

Acknowledgements

We are grateful for technical assistance by Ms. Kristina I Moody (Summer Intern) and Ms. Aparna Chhibber at Applied Biosystems.

References

[1] Jones, P. A., and Takai, D. (2001) *Science 293,* 1068-70.
[2] Issa, J. P. (2003) *Clin Cancer Res 9,* 2879-81.
[3] Novik, K. L., Nimmrich, I., Genc, B., Maier, S., Piepenbrock, C., Olek, A., and Beck, S. (2002) *Curr Issues Mol Biol 4,* 111-28.
[4] Garinis, G. A., Patrinos, G. P., Spanakis, N. E., and Menounos, P. G. (2002) *Hum Genet 111,* 115-27.
[5] Widschwendter, M., and Jones, P. A. (2002) *Oncogene 21,* 5462-82.
[6] Widschwendter, M., and Jones, P. A. (2002) *Clin Cancer Res 8,* 17-21.
[7] Oakeley, E. J. (1999) *Pharmacol Ther 84,* 389-400.
[8] Hayatsu, H., Wataya, Y., Kai, K., and Iida, S. (1970) *Biochemistry 9,* 2858-65.
[9] Aranyi, T., Varadi, A., Simon, I., and Tusnady, G. E. (2006) *BMC Bioinformatics 7,* 431.
[10] Tusnady, G. E., Simon, I., Varadi, A., and Aranyi, T. (2005) *Nucleic Acids Res 33,* e9.
[11] Hayatsu, H., Negishi, K., and Shiraishi, M. (2004) *Nucleic Acids Symp Ser (Oxf),* 261-2.
[12] Mill, J., Yazdanpanah, S., Guckel, E., Ziegler, S., Kaminsky, Z., and Petronis, A. (2006) *Biotechniques 41,* 603-7.
[13] Bianchi, D. W. (2006) *Clin Chem 52,* 2182-3.
[14] Tong, Y. K., Ding, C., Chiu, R. W., Gerovassili, A., Chim, S. S., Leung, T. Y., Leung, T. N., Lau, T. K., Nicolaides, K. H., and Lo, Y. M. (2006) *Clin Chem 52,* 2194-202.
[15] Grunau, C., Clark, S. J., and Rosenthal, A. (2001) *Nucleic Acids Res 29,* E65-5.
[16] Boyd, V. L., and Zon, G. (2004) *Anal Biochem 326,* 278-80.
[17] Reinhart, B., Paoloni-Giacobino, A., and Chaillet, J. R. (2006) *Mol Cell Biol 26,* 8347-56.
[18] Kress, C., Thomassin, H., and Grange, T. (2006) *Proc Natl Acad Sci U S A 103,* 11112-7.

[19] Jeronimo, C., Henrique, R., Hoque, M. O., Mambo, E., Ribeiro, F. R., Varzim, G., Oliveira, J., Teixeira, M. R., Lopes, C., and Sidransky, D. (2004) *Clin Cancer Res 10,* 8472-8.

[20] Saxonov, S., Berg, P., and Brutlag, D. L. (2006) *Proc Natl Acad Sci U S A 103,* 1412-7.

[21] Herman, J. G., Graff, J. R., Myohanen, S., Nelkin, B. D., and Baylin, S. B. (1996) *Proc Natl Acad Sci U S A 93,* 9821-6.

[22] Rakyan, V. K., Hildmann, T., Novik, K. L., Lewin, J., Tost, J., Cox, A. V., Andrews, T. D., Howe, K. L., Otto, T., Olek, A., Fischer, J., Gut, I. G., Berlin, K., and Beck, S. (2004) *PLoS Biol 2,* e405.

[23] Boyd, V. L., Moody, K. I., Karger, A. E., Livak, K. J., Zon, G., and Burns, J. W. (2006) *Anal Biochem 354,* 266-73.

[24] Shinde, D., Lai, Y., Sun, F., and Arnheim, N. (2003) *Nucleic Acids Res 31,* 974-80.

[25] Mehra, S., and Hu, W. S. (2005) *Biotechnol Bioeng 91,* 848-60.

[26] Eads, C. A., Danenberg, K. D., Kawakami, K., Saltz, L. B., Blake, C., Shibata, D., Danenberg, P. V., and Laird, P. W. (2000) *Nucleic Acids Res 28,* E32.

[27] Hoheisel, J. D., Craig, A. G., and Lehrach, H. (1990) *J Biol Chem 265,* 16656-60.

[28] Lasken, R. S., Schuster, D. M., and Rashtchian, A. (1996) *J Biol Chem 271,* 17692-6.

[29] Warnecke, P. M., Stirzaker, C., Melki, J. R., Millar, D. S., Paul, C. L., and Clark, S. J. (1997) *Nucleic Acids Res 25,* 4422-6.

[30] Oetting, W. S., Lee, H. K., Flanders, D. J., Wiesner, G. L., Sellers, T. A., and King, R. A. (1995) *Genomics 30,* 450-8.

[31] Don, R. H., Cox, P. T., Wainwright, B. J., Baker, K., and Mattick, J. S. (1991) *Nucleic Acids Res 19,* 4008.

[32] Voss, K. O., Roos, K. P., Nonay, R. L., and Dovichi, N. J. (1998) *Anal Chem 70,* 3818-3823.

[33] Hu, G. (1993) *DNA Cell Biol 12,* 763-70.

[34] Clark, J. M., Joyce, C. M., and Beardsley, G. P. (1987) *J Mol Biol 198,* 123-7.

[35] Clark, J. M. (1988) *Nucleic Acids Res 16,* 9677-9686.

[36] Dupont, J. M., Tost, J., Jammes, H., and Gut, I. G. (2004) *Anal Biochem 333,* 119-27.

[37] Albert, J., and Fenyo, E. M. (1990) *J Clin Microbiol 28,* 1560-4.

[38] Nakano, M., Komatsu, J., Matsuura, S., Takashima, K., Katsura, S., and Mizuno, A. (2003) *J Biotechnol 102,* 117-24.

[39] Hori, M., Fukano, H., and Suzuki, Y. (2007) *Biochem Biophys Res Commun 352,* 323-8.

[40] Romero, R. L., Juston, A. C., Ballantyne, J., and Henry, B. E. (1997) *J Forensic Sci 42,* 708-14.

[41] Kallakury, B. V., Sheehan, C. E., Winn-Deen, E., Oliver, J., Fisher, H. A., Kaufman, R. P., Jr., and Ross, J. S. (2001) *Cancer 92,* 2786-95.

[42] Olaussen, K. A., Soria, J. C., Park, Y. W., Kim, H. J., Kim, S. H., Ro, J. Y., Andre, F., and Jang, S. J. (2005) *Eur J Cancer 41,* 2112-9.

In: Progress in DNA Methylation Research
Editor: Hans P. Neumann, pp. 117-132
ISBN: 978-1-60021-722-7

Chapter III

ENVIRONMENTAL INFLUENCES ON DNA METHYLATION

***Chandrika J. Piyathilake*[1] *and Gary L. Johanning*[2,*]**
[1]Division of Nutritional Biochemistry and Genomics, Department of Nutrition Sciences, University of Alabama at Birmingham, USA;
[2]Department of Veterinary Sciences, The University of Texas M. D. Anderson Cancer Center, USA.

ABSTRACT

DNA methylation can alter the genetic code without changing the base sequence of DNA. Changes in DNA methylation accompany the development and progression of diseases including cancer, aging, imprinting, instability of DNA repeat elements, neurodevelopment and other disorders. Environmental factors, including diet, are known to play a role in DNA methylation and other associated epigenetic changes in the chromatin. This chapter focuses on the contribution of the environment toward DNA methylation, with an emphasis on diet and nutrition. The human diet contains several nutrients that have the potential to influence DNA methylation. Vitamins such as folic acid, vitamin B6 and vitamin B12, and the amino acids choline and methionine are capable of providing methyl groups for DNA methylation. These and other nutrients can impact DNA methylation at several key points. Dietary methyl donors may regulate the methyl supply for formation of S-adenosylmethionine (SAM), a key intermediate required for DNA methylation. Dietary components also modify the activity of DNA methyltransferases that methylate DNA and methyl CpG binding proteins that bind to CpG groups of methylated DNA. One area of DNA methylation that is receiving increase attention is DNA demethylation. There is little information available about how environmental factors influence the removal of methyl groups from DNA. We will discuss the effect of environmental factors on these aspects of DNA methylation. Other topics to be discussed include the relationship between diet, global DNA

* Supported in part by R01 CA105448, R01CA102489.

hypomethylation and gene-specific DNA hypermethylation, which is not clear, especially in disease states such as cancer. In addition, the permissive effect of the environment on the expression of polymorphic genes present in pathways leading to DNA methylation will be covered in this chapter. Finally, we will review how early environmental exposures might leave a DNA methylation imprint that affects disease risk in later life. Changes in DNA methylation do not occur in the absence of environmental influences, and in some cases the environment can play a key role in the process of DNA methylation.

Introduction

Diet and the environment do not generally alter the DNA sequence of genes. However, diet is believed to play a role in the etiology of cancer, possibly via effects on gene expression, but the mechanism is not clear. Environmental factors (diet, chemical/drug exposure, and other features of a person's environment) are known to play a role in DNA methylation and associated epigenetic changes in the chromatin. DNA methylation or histone modifications are therefore potential mediators of the effect of diet or other environmental factors on the etiology of disease.

In this chapter, we will explore the association between the environment and DNA methylation and associated epigenetic changes. Some environmental factors affect DNA methylation in a positive manner to promote health and well-being, whereas others alter methylation to increase disease risk. We will discuss the effect of nutrition, drug exposure and exposure to other environmental agents on DNA methylation and epigenetic changes other than DNA methylation, such as histone modification. Enzymes that methylate DNA are termed DNA methyltransferases (DNMTs), and the environment is capable of activating or repressing these enzymes. Diet and the environment also influence gene polymorphisms in metabolic pathways that supply methyl groups. We will include recent studies in our laboratory of vitamin status and DNA methylation in cancer as well as associations between vitamin status and polymorphisms of genes involved in folic acid metabolism. The possibility of employing environmental agents for epigenetic disease therapy will be discussed at the end of the chapter.

Metabolism of Methyl Donors Present in the Diet

DNA and histones are subject to changes in methylation both during normal development and in disease states such as cancer. When a cytosine residue of one or more CpG sequences in DNA becomes methylated at a promoter/regulatory region of the gene, expression of the gene in many instances is silenced. Methylation of DNA and histones requires a source of methyl groups, and these are commonly referred to as methyl donors. Dietary constituents that are capable of serving as methyl donors include folic acid, vitamin B6, vitamin B12, methionine, betaine, choline and S-adenosylmethionine (SAM). Zinc is a cofactor for DNMT, and while not a methyl donor itself, it does influence DNA methylation in its cofactor role, and is sometimes added as a methyl donor to experimental diets. The primary metabolic pathway that produces methyl donors is shown in Figure 1. Dietary folate (all of the various forms of the vitamin folic acid) is converted in a series of steps to SAM, which in turn is the donor of methyl groups for biological methylation reactions, including DNA and histone methylation. S-adenosylhomocysteine, the product formed after a methyl group is removed from SAM, is converted to homocysteine, which is a substrate for formation of additional methionine in this cycle. Note that when folate or vitamin B12 is deficient, homocysteine cannot be converted to methionine, and will accumulate.

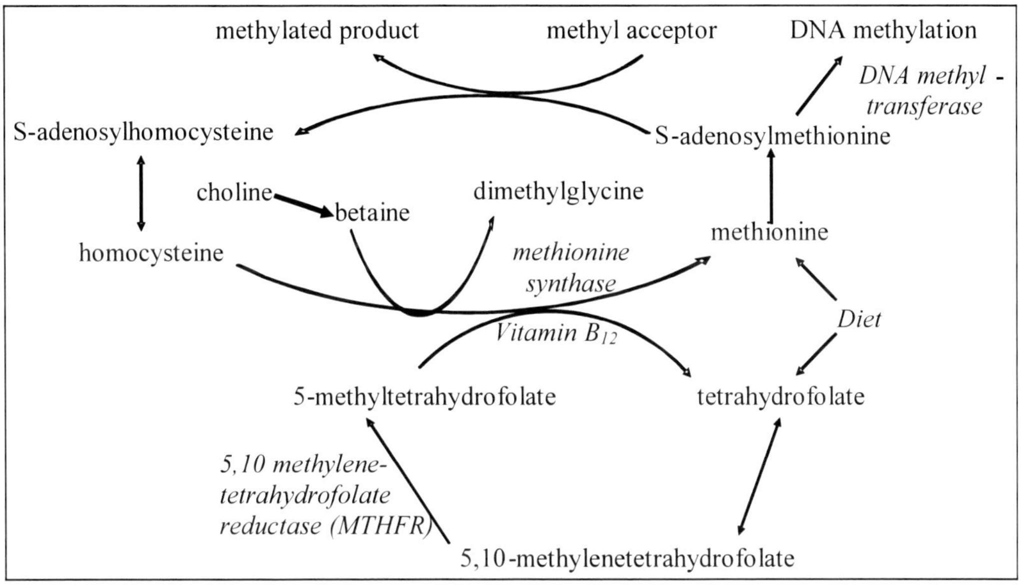

Figure 1. Metabolism of dietary methyl groups.

Does Diet Alter DNA Methylation?

Several reports in the last few years provide convincing experimental evidence that diet plays an important role in DNA methylation. In an early study, the expression of the *Agouti* gene was altered in the progeny by feeding mice dams a diet high in methyl group donors [1]. The *Agouti* gene regulates the production of black and yellow pigments in the hair follicles of mice. The *Avy* allele of the *Agouti* gene contains an insertion of a murine intracisternal A-particle long terminal repeat (LTR), which operates as a promoter/enhancer upstream of the transcription start site of the *Agouti* gene [2]. This LTR contains CpG sites, and expression of the *Avy* allele is regulated by methylation of the LTR. When the LTR is methylated, expression of the *Avy* allele is decreased and the mice show a pseudoagouti phenotype, or a predominately brown coat color. When the LTR is hypomethylated at various sites, the coat color becomes progressively more yellow. In addition to a yellow coat color, expression of the *Avy* allele is associated with several disorders, including obesity, diabetes, increased susceptibility to cancer and a short life. When the diet of the dams was supplemented with methyl donors, the Agouti coat phenotype became more prominent [1]. Moreover, the percentage of methylated cytosines was directly related to the Agouti coat color phenotype.

These studies provided the first suggestion that diet can modify DNA methylation, and in so doing can alter gene expression.

Additional investigations using the Agouti mouse model have shown that other nutrients are capable of modulating the expression of the *Agouti* gene via effects on LTR DNA methylation. When mouse dams received a diet supplemented with 250 mg/kg diet of genistein, the major phytoestrogen in soy, the coat color phenotype and adult body weight distribution of their offspring shifted toward the pseudoagouti phenotype, i.e., a darker coat color and a decreased body weight [3]. The *Avy* allele of the *Agouti* gene was hypermethylated at CpG sites in the LTR of this allele, which indicates that genistein protects against adult onset obesity in the Agouti model.

A recent extension of investigations with the Agouti mouse model provides evidence that the epigenetic mark induced by methylation diets generates an epigenetic change in the germ line. Cropley et al. [4] supplemented the diet of mouse dams with choline, betaine, L-methionine, ZnSO4, folic acid and vitamin B12 throughout pregnancy and during lactation. They hypothesized that a germ-line change in the *Avy* allele induced by a methylation diet would be reflected in phenotypes in a later generation that had not been exposed to methyl donor supplementation. To test this hypothesis, they supplemented pregnant dams with diets containing a methyl donor and the dams produced F1 offspring. The F1 offspring, who did not receive methyl donor supplementation, were bred, and phenotypes in the F2 generation were noted. The F2 generation had a pseudoagouti phenotype, which suggests that *Avy* allele methylation in germ cells of F1 mice is retained during the fertilization and development of the F2 embryo in the absence of methyl donor supplements. These results provide the first evidence that the diet of the mother can have lasting effects on the phenotype of succeeding generations, even if the offspring in subsequent generations consume a diet different from the mother.

Some types of dietary modification are known to initiate biochemical and metabolic changes that persist throughout adulthood and DNA methylation may play a role in these changes. Lillycrop et al. reported that dietary protein restriction of rat dams during pregnancy resulted in decreased methylation of the peroxisomal proliferator-activated receptor α (PPARα) and glucocorticoid receptor genes in the liver of the weaned offspring [5]. The observation that methylation of promoter regions of related genes such as PPARγ1 was not altered indicated that the hypomethylation was specific for only certain genes that were responsive to protein restriction. These epigenetic changes persisted after weaning, suggesting that the alterations in DNA methylation were stable. Another group demonstrated that experimental reduction of uterine blood flow in pregnant rats led to changes in the methylation status of the p53 gene in the kidney of the adult offspring [6]. It is very likely that future studies will show the diet and other environmental agents to be important inducers of epigenetic modifications in the fetus or neonate that persist into adulthood. There is potential for either beneficial or detrimental epigenetic events that could lead, respectively, to improved health or disease in later life.

Does the Environment Influence DNA Methylation?

The answer to this question is also yes. In a very interesting study, Manel Esteller's lab group reported that monozygotic twins are similar in their pattern of DNA methylation and histone modification in early life. As these monozygotic twins began to grow older and differ in their routine behavior, their genomic DNA methylation and histone acetylation degrees and patterns, both in lymphocytes and tissues, began to show notable differences [7]. Their results demonstrate conclusively that DNA methylation does not occur in a vacuum; if genetic factors were the sole determinant of DNA methylation, then the epigenetic changes occurring as the twins grew older would remain similar for each twin. However, their results indicate that some factors in the environment were likely responsible for causing epigenetic modifications in one twin that did not occur in the other.

Dietary Methyl Donors: Important Determinants of DNA Methylation in Health and Disease

The dietary methyl donors in Figure 1 contribute to methylation of DNA in healthy individuals, and are also capable of modifying DNA methylation in disease states such as cancer. The promoter/regulatory regions of cancer preventive genes such as tumor suppressor genes become hypermethylated in cancer, whereas DNA becomes globally hypomethylated in cancer. Dietary methyl deficiency is capable of influencing both of these processes. Methyl deficiency leads to global DNA hypomethylation as well as gene-specific changes in methylation.

A recent review has summarized the essentiality of choline, especially for development of the fetus [8]. Dietary choline deficiency removes a substrate for the biosynthesis of SAM, the methyl donor for DNMTs, and hence there is decreased DNA methylation in choline deficiency [9,10]. Methylation of specific genes is also affected by choline deficiency. For example, in neuroblastoma cells, deficiency of choline was associated with inhibition of cell proliferation, and this decreased proliferation was accompanied by incomplete methylation of most of the CpG sites in the promoter and first exon of the CDKN3 gene, a member of the cyclin-dependent kinase inhibitor group [11]. The promoter hypomethylation, in turn, resulted in overexpression of CDKN3 and inhibition of cell proliferation. Dietary choline deficiency in rats also led to gene-specific hypomethylation of the oncogenes c-myc, c-fos, and c-Ha-ras [12], and c-myc hypomethylation induced by dietary choline deficiency resulted in c-myc gene overexpression in hepatocellular carcinomas [9]. These results suggest that choline deficiency might play a role in hepatocarcinogenesis.

Dietary folic acid was also demonstrated to affect methylation of lymphocyte DNA in healthy elderly [13] and postmenopausal [14] women. In these studies, a low to moderate folate diet was provided for 5 or 7 weeks, and leukocyte or lymphocyte DNA methylation was determined. DNA hypomethylation was observed during moderate folate depletion, but clinical signs of folate deficiency were absent in one study [13]. In the other study [14], a subclinical folate deficiency with decreased plasma folate developed, which was accompanied by lymphocyte DNA hypomethylation [14]. These similar results from two independent laboratory groups indicate that low folic acid levels in the diet are capable of altering DNA methylation in normal, healthy humans.

As noted in other sections, depletion of methyl donors from the diet has been reported to hypomethylate specific genes, notably oncogenes. Other studies, however, have shown the opposite effect: a lower concentration of methyl donors induced gene-specific DNA hypermethylation. In an early study, folate deficiency induced hypermethylation of the *H-cadherin* gene in human nasopharyngeal carcinoma KB cells [15]. The promoter region of the *p16INK4A* gene was hypermethylated in liver tumors induced by folate/methyl deficiency [16]. A population-based study of neck squamous cell carcinoma also showed that *p16INK4A* promoter hypermethylation was associated with a low dietary folate intake [17]. An interesting aspect of the latter study is that both the 677 polymorphism and the 1298 polymorphism of the folate metabolism enzyme 5, 10 methylene tetrahydrofolate reductase (*MTHFR*) (Figure 1) increased the risk of cancer in subjects whose folate intake was low. The 677 and 1298 polymorphisms are common polymorphic variants of *MTHFR.* The T variant sequence is located at nucleotide 677 (*MTHFR* C677T) and the C variant is at nucleotide 1298 (*MTHFR* A1298C). The C677T variant is associated with higher serum homocysteine levels and with an increased risk of vascular disease and neural tube defects. Hypermethylation accompanying low folate intake might seem counterintuitive, since folate is a donor for SAM, the source of methyl groups for DNA. However, the activity of DNMT is increased in liver tumors of rats fed a diet low in methyl donors [18], and DNMT activation may be responsible for hypermethylation of specific gene promoters in the face of low methyl availability. Global DNA hypomethylation coupled with gene-specific hypermethylation is a yet unexplained phenomenon that has been observed in both folate deficiency and in carcinogenesis.

Studies from our laboratory have focused on vitamin depletion (primarily folate and vitamin B12) from cancer tissue or other easily accessible tissue, and the effect of this vitamin depletion on global DNA hypomethylation. A related focus is associations between vitamin status, polymorphisms in methylation pathway genes such as *MTHFR*, and cancer risk. Our studies of folate and DNA methylation in cancer began with the observation that exposure to cigarette smoke results in folic acid deficiency, rendering the bronchial epithelium more susceptible to neoplastic transformation by the carcinogenic hydrocarbons of tobacco smoke. Global DNA methylation was chosen as a marker of importance for malignant transformation because the vitamins of interest, folate and vitamin B12, are involved in generation of methyl groups for DNA methylation. The association between vitamin concentrations at the cell and tissue level and cancer risk has received little emphasis; most studies evaluate associations between blood vitamins and cancer risk.

Two studies by our group illustrate the concept that cellular vitamin concentrations in a given tissue may be associated with global DNA methylation, a marker of increased risk for malignancy. In the first study, we determined that folate and vitamin B-12 concentrations were significantly lower in squamous cell carcinoma (SCC) of the lung than in adjacent uninvolved bronchial mucosa [19]. Global DNA methylation was decreased in cancerous SCC tissues, in comparison to adjacent uninvolved tissues from the same subjects. We found a significant inverse correlation between cellular folate and DNA methylation, and between cellular vitamin B-12 and DNA methylation in cancerous tissues. However, in uninvolved tissues there was only a weak inverse association between folate and DNA methylation and no association between vitamin B-12 and DNA methylation. The significant inverse correlation between folate and vitamin B-12 concentrations and global DNA methylation in SCC tissues indicates the importance of both vitamins for DNA methylation.

A second investigation of SCC of the lung by our group found that ascorbic acid (vitamin C) concentrations of SCC tissues of the lung were significantly higher than matched uninvolved lung tissues [20]. This is in contrast folate and vitamin B-12 concentrations in the previous study, which were lower in lung SCC tissues. As in the previous study, cancerous tissues of the lung were hypomethylated relative to matched uninvolved tissues. There was an inverse correlation between ascorbic acid concentrations and DNA hypomethylation in cancerous tissues of the lung, such that high ascorbic acid was associated with prevention of the DNA hypomethylation that is frequently found in cancer tissues.

We have also examined the role of folate status and *MTHFR* polymorphisms in relation to cancer risk. In a recent study, we reported that the presence of any *MTHFR* polymorphism was inversely associated with cervical intraepithelial neoplasia 2 (moderate) or 3 (severe), even after adjustment for age, ethnic origin, smoking, age at first intercourse, use of hormonal contraception and plasma concentrations of folate, B12, vitamin C, vitamin A, vitamin E and total carotene [21]. Our results indicated that *MTHFR* polymorphism had an impact independent of blood folate, and we concluded that this may have been due to the fact that our study was conducted after the folate fortification of cereal grains in the United States. Colorectal cancer investigations have found no protective effects of *MTHFR* polymorphisms on cancer risk among individuals with low systemic folate [22], whereas protective effects, similar to what we found in our study, were observed among those with normal folate levels. Rather than an effect on DNA methylation, the more important consequence of *MTHFR* polymorphism in our study may have been that this polymorphism increases the availability of 5,10-methylenetetrahydrofolate for *de novo* purine synthesis, because a less active *MTHFR* would block the conversion of 5,10-methylenetetrahydrofolate to 5-methyltetrahydrofolate (Figure 1). Increased availability of 5,10-methylenetetrahydrofolate would conserve methyl groups for conversion of uracil to thymidine, whereas low thymidine levels would lead to uracil misincorporation into DNA, resulting in DNA strand breaks that lead to chromosome translocations and deletions. The *MTHFR* polymorphism might thus prevent DNA damage that could promote cervical cancer, and play a protective role in preventing cervical cancer.

Other Nutrients and Interactions

As already discussed, genistein, a component of soy and a dietary polyphenol, is capable of altering DNA methylation to such an extent that it influences phenotypes in vivo. Micronutrients and drugs not in the one-carbon metabolic pathway (Figure 1) can alter DNA methylation. Another dietary polyphenol, (2)-epigallocatechin 3-gallate (EGCG) from green tea, was recently found to competitively inhibit DNMT, and to reverse hypermethylation and reactivate expression of RARβ, MGMT, p16INK4a, and hMLH1, which were silenced in the esophageal cancer cell line KYSE 510 [23].

Selenium and arsenic supplementation were reported to alter DNA methylation in vivo and in vitro in colon cells [24]. More recent studies suggest possible interactions of selenium and folate [25] as well as selenium and arsenic [26] in methyl metabolism. Rats fed a diet deficient in zinc had a slower turnover of the methyl group of the SAM in the liver, which was accompanied by depressed rates of methylation of DNA and histones, suggesting that zinc is required for these methylation processes [27].

DNA and Histone Metyltransferases

The DNMTs occupy a critical position linking the methyl donor SAM with the target DNA that becomes methylated. Methyl groups are transferred to a cytosine residue of the DNA, present in most cases as a CpG dinucleotide. The effect of environment on DNMT expression has been evaluated in several studies. Mammalian genomes contain three DNMTs. DNMT1 methylates cytosine residues in hemimethylated DNA, which is DNA that is methylated on one of the two strands of DNA. This enzyme activity preserves the DNA methylation on the daughter strand when DNA is replicated. In contrast, DNMT3A and DNMT3B are *de novo* DNMTs, which can initiate DNA methylation on substrates that do not contain a methylated cytosine on either strand, and in fact show a preference for unmethylated DNA [28,29]. DNMT3A and DNMT3B are responsible for DNA methylation in the early embryo, during early embryonic development [30]. Dietary availability of methyl donors and gamma-radiation stress were reported to alter the maintenance DNMT1 profile [31]. Mice were fed a folate free diet, a folate and choline free diet, or a folate, choline and methionine free diet for two months. The methyl deficient diets led to a slight but nonsignificant decrease in DNMT1 activity. But exposure to γ radiation stress (1-3 Gy) caused significant decreases in DNMT1 activity in both the cytosol and nucleus. DNMT3 activity was not affected by diet or γ radiation stress. The radiation-induced loss of DNMT1 activity was greatest when all three of the methyl donors were deficient.

Ghoshal et al. recently reported differential expression of *de novo* DNA methyltransferases (DNMT3A and DNMT3B), DNMT1, and methyl CpG binding proteins during early stages of tumorigenesis in the livers of rats fed a diet deficient in L-methionine and devoid of folic acid and choline [32]. This methyl deficient diet leads to hepatocarcinogenesis in the absence of carcinogens, and is correlated with a decrease in SAM levels and an increase in SAH levels in the liver [33]. The levels of both DNMT1 mRNA and protein were increased in rats fed a methyl deficient diet, and the changes in DNMT levels were observed as a state of preneoplasia developed in the liver [32]. These authors found that DNMT3a mRNA levels were upregulated at early stages of methyl deficiency, but the DNMT3a protein level was not altered by methyl deficiency. Neither the mRNA nor the protein levels of DNMT3b were affected by the methyl deficient diets. The amounts of mRNA and protein of the methyl CpG binding protein MBD2 were also upregulated in the liver of methyl-deficient rats. The expression of two other methyl binding proteins, MBD1 and MBD4, were increased during methyl deficiency in this study. The authors hypothesized that the global DNA hypomethylation that occurs specifically in the livers of methyl deficient rats may induce expression of DNMTs and methyl CpG binding proteins.

As discussed in an earlier section, agents other than methyl donors can alter DNA methylation; similarly, other nutrients are capable of influencing DNMT expression. For example, supplemental selenium inhibits DNMT1 protein expression and enzyme activity; conversely, a deficiency of this trace mineral results in diminished DNMT1 expression [34,35]. Higher dietary arsenic levels were also reported to increase the activity of DNMT in the liver of rats [36].

Evidence for the presence of an enzyme that demethylates DNA [37] suggests that extrinsic factors that modulate demethylation of DNA or histones may exist. Few studies have been done to identify factors that might affect DNA demethylation. Acute homocysteine treatment caused transient demethylation of the homocysteine-induced endoplasmic reticulum protein promoter [38]. In addition to methylation on DNA, histones can also be methylated, and the methylation marks on histones made by histone methyltransferases alter transcriptional activity of genes. The source of methyl groups for histone methylation is also SAM. There have been recent reports of how environmental factors affect methylation and demethylation of histones. Hypoxia was shown to increase dimethylated histone H3 lysine 9 (H3K9me2) by inhibiting H3K9 demethylation [39]. Nickel ions directly decreased the activity of a histone H3K9 demethylase in nuclear extracts in vitro [40]. In another investigation, rats exposed to 420 p.p.m. tamoxifen in their diet for up to 24 weeks exhibited decreased trimethylation of histone H4K20, as well as decreased expression of DNMT1, DNMT3a and DNMT3b protein, in the liver [41]. The authors proposed that tamoxifen-induced hepatocarcinogenesis might result from both epigenetic changes in histone and DNA methylation as well as a genotoxic change stemming from the formation of tamoxifen–DNA adducts. In a recent study by this group, a methyl deficient diet fed to rats led not only to the already discussed global and gene-specific DNA hypomethylation, but also to a gradual decrease in H3K9 trimethylation and acetylation, and to decreased H4K20 trimethylation [42]. In addition, expression of associated histone-modifying enzymes was also decreased. They indicated that decreased methylation of these histones might lead to hepatocarcinogenesis via effects on heterochromatin organization and resultant impaired genome stability, damage checkpoint control, cell cycle arrest and/or terminal differentiation. The current interest in this topic suggests that the influence of environmental agents on DNA demethylation and histone methylation and acetylation will be an important subject for research in the next several years.

Environmental Agents and Epigenetic Therapy

There is a growing interest in cancer therapy with pharmacologic agents that inhibit the methylation of DNA and the methylation or acetylation of histones. The promoter/regulatory region of numerous tumor suppressor, cell cycle and DNA repair genes become hypermethylated in cancer, and expression of these genes is silenced. DNA hypomethylating agents that prevent or reverse this epigenetic silencing are being used clinically to restore expression of tumor suppressor genes. The appeal of this type of therapy is that it does not involve alterations in the base sequence of DNA, and that these epigenetic therapies can functionally reverse the gene-specific DNA hypermethylation that develops in many cancers.

Two major demethylating agents used in preclinical and clinical trials are azacitidine and its deoxy derivative, 5-aza-2'-deoxycytidine (decitabine; DAC). Azacitidine is metabolized to 5-aza-2'-deoxycytidine triphosphate, and is incorporated into DNA instead of cytosine, which is normally incorporated. The cellular DNMTs are unable to methylate this substituted base, and thus the DNA becomes demethylated upon subsequent DNA replication. These agents have shown promising results in randomized clinical trials involving patients with myelodysplastic syndrome [43] and chronic myelomonocytic leukemia [44], and they have received Food and Drug Administration approval for treatment of these diseases. The next step, already in progress, is to expand the clinical trials to solid tumors. In a Phase I trial, DAC administration induced expression of the target genes NY-ESO-1, MAGE-3, and p16 in non–small cell lung cancer and small-cell lung cancer biopsies [45]. In another Phase I trial, 19 patients with metastatic solid tumors (twelve of whom had metastatic melanoma) were infused continuously with DAC from days 1–3 of a 28-day cycle [46]. Although a posttreatment decrease in methylation was observed in 11 of the 19 genes studied, there was no relationship between the dose of DAC and the effect on methylation. A third Phase I clinical trial found that continuous infusion of DAC for 168 hours led to *MAGE-1* promoter hypomethylation and to significant genomic DNA hypomethylation in peripheral blood cells within 14 days after the start of treatment. Genomic DNA methylation returned to baseline levels between 28 and 35 days after the start of treatment, which provides an idea of the length of time that functional effects of DAC infusion persist [47].

There is no hard evidence to date that folate or other dietary component can prevent or treat cancer via effects on the epigenome. In fact, since folate and vitamins in general promote cell viability, there is a possibility that folate supplementation at later stages of cancer could stimulate cancer cell growth [48]. However, hypomethylating agents like decitabine and 5-azacitidine are showing efficacy in clinical trials for cancer. Are there other environmental agents that alter DNA methylation that are useful for cancer prevention or therapy? Global DNA hypomethylation is a hallmark of cancer. Would reversing the global hypomethylation serve to prevent or treat cancer? Pereira's lab group has reported in several cases that cancer-preventive agents reverse DNA hypomethylation in tumors, while agents that do not prevent cancer have no effect on DNA hypomethylation [49-53]. This group determined that the anti-inflammatory glucocorticoid budesonide modulates DNA methylation and prevents tumor formation in animal models. In a recent study by this group [54], combinations of budesonide and R115777 (a farnesyl transferase inhibitor that prevents lung tumors in mice, Zarnestra, Tipifarnib) prevented lung tumors and prevented and reversed DNA hypomethylation in lung tumors.

S-adenosylmethionine (SAM) also hypermethylates DNA by inhibiting DNA demethylation. Administration of SAM (but not SAH) to MDA-MB-231 breast cancer cells inhibited urokinase (uPA) expression as well as tumor cell invasion through Matrigel and metastasis *in vivo* [55-58]. Similar results were obtained in studies of SAM hypermethylation in prostate cancer [59].

Conclusion

There is convincing evidence that agents in the environment, including diet, are capable of influencing DNA methylation. Nutrients that are known to donate methyl groups for DNA and histone methylation influence methylation, but other nutrients that do not play a role as methyl donors are also capable of contributing to DNA methylation. The environment affects the expression and activity of enzymes that methylate DNA and histones, and also of proteins that bind to methyl groups on DNA. There is an escalating use of inhibitors of DNA methylation as therapeutic agents for diseases such as cancer, and the demethylating agents have been effective in some types of cancer. These agents are now being tested for efficacy against solid tumors in cancer patients. There is parallel interest in DNA hypermethylating agents for cancer prevention and therapy, because these agents reverse the progressive global DNA hypomethylation characteristic of many cancers.

Dietary methyl donors modulate DNA methylation, both in health and disease. Methyl restriction and DNMT modulators paradoxically cause global DNA hypomethylation alongside of gene-specific DNA hypermethylation, in some instances. Deficiency of methyl donors such as folate or vitamin B12 induces global DNA hypomethylation in tumor tissues, and the hypomethylation is accompanied by decreased tissue levels of the vitamins.

There are several remaining unanswered questions related to the environment and DNA methylation. Which dietary components play the most important roles in effecting changes in DNA methylation? Dietary methyl donors are known to alter DNA methylation, but a host of other nutritional and non-nutritional components of the diet have the potential to modulate methylation. Both hypermethylating and hypomethylating agents are currently under investigation for epigenetic therapy of cancer. Are there advantages of using hypomethylating agents (e.g., decitabine and azacitidine) in some types of cancer, and hypermethylating agents (e.g., budesonide) in other types? Are hypomethylating agents effective in patients with a pre-existing cancer, and hypermethylating agents effective in cancer prevention and early-stage cancer? Since DNA becomes globally hypomethylated in some cancers, will the demethylating agents that are being increasingly used in epigenetic therapy lead to long-term recurrence of cancer? Are there additional dietary components or natural products that alter DNA methylation? These might be less toxic and less expensive than drugs in current use. How important is DNA methylation in diseases other than cancer? There is emerging evidence that DNA methylation might play an important role in cardiovascular disease, arthritis, Alzheimer's disease and perhaps many other diseases. These questions cannot be answered with certainty now, but future research will almost certainly help to resolve these and many other issues related to the influence of diet and other environmental agents on DNA methylation.

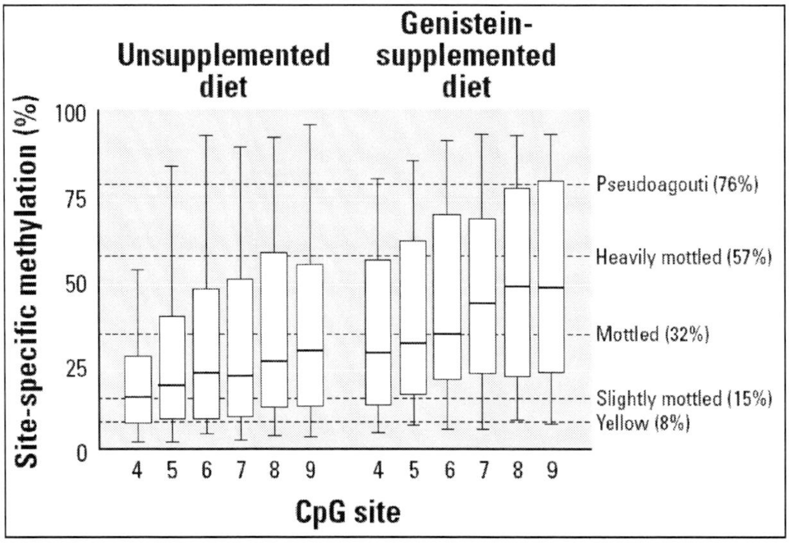

Figure 2. Box plots representing the percentage of cells methylated at sites 4–9 in unsupplemented (n = 52) and genistein-supplemented (n = 44) Avy/a offspring. Ends of the boxes indicate the interquartile range representing the 25th to 75th percentiles of the data; horizontal lines within each box indicate median; and dashed horizontal lines represent average percent methylation of CpG sites 4–9 according to coat-color phenotype. Reproduced with permission from *Environmental Health Perspectives,* Dolinoy et al., volume 114, number 4, April 2006.

References

[1] Cooney, CA; Dave, AA; Wolff, GL. Maternal methyl supplements in mice affect epigenetic variation and DNA methylation of offspring. *The Journal of Nutrition,* 2002;132:2393S-2400S.

[2] Wolff, GL; Kodell, RL; Moore, SR; Cooney, CA. Maternal epigenetics and methyl supplements affect agouti gene expression in Avy/a mice. *FASEB Journal,* 1998;12:949-57.

[3] Dolinoy, DC; Weidman, JR; Waterland, RA; Jirtle, RL. Maternal genistein alters coat color and protects Avy mouse offspring from obesity by modifying the fetal epigenome. *Environmental Health Perspectives,* 2006;114:567-72.

[4] Cropley, JE; Suter, CM; Beckman, KB; Martin, DI. Germ-line epigenetic modification of the murine A vy allele by nutritional supplementation. *Proceedings of the National Academy of Sciences U S A,* 2006;103:17308-12.

[5] Lillycrop, KA; Phillips, ES; Jackson, AA; Hanson, MA; Burdge, GC. Dietary protein restriction of pregnant rats induces and folic acid supplementation prevents epigenetic modification of hepatic gene expression in the offspring. *The Journal of Nutrition,* 2005;135:1382-6.

[6] Pham, TD; MacLennan, NK; Chiu, CT; Laksana, GS; Hsu, JL; Lane, RH. Uteroplacental insufficiency increases apoptosis and alters p53 gene methylation in the full-term IUGR rat kidney. *The American Journal of Physiology - Regulatory, Integrative and Comparative Physiology,* 2003;285:R962-70.

[7] Fraga, MF; Ballestar, E; Paz, MF; Ropero, S; Setien, F; Ballestar, ML; Heine-Suner, D; Cigudosa, JC; Urioste, M; Benitez, J; Boix-Chornet, M; Sanchez-Aguilera, A; Ling, C; Carlsson, E; Poulsen, P; Vaag, A; Stephan, Z; Spector, TD; Wu, YZ; Plass, C; Esteller, M. Epigenetic differences arise during the lifetime of monozygotic twins. *Proceedings of the National Academy of Sciences U S A,* 2005;102:10604-9.

[8] Zeisel, SH. Choline: critical role during fetal development and dietary requirements in adults. *Annual Review of Nutrition,* 2006;26:229-50.

[9] Tsujiuchi, T; Tsutsumi, M; Sasaki, Y; Takahama, M; Konishi, Y. 1999. Hypomethylation of CpG sites and c-myc gene overexpression in hepatocellular carcinomas, but not hyperplastic nodules, induced by a choline-deficient L-amino acid–defined diet in rats. *Japanese Journal of Cancer Research,* 1999;90:909–13.

[10] Locker, J; Reddy, TV; Lombardi, B. DNA methylation and hepatocarcinogenesis in rats fed a choline devoid diet. *Carcinogenesis,* 1986;7:1309–12.

[11] Niculescu, MD; Yamamuro, Y; Zeisel, SH. Choline availability modulates human neuroblastoma cell proliferation and alters the methylation of the promoter region of the cyclin-dependent kinase inhibitor 3 gene. *Journal of Neurochemistry,* 2004;89:1252-9.

[12] Christman, JK; Sheikhnejad, G; Dizik, M; Abileah, S; Wainfan, E. Reversibility of changes in nucleic acid methylation and gene expression induced in rat liver by severe dietary methyl deficiency. *Carcinogenesis,* 1993;14:551–57.

[13] Rampersaud, GC; Kauwell, GP; Hutson, AD; Cerda, JJ; Bailey, LB. Genomic DNA methylation decreases in response to moderate folate depletion in elderly women. *American Journal of Clinical Nutrition,* 2000;72:998-1003.

[14] Jacob, RA; Gretz, DM; Taylor, PC; James, SJ; Pogribny, IP; Miller, BJ; Henning, SM; Swendseid, ME. Moderate folate depletion increases plasma homocysteine and decreases lymphocyte DNA methylation in postmenopausal women. *The Journal of Nutrition,* 1998;128:1204-12.

[15] Jhaveri MS, Wagner C, Trepel JB. Impact of extracellular folate levels on global gene expression.
Molecular Pharmacology, 2001;60:1288-95.

[16] Pogribny, IP; James, SJ. De novo methylation of the p16INK4A gene in early preneoplastic liver and tumors induced by folate/methyl deficiency in rats. *Cancer Letters,* 2002;187:69-75.

[17] Kraunz, KS; Hsiung, D; McClean, MD; Liu, M; Osanyingbemi, J; Nelson, HH; Kelsey, KT. Dietary folate is associated with p16(INK4A) methylation in head and neck squamous cell carcinoma. *International Journal of Cance,r* 2006;119:1553-7.

[18] Lopatina, NG; Vanyushin, BF; Cronin, GM; Poirier, LA. Elevated expression and altered pattern of activity of DNA methyltransferase in liver tumors of rats fed methyl-deficient diets. *Carcinogenesis,* 1998;19:1777-81.

[19] Piyathilake, CJ; Johanning, GL; Macaluso, M; Whiteside, M; Oelschlager, DK; Heimburger, DC; Grizzle, WE. Localized folate and vitamin B-12 deficiency in squamous cell lung cancer is associated with global DNA hypomethylation. *Nutrition and Cancer,* 2000;37:99-107.

[20] Piyathilake, CJ; Bell, WC; Johanning, GL; Cornwell, PE; Heimburger, DC; Grizzle, WE. The accumulation of ascorbic acid by squamous cell carcinomas of the lung and larynx is associated with global methylation of DNA. *Cancer,* 2000;89:171-6.

[21] Henao, OL; Piyathilake, CJ; Waterbor, JW; Funkhouser, E; Johanning, GL; Heimburger, DC; Partridge, EE. Women with polymorphisms of methylenetetrahydrofolate reductase (MTHFR) and methionine synthase (MS) are less likely to have cervical intraepithelial neoplasia (CIN) 2 or 3. *International Journal of Cancer,* 2005;113:991-7.

[22] Chen, J; Giovannucci, E; Kelsey, K; Rimm, EB; Stampfer, MJ; Colditz, GA; Spiegelman, D; Willett, WC; Hunter, DJ. A methylenetetrahydrofolate reductase polymorphism and the risk of colorectal cancer. *Cancer Research,* 1996;56:4862-4.

[23] Fang, M; Chen, D; Yang, CS. Dietary polyphenols may affect DNA methylation. *The Journal of Nutrition,* 2007;137:223S-8S.

[24] Davis, CD; Uthus, EO; Finley, JW. Dietary selenium and arsenic affect DNA methylation in vitro in Caco-2 cells and in vivo in rat liver and colon. *The Journal of Nutrition,* 2000;130:2903-9.

[25] Uthus, EO; Ross, SA; Davis, CD. Differential effects of dietary selenium (se) and folate on methyl metabolism in liver and colon of rats. *Biological Trace Element Research,* 2006;109:201-14.

[26] Zeng, H; Uthus, EO; Combs, GF Jr. Mechanistic aspects of the interaction between selenium and arsenic. *Journal of Inorganic Biochemistry,* 2005;99:1269-74.

[27] Wallwork, JC; Duerre, JA. Effect of zinc deficiency on methionine metabolism, methylation reactions and protein synthesis in isolated perfused rat liver. *The Journal of Nutrition,* 1985;115:252-62.

[28] Okano, M; Xie, S; Li, E. Cloning and characterization of a family of novel mammalian DNA (cytosine-5) methyltransferases. *Nature Genetics,* 1998;19:219-20.

[29] Yokochi, T; Robertson, KD. Preferential methylation of unmethylated DNA by Mammalian de novo DNA methyltransferase Dnmt3a. *The Journal of Biological Chemistry,* 2002;277:11735-45.

[30] Okano, M; Bell, DW; Haber, DA; Li, E. DNA methyltransferases Dnmt3a and Dnmt3b are essential for de novo methylation and mammalian development. *Cell,* 1999;99:247-57.

[31] Batra, V; Mishra, KP. Modulation of DNA methyltransferase profile by methyl donor starvation followed by gamma irradiation. *Molecular and Cellular Biochemistry,* 2006;Jul 20, DOI 10.1007/s11010-006-9258-8.

[32] Ghoshal, K; Li, X; Datta, J; Bai, S; Pogribny, I; Pogribny, M; Huang, Y; Young, D; Jacob, ST. A folate- and methyl-deficient diet alters the expression of DNA methyltransferases and methyl CpG binding proteins involved in epigenetic gene silencing in livers of F344 rats. *The Journal of Nutrition,* 2006;136:1522-7.

[33] Shivapurkar, N; Poirier, LA. Tissue levels of *S*-adenosylmethionine and *S*-adenosylhomocysteine in rats fed methyl-deficient, amino acid-defined diets for one to five weeks. *Carcinogenesis,* 1983;4:1051-1057.

[34] Davis, CD; Uthus, EO. Dietary selenite and azadeoxycytidine treatments affect dimethylhydrazine-induced aberrant crypt formation in rat colon and DNA methylation in HT-29 cells. *The Journal of Nutrition,* 2002;132:292-7.
[35] Fiala, ES; Staretz, ME; Pandya, GA; El-Bayoumy, K; Hamilton, SR. Inhibition of DNA cytosine methyltransferase by chemopreventive selenium compounds, determined by an improved assay for DNA cytosine methyltransferase and DNA cytosine methylation. *Carcinogenesis,* 1998;19:597-604.
[36] Uthus, EO; Davis, C. Dietary arsenic affects dimethylhydrazine-induced aberrant crypt formation and hepatic global DNA methylation and DNA methyltransferase activity in rats. *Biological Trace Element Research,* 2005;103:133-45.
[37] Ramchandani, S; Bhattacharya, SK; Cervoni, N; Szyf, M. DNA methylation is a reversible biological signal. *Proceedings of the National Academy of Sciences U S A,* 1999;96:6107-12.
[38] Lenz, B; Bleich, S; Beutler, S; Schlierf, B; Schwager, K; Reulbach, U; Kornhuber, J; Bonsch, D. Homocysteine regulates expression of Herp by DNA methylation involving the AARE and CREB binding sites. *Experimental Cell Research,* 2006;312:4049-55.
[39] Chen, H; Yan, Y; Davidson, TL; Shinkai, Y; Costa, M. Hypoxic stress induces dimethylated histone H3 lysine 9 through histone methyltransferase G9a in mammalian cells. *Cancer Research,* 2006;66:9009-16.
[40] Chen, H; Ke, Q; Kluz, T; Yan, Y; Costa, M. Nickel ions increase histone H3 lysine 9 dimethylation and induce transgene silencing. *Molecular and Cellular Biology,* 2006;26:3728-37.
[41] Tryndyak, VP; Muskhelishvili, L; Kovalchuk, O; Rodriguez-Juarez, R; Montgomery, B; Churchwell, MI; Ross, SA; Beland, FA; Pogribny, IP. Effect of long-term tamoxifen exposure on genotoxic and epigenetic changes in rat liver: implications for tamoxifen-induced hepatocarcinogenesis. *Carcinogenesis,* 2006;27:1713-20.
[42] Pogribny, IP; Tryndyak, VP; Muskhelishvili, L; Rusyn, I; Ross, SA. Methyl deficiency, alterations in global histone modifications, and carcinogenesis. *The Journal of Nutrition,* 2007;137:216S-22S.
[43] Silverman, LR; Demakos, EP; Peterson, BL; Kornblith, AB; Holland, JC; Odchimar-Reissig, R; Stone, RM; Nelson, D; Powell, BL; DeCastro, CM; Ellerton, J; Larson, RA; Schiffer, CA; Holland, JF. Randomized controlled trial of azacitidine in patients with the myelodysplastic syndrome: a study of the cancer and leukemia group B. *Journal of Clinical Oncology,* 2002;20:2429-40.
[44] Kantarjian, H; Oki, Y; Garcia-Manero, G; Huang, X; O'brien, S; Cortes, J; Faderl, S; Bueso-Ramos, C; Ravandi, F; Estrov, Z; Ferrajoli, A; Wierda, W; Shan, J; Davis, J; Giles, F; Saba, HI; Issa, JP. Results of a randomized study of three schedules of low-dose decitabine in higher risk myelodysplastic syndrome and chronic myelomonocytic leukemia. *Blood,* 2006 Aug 1.
[45] Schrump, DS; Fischette, MR; Nguyen, DM; Zhao, M; Li, X; Kunst, TF; Hancox, A; Hong, JA; Chen, GA; Pishchik, V; Figg, WD; Murgo, AJ; Steinberg, SM. Phase I study of decitabine-mediated gene expression in patients with cancers involving the lungs, esophagus, or pleura. *Clinical Cancer Research,* 2006;12:5777-85.

[46] Aparicio, A; Eads, CA; Leong, LA; Laird, PW; Newman, EM; Synold, TW; Baker, SD; Zhao, M; Weber, JS. Phase I trial of continuous infusion 5-aza-2'-deoxycytidine. *Cancer Chemotherapy and Pharmacology,* 2003;51:231-9.

[47] Samlowski, WE; Leachman, SA; Wade, M; Cassidy, P; Porter-Gill, P; Busby, L; Wheeler, R; Boucher, K; Fitzpatrick, F; Jones, DA; Karpf, AR. Evaluation of a 7-day continuous intravenous infusion of decitabine: inhibition of promoter-specific and global genomic DNA methylation. *Journal of Clinical Oncology,* 2005;23:3897–905.

[48] Kim, YI. Folate: a magic bullet or a double edged sword for colorectal cancer prevention? *Gut,* 2006;55:1387-1389.

[49] Pereira, MA; Tao, L; Liu, Y; Li, L; Steele, VE; Lubet, RA. Modulation by budesonide of DNA methylation and mRNA expression in mouse lung tumors. *International Journal of Cancer,* 2006, Dec 12.

[50] Alyaqoub, FS; Tao, L; Kramer, PM; Steele, VE; Lubet, RA; Gunning, WT; Pereira MA. Prevention of mouse lung tumors and modulation of DNA methylation by combined treatment with budesonide and R115777 (ZarnestraMT). *Carcinogenesis,* 2007;28:124-9.

[51] Pereira, MA; Tao, LH; Wang, W; Gunning, WT; Lubet, R. Chemoprevention: mouse colon and lung tumor bioassay and modulation of DNA methylation as a biomarker. *Experimental Lung Research,* 2005;31:145-63.

[52] Pereira, MA; Tao, L; Wang, W; Li, Y; Umar, A; Steele, VE; Lubet, RA. Modulation by celecoxib and difluoromethylornithine of the methylation of DNA and the estrogen receptor-alpha gene in rat colon tumors. *Carcinogenesis,* 2004;25:1917-23.

[53] Tao, L; Wang, W; Kramer, PM; Lubet, RA; Steele, VE; Pereira, MA. Modulation of DNA hypomethylation as a surrogate endpoint biomarker for chemoprevention of colon cancer. *Molecular Carcinogenesis,* 2004;39:79-84.

[54] Pereira, MA; Wang, W; Kramer, PM; Tao, L. Prevention by methionine of dichloroacetic acid-induced liver cancer and DNA hypomethylation in mice. *Toxicological Science* 2004;77:243-8.

[55] Pakneshan, P; Szyf, M; Farias-Eisner, R; Rabbani, SA. Reversal of the hypomethylation status of urokinase (uPA) promoter blocks breast cancer growth and metastasis. *The Journal of Biological Chemistry,* 2004;279:31735-44.

[56] Pakneshan, P; Szyf, M; Rabbani, SA. Hypomethylation of urokinase (uPA) promoter in breast and prostate cancer: prognostic and therapeutic implications. *Current Cancer Drug Targets,* 2005;5:471-88.

[57] Szyf, M; Pakneshan, P; Rabbani, SA. DNA methylation and breast cancer. *Biochemical Pharmacology,* 2004;68:1187-97.

[58] Szyf, M; Pakneshan, P; Rabbani, SA. DNA demethylation and cancer: therapeutic implications. *Cancer Letters,* 2004;211:133-43.

[59] Shukeir, N; Pakneshan, P; Chen, G; Szyf, M; Rabbani, SA. Alteration of the methylation status of tumor-promoting genes decreases prostate cancer cell invasiveness and tumorigenesis in vitro and in vivo. *Cancer Research,* 2006;66:9202-10.

In: Progress in DNA Methylation Research
Editor: Hans P. Neumann, pp. 133-152
ISBN: 978-1-60021-722-7

Chapter IV

DNA METHYLATION IN ZEBRAFISH

Nobuyoshi Shimoda*
Department of Regenerative Medicine, National Institute for Longevity Sciences, 36-3 Gengo, Morioka-cho, Ohbu, Aichi 474-8522, Japan.

ABSTRACT

In contrast to a large number of publications as to zebrafish mutants with defects in early morphogenesis, no zebrafish mutants affecting DNA methylation have been reported, hence little is known about DNA methylation in the fish. But I believe that zebrafish will be recognized as an excellent model organism for studying DNA methylation machinery in vertebrates, considering shared characteristics of cytosine methylation between mammals and the fish. Here, I first introduce approaches and methodologies established for zebrafish developmental biology, which can be applicable to the analyses of roles and mechanisms of DNA methylation in zebrafish. Next, I show a few examples of *de novo* methylation and demethylation that are observed in zebrafish embryos. I finally describe two methyltransferase genes in the zebrafish genome, *dnmt1* and *dnmt7*, whose expression patterns and functions are already demonstrated.

INTRODUCTION

Cytosine methylation in the zebrafish genome shares some characteristics with that in mammalian genomes. First, it predominantly occurs at cytosine residues in CpG dinucleotides. Second, the occurrence of CpG sequences is lower than that expected from the nucleotide composition of the genome. The observed/expected ratio is 0.55, although it is still greater than that in mammals (0.23 in human and 0.22 in mouse) (Jabbari and Bernardi,

* Correspondence concerning this article should be addressed to Nobuyoshi Shimoda, Department of Regenerative Medicine, National Institute for Longevity Sciences, 36-3 Gengo, Morioka-cho, Ohbu, Aichi 474-8522, Japan. E-mail: shimoda@nils.go.jp; Tel and fax: +81-562-46-8629.

2004). Third, most genes have CpG islands (~60% in mammals) and they are unmethylated irrespective of expression (Shimoda, unpublished result). Although the numbers of zebrafish CpG islands reported to date are limited (Yamakoshi and Shimoda, 2003a), they fulfill the criteria defined for mammalian CpG island: a stretch of nonmethylated DNA longer than 200 bp with over 50% G+C content and a CpG frequency at least 0.6 of that statistically expected (Gardiner-Garden and Frommer, 1987). However, CpG islands in fish species including zebrafish differ from those in mammals in that they are no GC-rich (Cross *et al.*, 1991, Shimoda, unpublished). This lack of GC-richness in zebrafish CpG islands may limit the application of some techniques for the analysis of CpG island to zebrafish (Toyota *et al.*, 1999; Ushijima *et al.*, 1997). Fourth, CpG dinucleotides in genic regions excluding CpG islands and in transposons are densely methylated (Macleod *et al.*, 1999; Yamakoshi and Shimoda, 2003a; Shimoda *et al.*, 2005).

Unlike in mammals, it is not likely that parental imprinting and X inactivation, two phenomena associated with DNA methylation, occur in zebrafish. Neither of sex chromosome or mutations indicative of parental imprinting has been found in zebrafish (Sola and Gornung, 2001). Furthermore, zebrafish do not have an ortholog of *Dnmt3l*, which is essential for imprint establishment in the parental germline in mouse (Yokomine *et al.*, 2006). Even if imprinting genes were to exist, they would not be essential genes since androgenetic and gynogenetic zebrafish are viable (Corley-Smith *et al.*, 1996).

Do global Methylation Changes occur in the Zebrafish Genome?

The existence of global methylation changes in the zebrafish genome DNA has been a contentious issue. There are three groups that surveyed genome-wide methylation changes in zebrafish (Macleod *et al.*, 1999; Martin *et al.*, 1999; Mhanni and McGowan, 2004). For the purpose they all performed Southern hybridization under very similar conditions: using the same repetitive sequence DANA (*mermaid*) for the probe and using the same combination of restriction enzymes for digestion of genomic DNA. Nevertheless, two of the three groups detected the changes but the other did not. Moreover, the developmental stages at which the changes were observed are different between the two groups. One of the two groups showed DNA hypomethylation at the blastula and gastrula stages, which correspond to 4 and 6 hpf (hours post fertilization), respectively (Martin *et al.*, 1999), while the other group did not observe hypomethylation at these stages but at earlier stages, 1-2 cell and 128 cell (2.2 hpf) stages (Mhanni and McGowan, 2004). It may be possible that the changes the two groups observed are due to a slight difference in quality and/or quantity of the DNA used for Southern analysis. The third group that did not detect global-methylation changes in zebrafish by Southern analysis confirmed the absence of methylation changes in three zebrafish genes at any stages examined by an alternative method, bisulfite sequencing (Macleod *et al.*, 1999). In agreement with their data, using bisulfite sequencing I did not observe any hypomethylation in a DANA or type I centromeric repeats at 128-cell or gastrula stage (Shimoda, unpublished). Thus, I suggest that, unlike in mammals, global changes in DNA methylation do not occur in the zebrafish genome during development. Absence of the genome-wide changes in DNA methylation is also reported for another fish, medaka (Walter *et al.*, 2002).

Forward and Reverse Genetic Approach used for Zebrafish Research

Here I introduce three major approaches that have been successfully used to study gene function in zebrafish. Another approach, chemical genetics, that has just been exploited and seems very promising, is also described. Knowing these approaches will help those unfamiliar to zebrafish genetics to understand the description below and also to design experiments for epigenetic studies using zebrafish.

Forward Genetic Approach 1: Chemical Mutagenesis

Zebrafish emerged as a vertebrate model system in which genome-wide random mutagenesis can be performed. By using chemical mutagen, ethylnitrosourea (ENU), that introduces base-substitutions in DNA, two large-scale screenings have been applied to zebrafish (Driever *et al.*, 1996; Haffter *et al.*, 1996), which were followed by numerous medium- and small-scale ones. Possibly more than tens of thousands of mutants were isolated in total from such screens, but most of the screens were intended to find developmental anomalies, and none affecting DNA methylation was isolated or even searched. However, thanks to those screenings, methods for ENU mutagenesis and screening strategies are empirically well established. Further, finding point mutations in mutants, which had been a time-consuming step, is greatly facilitated by the ongoing zebrafish genome project.

Forward Genetic Approach 2: Insertional Mutagenesis

Retroviral-mediated insertional mutagenesis in zebrafish that was established by Nancy Hopkins' group is an alternative to chemical mutagenesis. Although insertional mutagenesis is usually less efficient (1/7~1/10) than ENU, insertions serve as a molecular tag to aid in the isolation of the mutated genes (Amsterdam and Hopkins, 2006). The methodology is currently being developed by Znomics Inc. (http://www.znomics.com), who are creating a library of frozen sperm samples containing multiple insertions each with a parallel library of genomic DNA samples; the DNA adjacent to each insert will be cloned and sequenced, thus identifying the disrupted genes in each sperm sample. Fish harboring the desired insertions can then be recovered by *in vitro* fertilization. Znomics expects to have located the insertion sites of half a million insertions by early 2007, which should include at least one insertion in every gene. In fact, as of the time of writing, all of the eight-zebrafish methyltransferase genes identified except *dnmt2* are tagged by retroviral insertions, indicating a very good coverage of the genome.

Reverse Genetic Approach: Knockdown of Gene Function by Morpholinos

Gene targeting by homologous recombination in zebrafish ES-like cells is reported (Fan *et al.*, 2006), but no knockout fish has been generated. On the other hand, inhibition of gene function *in vivo* is easily accomplished by morpholino phosphorodiamidate oligonucleotides (morpholinos; MOs) (Nasevicius & Ekker, 2000). Successful uses of MOs to knockdown some zebrafish DNA methyltransferase genes and other methylation-related genes are described below. MOs are synthetic DNA analogues and have a morpholine ring instead of the standard ribose sugar moiety and contain a neutral charge backbone that minimizes interaction with proteins, eliminating non-antisense effects. With high affinity for RNA, MOs serve as excellent antisense reagents through blocking translation or splicing. By injecting a MO into fertilized eggs from one to eight-cell stage, the product of the target gene often decrease to levels undetectable by Western blot (Nasevicius & Ekker, 2000).

A major strategy for MO gene targeting is translational inhibition. In this approach, an antisense MO selected against the 5'cap to about 25 bases 3' to the AUG translational start site of a target mRNA inhibits scanning of the mRNA by the 40S ribosomal subunit, resulting in the prevention of protein synthesis.

Another approach is the alteration of transcriptional processing of newly synthesized (i.e., zygotic) transcripts. Binding of an antisense MO to one potential intron/exon junction of a pre-mRNA can block pre-mRNA processing and results in exon deletions (Draper *et al.*, 2001). One significant advantage of this approach is that the aberrant mRNA can be readily detected using RT-PCR.

The most frequently observed problems with the use of MOs are mistargeting, which is presumably caused by the simultaneous inactivation of an essential gene with serendipitous homology (off-target). When mistargeting is suspected, use of shorter MO or another antisense reagent HypNA-pPNA should be considered (Wickstrom *et al.*, 2004). Compared with other oligonucleotide derivatives, HypNA-pPNA display the highest T_m for duplexes formed with single-stranded DNA or RNA, allowing it stringent mismatch discrimination. HypNA-pPNA is commercially available as gripNA (Active Motif). When either MO or HypNA-pPNA is used, conclusive determination of any phenotype should be made through either the targeting of the same gene by a second, non-overlapping sequence oligo or RNA rescue.

Morpholino-based gene targeting is completely penetrant throughout the first two to three days of development, which include the critical vertebrate processes of somitogenesis and organogenesis in the zebrafish embryo. However, it begins to be compromised thereafter (van der Sar *et al.*, 2002; Shimoda, unpublished).

Chemical Approach: Knockdown of Gene Function by small Molecules

Chemical genetics is screening small molecule compounds for specific phenotypes and identification of their targets (den Hertog, 2005). It resembles classical forward genetics in that both are aimed at elucidation of gene function through analysis of phenotypes. An important advantage of the use of small molecule compounds over genetic mutations or MOs is that the compounds may be administered at different times, allowing pinpointing of the exact timing of the requirement for the target protein. If a gene is essential for distinct early and late functions, a mutation or MO knocking out the early function might hide the late function, but administering the compounds after the early essential period would reveal it. In fact, it has been shown that temporal control of gene inactivation is possible with a small molecule (Yang and Johnson, 2006). Until recently, high throughput screens for biological activities of compound libraries were carried out with purified proteins or with cell cultures, focusing on particular cell biological phenomena. However, many processes that occur in an organism cannot be reproduced in cell cultures and should be investigated in the whole organism. It is considered that compounds can be screened in zebrafish embryos, using the system to its full potential. Zebrafish embryos are about 1mm in diameter, facilitating scaling up to 96-well plates and the inhibitors can just be added to the aqueous embryo medium. Currently, there are no vertebrate alternatives, because due to its unique features, only the zebrafish system is amenable to such large-scale drug screens.

A medium-scale screen with 1100 randomly selected small molecules, focusing on morphological defects after 1, 2 or 3 days has been done, leading to the identification of compounds that are involved in development of specific tissues in zebrafish (Peterson *et al.*, 2000). This approach should be also useful for the isolation of DNA methylation mutants.

De Novo Methylation and Demethylation in Zebrafish

De novo Methylation of the no Tail Gene

A portion of CpG island in mammals undergoes cytosine methylation during development and differentiation, although most CpG dinucleotides in CpG islands are normally constitutively unmethylated, irrespective of expression. CpG methylation leads to long-term shutdown of the associated genes by blocking the binding of transcription factors (Takizawa *et al.*, 2001) and by promoting the formation of inactive chromatin (Futscher *et al.*, 2002). Thus, cytosine methylation of CpG island is considered to be involved in the establishment and maintenance of cell type-specific expression of developmentally regulated genes (Ehrlich, 2003; Futscher et *al.*, 2002; Takizawa *et al.*, 2001).

The zebrafish gene *no tail* (*ntl*) is a mouse *Brachyury* (*T*) homolog that encodes a transcription factor with a DNA-binding domain (Kispert and Herrmann, 1993; Schulte-Merker *et al.*, 1994). *ntl* is transiently expressed in prospective mesoendoderm, with a notably consistent expression pattern in presumptive notochord cells (Schulte-Merker *et al.*, 1992; Thisse *et al.*, 2001), and is required for the formation of a differentiated notochord and a proper posterior structure (Halpern *et al.*, 1993; Odenthal *et al.*, 1996; Stemple *et al.*, 1996). However, how *ntl* expression is controlled remains unknown.

In *Xenopus laevis*, the expression of a *Brachyury* homolog, *Xbra*, is regulated temporally and spatially, like that of *ntl* (Smith *et al.*, 1991). It has been suggested that DNA methylation of CpG islands has a role in regulating the expression timing of immediate early genes including *Xbra* in *Xenopus* embryos (Stancheva *et al.*, 2002). A loss of methylation at CpG islands of immediate early genes occurs before they were activated at the "midblastula transition", the period of development when zygotic genes start expression, whereas no changes are observed in the methylation patterns of repeated sequences, genes that are inactive at the midblastula transition, or the coding regions of individual genes (Stancheva *et al.*, 2002).

To examine if a structurally and functionary conserved gene is similarly regulated in terms of DNA methylation between evolutionary diverged organisms, bisulfite sequencing analysis was done for the zebrafish *ntl* (Yamakoshi and Shimoda, 2003a).

As shown in Figure 1A,B, there is a dense cluster of CpG dinucleotides that covers the 5' promoter region and the first exon of *ntl*. The cluster meets the criteria of a CpG island (Gardiner-Garden and Frommer, 1987) on the basis of its size (335 bp), G C content (54%), and CpG dinucleotide frequency (1.31), as well its location with respect to the transcriptional unit (–132 to 203). Two discrete regions of the CpG island were amplified by PCR: one of the two regions, which is 222 bp long and located upstream of the other, covers the promoter region of *ntl* (region I in Figure 1B), and the other region, which is 290 bp long, covers a part of exon 1 (region II in Figure 1B). By sequencing, *de novo* DNA methylation was detected at all the CpG dinucleotides in the two regions (Figure 1C). It starts from the 10-somite stage and continued even after the larval stage (7 days old). CpG dinucleotides in the two regions in sperm and in early embryos, from the sphere to 6-somite stages, were all hypomethylated. The degree of the methylation was the heaviest in genomic DNA extracted from the whole body of an adult fish: on average 72% of CpG dinucleotides in the CpG island were methylated. It should be noted that *de novo* methylation occurred most markedly during a limited developmental period in zebrafish, between 14 and 48 hpf (Figure 1C). On average, 43% of CpG dinucleotides in the CpG island were methylated by 48 hpf, which corresponded to 60% of methylated cytosines observed in the region in the adult fish. In a zebrafish cell line, LFF (Chevrette *et al.*, 1997), CpG dinucleotides in the *ntl* CpG island were methylated as heavily as the ones in the adult fish. In contrast to CpG dinucleotides in the CpG island, CpG dinucleotides in the last exon of *ntl* were heavily methylated irrespective of the developmental stage (Figure 1C). This result was consistent with a previous report showing hypermethylation of non-CpG island regions in some other zebrafish genes at all developmental stages (Macleod *et al.*, 1999).

Global Methylation of the ntl CpG Island in the Zebrafish Body

It was somewhat curious that even in the adult fish, in which the highest degree of cytosine methylation at the *ntl* CpG island was observed, the degree was still lower than that observed at the non-CpG island region in *ntl* (Figure 1C). It was possible that this might be due to hypomethylation of the CpG island in some particular tissues in adult zebrafish because cell-type-specific hypomethylation of mammalian genes has been reported (Ehrlich, 2003; Futscher *et al.*, 2002; Takizawa *et al.*, 2001). As shown in Figure 1D, however, in all organs examined the *ntl* CpG island region was either moderately or heavily methylated. The degree of cytosine methylation in the region was the lowest in muscle (42%), but it was still more than half of that in spleen or brain, in which the highest degree of cytosine methylation (79%) occurred. These results showed that *de novo* methylation at the *ntl* CpG island occurred globally in zebrafish but it did not occur so extensively as at the non-CpG island region (Figure 1C). This explains why the methylation level at the CpG island was lower than that at the non-CpG island region in whole-body DNA preparations.

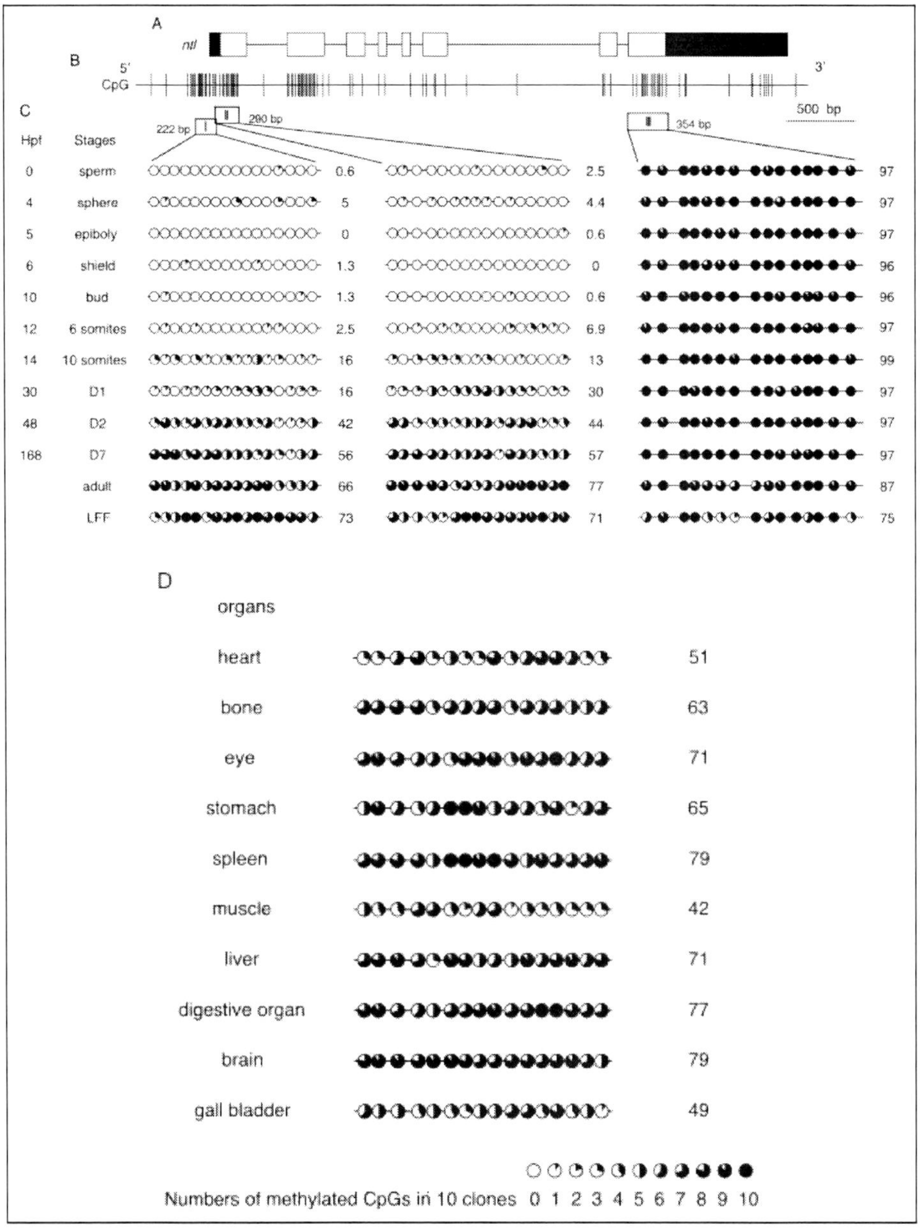

Figure 1. The *ntl* CpG island becomes methylated during development. (A) Structure of *ntl*. Boxes represent exons, white and black portions representing translated and untranslated regions, respectively. The asterisk indicates the position of a *Sau* 3AI site used for the chromatin accessibility assays. (B) CpG plot for *ntl*. Each vertical line indicates the position of a CG dinucleotide. The bracket indicates the *ntl* CpG island. (C) Summary of methylcytosine levels measured by sodium bisulfite sequencing of the *ntl* CpG island (region I and II) and a non-CpG island region (III). For each stage of development and for a cell line, LFF, 10 independent clones were sequenced. Each circle corresponds to a CpG, and is filled according to the number of methylated cytosines found in the 10 clones. Numbers on the right show the average number of methylated CpGs in the total CpGs checked. (D) Methylation status at the *ntl* CpG island (region II) in zebrafish adult tissues. Hpf, hours post fertilization. (De novo DNA methylation at the CpG island of the zebrafish *no tail* gene. Yamakoshi and Shimoda, Genesis, 37, Copyright ©[2003] John Wiley & Sons, Inc.)

De novo Methylation of Plasmid DNA

Collas found that a reporter gene construct injected into zebrafish embryos became gradually methylated during development, starting at ~12h post-injection (Collas, 1998). He showed that the methylation of injected plasmid occurred not only in CpG, but also in CpT, CpA, and CpC, whose cytosines are not methylated in the zebrafish protein-coding genes I have investigated to date (Shimoda, unpublished). *De novo* non-CpG methylation of episomal vectors is also observed in human ES cells, although it is minor compared to CpG methylation (Kameda *et al*., 2006). Further, non-CpG methylation in integrated virus DNA in mouse ES cells is reported (Dodge *et al*., 2002). These results suggest that non-CpG methylation might be part of a *de novo* methylator response to invading DNA.

Demethylation of Plasmid DNA in Zebrafish Embryos

Collas also noted the first instance of demethylation in zebrafish: when *in vitro*-methylated plasmid DNA was injected into zebrafish embryos, it was rapidly demethylated before being remethylated at 12-24 h (Collas, 1998). It appeared to be an active (replication-independent) mechanism since inhibition of DNA replication with topoisomerase I inhibitor camptothenin did not affect the plasmid demethylation. In contrast to the *in vivo* demethylation, *in vitro* plasmid demethylation was detected in nuclear extracts from 2 hpf to 4-day embryos. This difference suggests that zebrafish embryos contain substantial levels of demethylating activity up to at least 4 days of development and it is overridden by increasing levels of methyltransferase activity allowing *de novo* methylation of newly replicating DNA. Since demethylation activity was observed *in vivo* and *in vitro* before the onset of zygotic gene expression (~3 hpf), the activity must be maternally supplied. Neither any genomic regions subject to demethylation nor the biological significance of the demethylation activity in early zebrafish embryo is known.

He also found that incubation of plasmid-injected embryos with DNA methylation inhibitor 5-azazcytidine or the histone deacetylation inhibitor butyrate immediately after injection inhibited plasmid methylation and extends the period of reporter gene expression (Collas, 1998). When applied after *de novo* methylation has occurred, both inhibitors prevent methylation of newly replicated DNA and promote reporter gene expression. These data suggest that methylation of the injected construct during early development induces repression of the reporter gene, perhaps by converting the construct to a repressive chromatin structure.

Interestingly, the demethylating activity was heat-labile and sensitive to protease and RNase and requires ATP hydrolysis, suggesting the involvement of protein and RNA components. However, there is a contradictory report as to a role of RNA in demethylation (Swisher *et al*., 1998).

DNA Methyltransferase Genes in Zebrafish

Structures of Zebrafish Methyltransferase Genes

In mammals, three distinct families of DNA methyltransferase genes, termed *Dnmt1*, *Dnmt2*, and *Dnmt3*, have been identified. Dnmt1 shows marked preference for hemimethylated DNA and acts as a maintenance methyltransferase (Bestor, 1992). Dnmt2 contains all the highly conserved methyltransferase motifs, but lacks the N-terminal domain of Dnmt1. Dnmt2 had been an enigmatic protein since no methyltransferase activity against DNA is shown *in vitro* or *in vivo* (Okano *et al.*, 1998). But, Dnmt2 has recently shown to have a tRNA methyltransferase activity (Goll *et al.*, 2006). The Dnmt3 family consists of three related genes, termed Dnmt3a, Dnmt3b, and Dnmt3L; the first two of them act as *de novo* methyltransferases (Okano *et al.*, 1999). Dnmt3a and Dnmt3b show striking sequence similarity and have shorter N-terminal domains with low homology to that of Dnmt1.

In the zebrafish genome eight DNA methyltransferase genes are so far identified (Martin et al., 1999; Mhanni et al., 2001; Shimoda *et al.*, 2005). Two of the eight are the orthologues of *Dnmt1* and *Dnmt2*. The remaining six-methyltransferase genes encode in common a PWWP domain, involved in protein-protein interaction, and C-terminal catalytic domain. Since both of which are present in the murine *de novo* DNA methyltransferases Dnmt3a and Dnmt3b, we regard these as members of the Dnmt3 family. However, they differ in their N-terminal domains and none of them appear to be an ortholog of *Dnmt3a* or *Dnmt3b*. This is in contrast to the presence of the orthologs of the murine *Dnmt1* and *Dnmt2*.

It should be noted that Dnmt3 and Dnmt7 have a Calponin-homology domain at their N-terminal ends, which is known to be a microtubule-binding domain (Bu and Su, 2003). No other Dnmt proteins containing Calponin-homology domain have been reported and the function of the Calponin-homology domains in Dnmt3 and Dnmt7 remain unknown. Except the calponin-homology domain, zebrafish Dnmt3 and Dnmt7 are not similar at their N-terminal ends. Although there might be methyltransferase genes unidentified, the number should be small, if any, judging from the current status of the zebrafish genome project, which is close to the completion.

Genomic Positions of the Zebrafish Methyltransferase Genes

Why are so many possible *de novo* methyltransferase genes compared to three in mammals? Additional homologs are often found in the zebrafish genome, and this is best explained by a model proposed by the Postlethwait group: two polyploidization events occurred in a common ancestor of zebrafish and mammals, and either chromosome duplication or another tetraploidization event in the later zebrafish lineage after the divergence of the common ancestor (Amores *et al.*, 1998).

The ongoing zebrafish genome project mainly conducted by the Welcome Trust Sanger Institute has been providing invaluable information for the functional analyses and positional cloning of zebrafish genes (http://www.sanger.ac.uk/Projects/D_rerio/). However, due to the absence of inbred lines of zebrafish, misalignment of BAC clones in the assembly are sometimes observed at present. Thus, it is still necessary to confirm locations of zebrafish genes in the genome by an alternative method such as genetic linkage mapping (Shimoda *et al.*, 1999) or radiation hybrid mapping (Geisler *et al.*, 1999; Hukriede *et al.*, 1999). By using radiation mapping, I located the six *Dnmt3* family genes at five distinct loci (Figure 2). The positions of *dnmt3* and *dnmt5* were not separable by the method, meaning that they are very close to each other. This result is in agreement with information from the genome project (Smith *et al.* 2005). The tight clustering seen at *dnmt3* and *dnmt5* genes, together with their high similarity at the nucleotide level, suggest that, at least in this instance, tandem gene duplication has occurred.

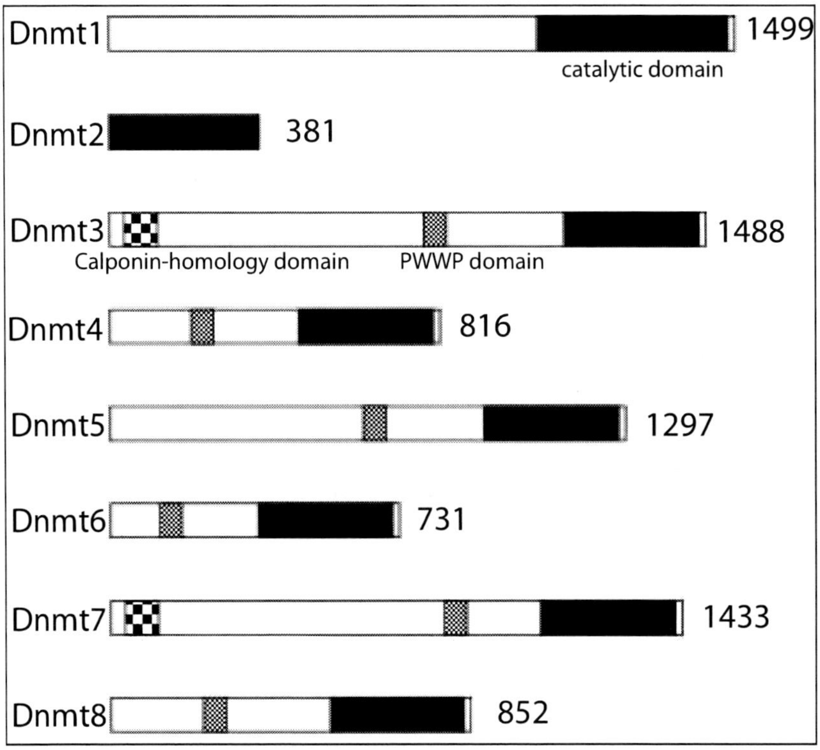

Figure 2. Structures of zebrafish methyltransferases. Schematic representations of the zebrafish DNA methyltransferase (Dnmt) family. The numbers shown are deduced amino-acid lengths. Note that Dnmt3 and Dnmt7 have a calponin-homology domain in their N-terminal regions. LG: Linkage Group. N.D.: Not Determined. The GenBank accession numbers for *dnmts* are as follows: AF483203 (*dnmt1*), AB196920 (*dnmt2*), AB196914 (*dnmt3*), AB196915 (*dnmt4*), AB196916 (*dnmt5*), AB196917 (*dnmt6*), AB196918 (*dnmt7*), and AB196919 (*dnmt8*).

Expression Patterns and Functions of two Zebrafish Methyltransferases

Since functional analyses are done only for *dnmt1* and *dnmt7* among the eight-zebrafish methyltransferase genes identified, I focus on the two genes in this section. Expression patterns of *dnmt2* and *dnmt4* can be seen on the zebrafish web server at http://zfin.org/cgi-bin/webdriver?MIval=aa-ZDB_home.apg. Splice variants are observed for *dnmt5*, (*dnmt3*-2 in Smith *et al.*, 2005) and for *dnmt6* (Shimoda, unpublished), reminiscent of what is seen with the human DNMT3s. Expression analyses of *dnmt5* variants by RT-PCR suggest they are independently controlled (Smith *et al.*, 2005)

Expression Pattern and Function of dnmt1

dnmt1 mRNA is present abundantly and ubiquitously in embryos from the time of fertilization to 6-h gastrula-stage, indicating that the transcripts are of maternal origin, since embryonic transcription, in zebrafish, starts at around the 1000-cell stage (3 hpf) (Martin *et al.*, 1999; Mhanni & McGown, 2002). At 16 hpf the expression is strongest in the brain, neural tube, and newly formed anterior somites. In tail-stage embryos (24 hpf), *dnmt1* mRNA expression is strongest in the eyes, central nervous system, and the somites of the tail bud.

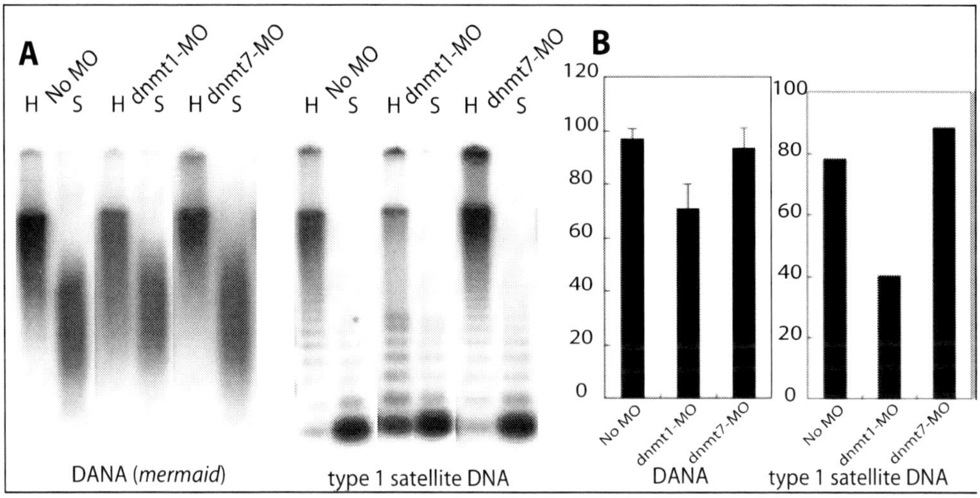

Figure 3. Methylation analysis of repetitive sequences. (A) Southern blot analysis of methylation status of DANA/*mermaid* repeat and Type I satellite sequence. The enzyme, *Hpy*CH4IV (H), which is present in both repeats, cannot cleave its recognition sequence of "ACGT" when its CpG is methylated. Compared to the control (No MO), increased cleavages of DANA (Shimoda *et al.*, 2005) and Type I (Ekker *et al.*, 1992; He *et al.*, 1992) with the enzyme were observed in the genomes of *dnmt1*-MO-injected embryos, reflecting substantial hypomethylation of CpGs. On the other hand, identical patterns were observed between the control and *dnmt7*-injected embryos, indicating no change in CpG methylation. Since isoschizomers of *Hpy*CH4IV, which are not sensitive to CpG methylation, were not available, we used *Sau*3AI (S) to see how binding patterns were changed when methylation in these repeats was reduced, since *Sau*3AI digestion is less affected by CpG methylation than *Hpy*CH4IV digestion. (B) Quantitative analysis of methylation status of DANA/*mermaid* and Type I sequences. There are fifteen CpG dinucleotides in the sequenced region in the DANA. Mean values and standard errors were determined based on ten independent sequences. Since there is only one CpG site in Type I repeat, only mean values are shown for Type I. Template DNA used for (A) and (B) were the same (Identification of a gene required for de novo DNA methylation of the zebrafish *no tail* gene. Shimoda, Yamakoshi, Miyake, and Takeda, Developmental Dynamics, 233, Copyright ©[2005] John Wiley & Sons, Inc.).

Injection of a *dnmt1*-MO into fertilized eggs reduced the methylation level over the entire genome (Figure 3A, B), a phenomenon observed in organisms in which the activity of maintenance methyltransferases is suppressed (Li *et al.*, 1992; Stancheva and Meehan, 2000). The hypomethylation was also confirmed at a particular locus: the *dnmt1*-MO reduced the methylation level at the CpG island and non-CpG island regions of *ntl* to 60%, compared to controls (Figure 4A, B). These results strongly suggest that the zebrafish Dnmt1 is a genuine maintenance methyltransferase.

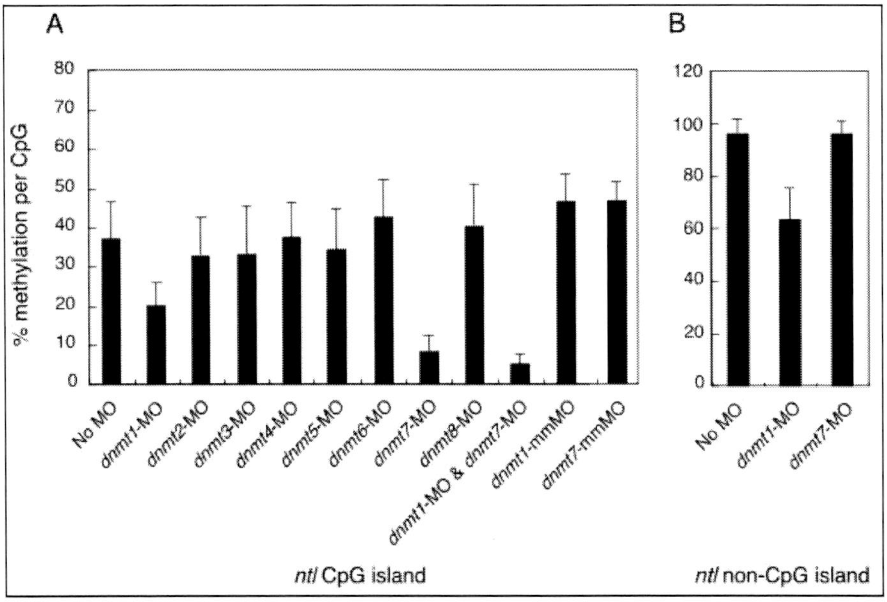

Figure 4. Suppression of Dnmt1 or Dnmt7 activity affects *ntl* methylation. (A) Methylation status of the *ntl* CpG island was analyzed 48 hours after the injection of morpholino oligo (MO) against each *dnmt*. *mm-dnmt1* and *mm-dnmt7* are negative-control MOs carrying mismatch nucleotides in their original *dnmt1*-MO and *dnmt7*-MO sequences, respectively. Part of the *ntl* CpG island containing 16 CpG dinucleotides was PCR amplified and 10 independent clones were sequenced. Each column represents the mean ± standard error of methylated cytosines. (B) Methylation status of an *ntl* 3' exon region. An *ntl* 3' region harboring 15 CpGs was amplified and 10 independent clones were sequenced (Identification of a gene required for de novo DNA methylation of the zebrafish *no tail* gene. Shimoda, Yamakoshi, Miyake, and Takeda, Developmental Dynamics, 233, Copyright ©[2005] John Wiley & Sons, Inc.).

Rai *et al.* also demonstrated the reduction of genome-methylation level by the injection of another *dnmt1*-MO at a semi-lethal dose (40% death in 24h) (Rai *et al.*, 2006). Interestingly, they reported that their *dnmt1*-MO-injected embryos had defects in terminal differentiation of the intestine, exocrine pancreas, and retina. Since zebrafish *dnmt1* or human *DNMT1* mRNA could rescue the defects, they reasoned that they were due to Dnmt1 deficiency. Furthermore, they reported that *dnmt1*-MO embryos exhibited dramatic reduction of genome-wide H3K9 trimethyl level and that embryos lacking Suv39h1 had organ-specific terminal differentiation defects that produced largely phenocopies of *dnmt1*-MO embryos. Based on these results, they suggest that Dnmt1 activity helps direct histone methylation by Suv39h1 and that, together, Dnmt1 and Suv39h1 help guide the terminal differentiation of particular tissues.

In some organisms, reduction in genomic DNA methylation level causes increase in the transposition frequency of many DNA-based transposable elements (Miura *et al.*, 2001, Iida *et al.*, 2006), but it remains unknown whether activation of transposons also occurs in zebrafish by genome-wide hypomethylation.

Expression Pattern and Function of dnmt7

dnmt7 was identified as a gene essential for *de novo* methylation of the *ntl* gene (Shimoda *et al.*, 2005). At 48 hours after injection, a MO against *dnmt7* reduced the methylation level at the *ntl* CpG island to 20% of that in wild-type (Figure 4A). The incomplete repression of *ntl* methylation by the *dnmt7*-MO may be due to the activity of the maternal Dnmt7 protein that MOs are unable to eliminate.

The *dnmt7*-MO did not induce genome-wide hypomethylation or lower the methylation level at the *ntl* 3' region (Figure 4B). These results indicated that *dnmt7* is a gene that selectively methylates the *ntl* CpG island *de novo*. Presumably Dnmt1 maintains the *ntl* methylation pattern established by Dnmt7, and the severest reduction of *ntl* methylation by coinjection of *dnmt1*-MO and *dnmt7*-MO (Figure 4A) is possibly caused by failure in the both process: *de novo* and maintenance methylation of the *ntl* CpG island.

The expression pattern of *dnmt7* is shown in Figure 5. *dnmt7* transcripts were detectable as early as the 4-cell stage (Figure 5A), indicating that they are maternally supplied. From blastula to mid-segmentation stages, *dnmt7* is expressed ubiquitously in embryos. At the end of the segmentation stage, 24 hpf, a slightly higher expression was observed in the brain (Figure 5I), and the highest expression was detected in the tectum (Figure 5H). The transcripts were almost undetectable by 2 days post fertilization (Figure 5J). The expression pattern shown in Figure 5 suggests that *ntl* methylation by Dnmt7 occurs globally in early zebrafish embryos. This is in agreement with a result showing that, apart from sperm, the *ntl* CpG island was methylated in the genomes of all adult organs and tissues examined (Yamakoshi and Shimoda, 2003a). In addition, the strong expression of *dnmt7* in the brain may be related to the highest degree of *ntl* methylation observed in the brain of an adult zebrafish (Yamakoshi and Shimoda, 2003a).

Although maternal *dnmt7* transcripts are supplied in cleavage-stage embryos and zygotic *dnmt7* is expressed throughout the blastula and gastrula stages, *de novo* methylation of *ntl* becomes evident ~14 hpf (Yamakoshi and Shimoda, 2003a), well after the end of gastrulation (~10 hpf). The absence of *ntl* methylation in spite of the presence of *dnmt7* mRNA may be caused by demethylation activity in early zebrafish embryos, which is detected for at least 6 hpf (Collas, 1998). The demethylation activity may be gradually overridden by increasing methyltransferase activity of Dnmt7.

Except for *ntl*, plasmid DNA microinjected into zebrafish eggs is the only other known example in which *de novo* methylation occurs in zebrafish (Collas, 1998). The *de novo* methylation of microinjected DNA starts around 12 hpf (Collas, 1998), as does that of *ntl* (Yamakoshi and Shimoda, 2003a). Although this fact suggested a possibility that Dnmt7 was also involved in methylation of plasmid DNA, *de novo* methylation occurred in plasmid DNA injected into fertilized eggs in the presence of *dnmt7*-MO (Shimoda *et al.*, 2005). Further, methylation occurred not only in CpG, but also in CpT, CpA, and CpC, as described above. Therefore, *de novo* methylation of injected exogenous DNA differs from that of *ntl* in terms of enzymes involved and cytosines targeted.

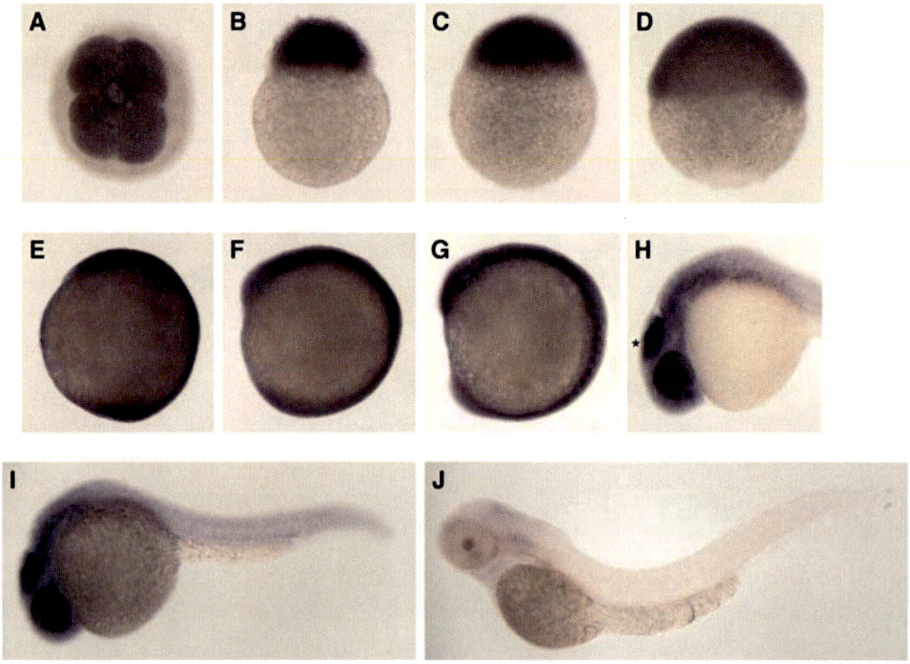

Figure 5. Expression pattern of *dnmt7*. The embryo stages used for in situ hybridization were (A) 4 cells, (B) 256 cells, (C) 1000 cells, (D) 50% epiboly, (E) tailbud, (F) 5 somite, (G) 10 somite, (H, I) day 1, (J), day 2. *dnmt7* mRNA could be detected before the mid-blastula stage (1,000 cells) of zebrafish embryos (A, B), indicating its maternal supply. During the somite stage (F, G), *dnmt7* was ubiquitously expressed in zebrafish embryos, and its expression persisted until day 2. In 1-day-old embryos, higher expression was observed in the brain, especially in the tectum (asterisk in H) (Identification of a gene required for de novo DNA methylation of the zebrafish *no tail* gene. Shimoda, Yamakoshi, Miyake, and Takeda, Developmental Dynamics, 233, Copyright ©[2005] John Wiley & Sons, Inc.).

The biological significance of *ntl* methylation by Dnmt7 remains unclear since *dnmt7*-MO-injected embryos normally develop and no difference in *ntl* expression was observed between wild-type and *dnmt7*-MO-injected embryos at any stages examined (data not shown). Further analyses await mutants in which no functional Dnmt7 protein is produced.

Also unknown is how Dnmt7 selectively methylates *ntl*. In plants correlation has been suggested between DNA methylation and short RNA (Chan *et al.*, 2004). DNA methylation may be targeted to identical sequences by short RNAs derived via Dicer cleavage of double-stranded RNA. To test if short RNA is involved in *ntl* methylation, knockdown of zebrafish Dicer1 using MO was conducted since *dicer1*-MO-injected zebrafish embryos are defective in the maturation of short RNA (Wienholds *et al.*, 2003). Although it cannot eliminate maternal Dicer activity (Giraldez *et al.*, 2005), the *de novo* methylation of *ntl* occurred normally in the embryos (data not shown). In addition, neither an antisense strand of *ntl* mRNA or short RNA homologous to *ntl* could be detected (data not shown). Thus, so far no data suggesting the involvement of RNA in *ntl* methylation is obtained.

An alternative possibility accounting for *ntl* methylation is DNA-DNA pairing. A precedent for DNA-DNA pairing as a trigger for DNA methylation is provided by the MIP (methylation induced premeiotically) phenomenon in filamentous fungi, *Ascobolus immerses* (Selker, 1997). MIP is pairing-dependent silencing and the methylation of duplicated DNA regions longer than ~400 bp. By analogy, it is tempting to speculate the possibility that DNA pairing between *ntl* and *cryptail*, a *ntl*-pseudogene-like DNA (Yamakoshi and Shimoda, 2003b), occurs via the transient formation of double-stranded DNA ends in an *ntl* intron (Yamakoshi *et al.*, 2005), and that the resultant heteroduplex becomes a target of Dnmt7.

CXXP Finger Protein 1 (CFP1): a Factor Necessary for DNA Methylation in Zebrafish

Human CXXP finger protein1 (CFP1) was serendipitously found as a protein that binds to unmethylated CpG dinucleotides (Voo *et al.*, 2000). CFP1 is a component of the Set1 histone methyltransferase complex and localizes euchromatic nuclear speckles and associates with the nuclear matrix (Lee & Skalnik, 2005). Mice lacking CFP1 suffer a peri-implantation lethal phenotype, and CFP1-deficient embryonic stem cells are viable but unable to differentiate and exhibit a 60-80% decrease in genomic cytosine methylation (Carlone *et al.*, 2005).

Two zebrafish homologs of CFP1 were cloned and designated as zCFP1 and zCFP1b (Young *et al.*, 2006). As in the CFP1-deficient ES cells, genomic DNA isolated from zebrafish embryos treated with a zCFP1 MO exhibited a 60% reduction of global genomic cytosine methylation. The zCFP1-MO-injected embryos had little or no circulating red blood cells and exhibited abnormal yolk sac morphology at 48 hpf, although hematopoietic markers were expressed normally in these embryos. It remains unknown how zCFP1 is involved in the maintenance of genome methylation.

CONCLUSION

Forward genetical approach that has been applied to *Arabidopsis* yielded a wealth of mutants affecting DNA methylation, and following isolation of responsible genes helped us to understand the molecular mechanism and the significance of DNA methylation in the plant. Since the zebrafish genome project is almost finished and positional cloning of ENU-mutated genes became much easier than before, I believe that it is now a good time to screen zebrafish methylation mutants, which will make zebrafish a complementary system to the mouse one to understand the significance of DNA methylation in vertebrates.

REFERENCES

Amores, A., Force, A., Yan, Y.-L., Joly, L., Amemiya, C., Fritz, A., Ho, R. K., Langeland, J., Prince, V., Wang, Y.-L., Westerfield, M., Ekker, M. & Postlethwait, J. H. (1998). Zebrafish *hox* clusters and vertebrate genome evolution. *Science, 282*, 1711-1714.

Amsterdam A. & Hopkins, N. (2006). Mutagenesis strategies in zebrafish for identifying genes involved in development and disease. *Trends in Genetics, 22*, 473-478.

Bestor, T. H. (1992). Activation of mammalian DNA methyltransferase by cleavage of a Zn binding regulatory domain. *EMBO Journal, 11*, 2611-2617.

Bu, W. & Su, L.-K. (2003). Characterization of functional domains of human EB1 family proteins. *Journal of Biological Chemistry, 278*, 49721-49731.

Carlone, D. L., Lee, J.-H., Young, S. R. L., Dobrota, E., Butler, J. S., Ruiz, J. & Skalnik, D. G. (2005). Reduced genomic cytosine methylation and defective cellular differentiation in embryonic stem cells lacking CpG binding protein. *Molecular and Cellular Biology, 25*, 4881-4891.

Chan, S. W.-L., Zilberman, D., Xie, Z., Johansen, L. K., Carrington, J. C. & Jacobsen, S. E. (2004). RNA silencing genes control *de novo* DNA methylation. *Science, 303*, 1336.

Chevrette, M., Joly, L., Tellis, P. & Ekker, M. (1997). Contribution of zebrafish-mouse cell hybrids to the mapping of the zebrafish genome. *Biochemistry and Cell Biology, 75*, 641–649.

Collas, P. (1998). Modulation of plasmid DNA methylation and expression in zebrafish embryos. *Nucleic Acids Research, 26*, 4454-4461.

Corley-Smith, G. E., Lim, C. J. & Brandhorst, B. P. (1996). Production of androgenetic zebrafish (*Danio rerio*). *Genetics, 142*, 1265-1276.

Cross, S., Kovarik, P., Schmidtke, J. & Bird, A. (1991). Non-methylated islands in fish genomes are GC-poor. *Nucleic Acids Research, 19*, 1469-1474.

den Hertog, J. (2005). Chemical genetics: drug screens in zebrafish. *Bioscience Reports, 25*, 289-297.

Dodge, J. E., Ramsahoye, B. H., Wo, Z. G., Okano, M. & Li, E. (2002). De novo methylation of MMLV provirus in embryonic stem cells: CpG versus non-CpG methylation. *Gene, 289*, 41-48.

Draper, B. W., Morcos, P. A. & Kimmel, C. B. (2001). Inhibition of zebrafish *fgf8* pre-mRNA splicing with morpholino oligos: a quantifiable method for gene knockdown. *Genesis, 30*, 154-156.

Driever, W., Solnica-Krezel, L., Schier, A. F., Neuhauss, S. C. F., Malicki, J., Stemple, D. L., Stainier, D. Y. R., Zwartkruis, F., Abdelilah, S., Rangini, Z., Belak, J. & Boggs, C. (1996). A genetic screen for mutations affecting embryogenesis in zebrafish. *Development, 123*, 37-46.

Ehrlich, M. (2003). Expression of various genes is controlled by DNA methylation during mammalian development. *Journal of Cellular Biochemistry, 88*, 899–910.

Ekker, M., Fritz, A. & Westerfield, M. (1992). Identification of two families of satellite-like repetitive DNA sequences from the zebrafish (*Brachydanio rerio*). *Genomics, 13*, 1169-1173.

Fan, L., Moon, J., Crodian, J. & Collodi, P. (2006). Homologous recombination in zebrafish ES cells. *Transgenic Research, 15*, 21-30.

Futscher, B. W., Oshiro, M. M., Wozniak, R. J., Holtan, N., Hanigan, C. L., Duan, H. & Domann, F. E. (2002). Role for DNA methylation in the control of cell type specific maspin expression. *Nature Genetics, 31*, 175-179.

Gardiner-Garden, M. & Frommer, M. (1987). CpG islands in vertebrate genomes. *Journal of Molecular Biology, 196*, 261–282.

Geisler, R., Rauch, G. J., Baier, H., van Bebber, F., Broß, L., Dekens, M. P., Finger, K., Ficke, C., Gates, M. A., Geiger, H., Geiger-Rudolph, S., Gilmour, D., Glaser, S., Gnügge, L., Habeck, H., Hingst, K., Holley, S., Keenan, J., Kirn, A., Knaut, H., Lashkari, D., Maderspacher, F., Martyn, U., Neuhauss, S., Neumann, C., Nicolson, T., Pelegri, F., Ray, R., Rick, J. M., Roehl, H., Roeser, T., Schauerte, H. E., Schier, A. F., Schönberger, U., Schönthaler, H. B., Schulte-Merker, S., Seydler, C., Talbot, W. S., Weiler, C., Nüsslein-Volhard, C. & Haffter, P. (1999). A radiation hybrid map of the zebrafish. *Nature Genetics, 23*, 86-89.

Giraldez, A. J., Cinalli, R. M., Glasner, M. E., Enright, A. J., Thomson, J. M., Baskerville, S., Hammond, S. M., Bartel, D. P. & Schier, A. F. (2005). MicroRNAs regulate brain morphogenesis in zebrafish. *Science, 308*, 833-838.

Goll, M. G., Kirpekar, F., Maggert K. A., Yoder, J. A., Hsieh, C.-L., Zhang, X., Golic, K. G., Jacobsen, S. E. & Bestor, T. H. (2006). Methylation of $tRNA^{Asp}$ by the DNA methyltransferase homolog Dnmt2. *Science, 311*, 395-398.

Haffter, P., Granato, M., Brand, M., Mullins, M. C., Hammerschmidt, M., Kane, D. A., Odenthal, J., van Eeden, F. J. M., Jiang, Y.-J., Heisenberg, C.-P., Kelsh, R. N., Furutani-Seiki, M., Vogelsang, E., Beuchle, D., Schach, U., Fabian, C. & Nüsslein-Volhard, C. (1996). The identification of genes with unique and essential functions in the development of the zebrafish, *Danio rerio. Development, 123*, 1-36.

Halpern, M. E., Ho, R. K., Walker, C. & Kimmel, C. B. (1993). Induction of muscle pioneers and floor plate is distinguished by the zebrafish *no tail* mutation. *Cell, 75*, 99 –111.

He, L., Zhu, Z., Faras, A. J., Guise, K. S., Hackett, P. B. & Kapuscinski, A. R. (1992). Characterization of *Alu*I repeats of zebrafish (*Brachydanio rerio*). *Molecular Marine Biology and Biotechnology, 1*, 125-135.

Hukriede, N. A., Joly, L., Tsang, M., Miles, J., Tellis, P., Epstein, J. A., Barbazuk, W. B., Li, F. N., Paw, B., Postlethwait, J. H., Hudson, T. J., Zon, L. I., McPherson, J. D., Chevrette, M., Dawid, I. B., Johnson, S. L. & Ekker, M. (1999). Radiation hybrid mapping of the zebrafish genome. *Proceedings of National Academy of Sciences of the United States of America, 96*, 9745-9750.

Iida, A., Shimada, A., Shima, A., Takamatsu, N., Hori, H., Takeuchi, K. & Koga, A. (2006). Targeted reduction of the DNA methylation level with 5-azacytidine promotes excision of the medaka fish *Tol2* transposable element. *Genetical Research,* 87, 187-193.

Izsvák, Z., Ivics, Z., Garcia-Estefania, D., Fahrenkrug, S. C. & Hackett, P. B. (1996). DANA elements: A family of composite, tRNA-derived short interspersed DNA elements associated with mutational activities in zebrafish (*Danio rerio*). *Proceedings of National Academy of Sciences of the United States of America. 93*, 1077-1081.

Jabbari, K. & Bernardi, G. (2004). Cytosine methylation and CpG, TpG (CpA) and TpA frequencies. *Gene, 333*, 143-149.

Kameda, T., Smuga-Otto, K. & Thomson, J. A. (2006). A severe *de novo* methylation of episomal vectors by human ES cells. *Biochemical and Biophysical Research Communications, 349*, 1269-1277.

Kispert, A. & Herrmann, B. G. (1993). The *Brachyury* gene encodes a novel DNA binding protein. *EMBO Journal, 12*, 3211–3220.

Lee, J.-H & Skalnik, D. G. (2005). CpG-binding protein (CXXC finger protein 1) is a component of the mammalian Set1 histone H3-Lys[4] methyltransferase complex, the analogue of the yeast Set1/COMPASS complex. *Journal of Biological Chemistry, 280*, 41725-41731.

Li, E., Bestor, T. H. & Jaenisch, R. (1992). Targeted mutation of the DNA methyltransferase gene results in embryonic lethality. *Cell, 69*, 915-926.

Macleod, D., Clark, V. H. & Bird, A. (1999). Absence of genome-wide changes in DNA methylation during development of the zebrafish. *Nature Genetics, 23*, 139-140.

Martin, C. C., Laforest, L., Akimenko, M.-A. & Ekker, M. (1999). A role for DNA methylation in gastrulation and somite patterning. *Developmental Biology, 206*, 189-205.

Mhanni, A. A. & McGowan, R. A. (2002). Variations in DNA (cytosine-5)-methyltransferase-1 expression during oogenesis and early development of the zebrafish. *Development, Genes and Evolution, 212*, 530-533.

Mhanni, A. A. & McGowan, R. A. (2004). Global changes in genomic methylation levels during early development of the zebrafish embryo. *Development, Genes and Evolution, 214*, 412-417.

Mhanni, A. A., Yoder, J. A., Dubesky, C. & McGowan, R. A. (2001). Cloning and sequence analysis of a zebrafish cDNA encoding DNA (cytosine-5)-methyltransferase-1. *Genesis, 30*, 213-219.

Miura, A., Yonebayashi, S., Watanabe, K., Toyama, T., Shimada, H. & Kakutani, T. (2001). Mobilization of transposons by a mutation abolishing full DNA methylation in *Arabidopsis*. *Nature, 411*, 212-214.

Nasevicius, A. & Ekker, S. C. (2000). Effective targeted gene 'knockdown' in zebrafish. *Nature Genetics, 26*, 216-220.

Odenthal, J., Haffter, P., Vogelsang, E., Brand, M., van Eeden, F. J. M., Furutani-Seiki, M., Granato, M., Hammerschmidt, M., Heisenberg, C.-P., Jiang, Y.-J., Kane, D. A., Kelsh, R. N., Mullins, M. C., Warga, R. M., Allende, M. L., Weinberg, E. S. & Nüsslein-Volhard, C. (1996). Mutations affecting the formation of the notochord in the zebrafish, *Danio rerio*. *Development, 123*, 103–115.

Okano, M., Bell, D. W., Haber, D. A. & Li, E. (1999). DNA methyltransferases Dnmt3a and Dnmt3b are essential for de novo methylation and mammalian development. *Cell, 99*, 247-257.

Okano, M., Xie, S. & Li, E. (1998). Dnmt2 is not required for *de novo* and maintenance methylation of viral DNA in embryonic stem cells. *Nucleic Acids Research, 26*, 2536-2540.

Peterson, R., Link, B. A., Dowling, J. E. & Schreiber, S. L. (2000). Small molecule developmental screens reveal the logic and timing of vertebrate development. *Proceedings of National Academy of Sciences of the United States of America, 97*, 12965-12969.

Rai, K., Nadauld, L. D., Chidester, S., Manos, E. J., James, S. R., Karpf, A. R., Cairns, B. R. & Jones, D. A. (2006). Zebra fish Dnmt1 and Suv39h1 regulate organ-specific terminal differentiation during development. *Molecular and Cellular Biology, 26*, 7077-7085.

Schulte-Merker, S., Ho, R. K., Herrmann, B. G. & Nüsslein-Volhard, C. (1992). The protein product of the zebrafish homologue of the mouse *T* gene is expressed in nuclei of the germ ring and the notochord of the early embryo. *Development, 116*, 1021–1032.

Schulte-Merker, S., van Eeden F. J. M., Halpern, M. E., Kimmel, C. B. & Nüsslein-Volhard, C. (1994). *no tail* (*ntl*) is the zebrafish homologue of the mouse *T* (*Brachyury*) gene. *Development, 120*, 1009–1015.

Selker, E. U. (1997). Epigenetic phenomena in filamentous fungi: useful paradigms or repeat-induced confusion? *Trends in Genetics, 13*, 296-301.

Shimoda, N., Chevrette, M., Ekker, M., Kikuchi, Y., Hotta, Y. & Okamoto, H. (1996). *Mermaid*: a family of short interspersed repetitive elements widespread in vertebrates. *Biochemical and Biophysical Research Communications, 220*, 226-232.

Shimoda, N., Knapik, E. W., Ziniti, J., Sim, C., Yamada, E., Kaplan, S., Jackson, D., de Sauvage, F., Jacob, H. & Fishman, M. C. (1999). Zebrafish genetic map with 2000 microsatellite markers. *Genomics, 58*, 219-232.

Shimoda, N., Yamakoshi, K., Miyake, A. & Takeda, H. (2005). Identification of a gene required for *de novo* DNA methylation of the zebrafish *no tail* gene. *Developmental Dynamics, 233,* 1509-1516.

Smith, T. H. L., Dueck, C. C., Mhanni, A. A. & McGowan, R. A. (2005). Novel splice variants associated with one of the zebrafish *dnmt3* genes. *BMC Developmental Biology, 5*, 1-7.

Smith, J. C., Price, B. M., Green, J. B., Weigel, D. & Herrmann, B. G. (1991). Expression of a *Xenopus* homolog of *Brachyury* (*T*) is an immediate-early response to mesoderm induction. *Cell, 67*, 79–87.

Sola, L. & Gornung, E. (2001). Classical and molecular cytogenetics of the zebrafish, *Danio rerio* (Cyprinidae, Cypriniformes): an overview. *Genetica, 111*, 397-412.

Stancheva, I., El-Maarri, O., Walter, J., Niveleau, A. & Meehan, R. R. (2002). DNA methylation at promoter regions regulates the timing of gene activation in *Xenopus laevis* embryos. *Developmental Biology, 243*, 155–165.

Stancheva, I. & Meehan, R. R. (2000). Transient depletion of xDnmt1 leads to premature gene activation in *Xenopus* embryos. *Genes & Development, 14*, 313-327.

Stemple, D. L., Solnica-Krezel, L., Zwartkruis, F., Neuhauss, S. C. F., Schier, A. F., Malicki, J., Stainier, D. Y. R., Abdelilah, S., Rangini, Z., Mountcastle-Shah, E. & Driever, W. 1996. Mutations affecting development of the notochord in zebrafish. *Development, 123*, 117–128.

Swisher, J. F. A., Rand, E., Cedar, H. & Pyle, A. M. (1998). Analysis of putative RNase sensitivity and protease insensitivity of demethylation activity in extracts from rat myoblasts. *Nucleic Acids Research, 26*, 5573-5580.

Takizawa, T., Nakashima, K., Namihara, M., Ochiai, W., Uemura, A., Yanagisawa, M., Fujita, N., Nakao, M. & Taga, T. (2001). DNA methylation is a critical cell-intrinsic determinant of astrocyte differentiation in the fetal brain. *Developmental Cell, 1*, 749-758.

Thisse, B., Pflumio, S., Fürthauer, M., Loppin, B., Heyer, V., Degrave, A., Woehl, R., Lux, A., Steffan, T., Charbonnier, X. Q. & Thisse, C. (2001). Expression of the zebrafish genome during embryogenesis. ZFIN on-line publication (http://zfin.org/).

Toyota, M., Ho, C., Ahuja, N., Jair, K-W., Ohe-Toyota, M., Baylin, S. B. & Issa, J. P. (1999). Identification of differentially methylated sequences in colorectal cancer by methylated CpG island amplification. *Cancer Research, 59*, 2307-2312.

Ushijima, T., Morimura, K., Hosoya, Y., Okonogi, H., Tatematsu, M., Sugimura, T. & Nagao, M. (1997). Establishment of methylation-sensitive-representational difference analysis and isolation of hypo- and hypermethylated genomic fragments in mouse liver tumors. *Proceedings of the National Academy of Sciences of the United States of America, 94*, 2284-2289.

van der Sar, A. M., Zivkovic, D. & Hertog, J. D. (2002). Eye defects in receptor-tyrosine phosphatase α knock-down zebrafish. *Developmental Dynamics, 223*, 292-297.

Voo, K. S., Carlone, D. L., Jacobsen, B. M., Flodin, A. & Skalnik, D. G. (2000). Cloning of a mammalian transcriptional activator that binds unmethylated CpG motifs and shares a CXXC domain with DNA methyltransferase, Human trithorax, and methyl-CpG binding domain protein 1. *Molecular and Cellular Biology, 20*, 2108-2121.

Walter, R. B., Li, H.-Y., Intano, G. W., Kazianis, S. & Walter, C. A. (2002). Absence of global genomic cytosine methylation pattern erasure during medaka (*Oryzias latipes*) early embryo development. *Comparative Biochemistry and Physiology Part B, 133*, 597-607.

Wickstrom, E., Urtishak, K. A., Choob, M., Tian, X., Sternheim, N., Cross, L. M., Rubinstein, A. & Farber, S. A. (2004). Downregulation of gene expression with negatively charged peptide nucleic acids (PNAs) in zebrafish embryos. *Methods in Cell Biology, 77*, 137-158.

Wienholds, E., Koudijs, M. J., van Eeden, F. J. M., Cuppen, E. & Plasterk, R. H. A. (2003). The microRNA-producing enzyme Dicer1 is essential for zebrafish development. *Nature Genetics, 35*, 217-218.

Yamakoshi, K. & Shimoda, N. (2003a). *De novo* DNA methylation at the CpG island of the zebrafish *no tail* gene. *Genesis, 37*, 195-202.

Yamakoshi, K. & Shimoda, N. (2003b). PCR-based cloning of an intronless zebrafish *no tail* gene. *Biochemical and Biophysical Research Communication, 306*, 598-602.

Yamakoshi, K., Shishido, Y. & Shimoda, N. (2005). Generation of aberrant transcripts of and free DNA ends in the zebrafish *no tail* gene. *Marine Biotechnology, 7*, 163-172.

Yang, C.-T. & Johnson, S. L. (2006). Small molecule-induced ablation and subsequent regeneration of larval zebrafish melanocytes. *Development, 133*, 3563-3573.

Yokomine, T., Hata, K., Tsudzuki, M. & Sasaki, H. (2006). Evolution of the vertebrate DNMT3 gene family: a possible link between existence of *DNMT3L* and genomic imprinting. *Cytogenetic and Genome Research, 113*, 75-80.

Young, S. R. L., Mumaw, C., Marrs, J. A. & Skalnik, D. G. (2006). Antisense targeting of CXXC finger protein 1 inhibits genomic cytosine methylation and primitive hematopoiesis in zebrafish. *Journal of Biological Chemistry, 281*, 37034-37044.

In: Progress in DNA Methylation Research
Editor: Hans P. Neumann, pp. 153-171

ISBN: 978-1-60021-722-7

Chapter V

Epigenetic Reprogramming and Developmental Potentials of Cloned Embryos after Somatic Cell Nuclear Transfer

Satoshi Kishigami
Laboratory for Genomic Reprogramming, Center for Developmental Biology, RIKEN Kobe, 2-2-3 Minatojima-minamimachi, Kobe 650-0047, Japan.

Abstract

The current nuclear transfer and microinsemination techniques enable us to produce animals even from adult somatic cells and immature sperm such as a round spermatid. On the other hand, the offspring production rates, when using those unnatural cells, are significantly lower than that of normally fertilized embryos. Even though somatic cell clones have been produced in many mammals including cow, mouse and pig after Dolly, the first cloned sheep, the success rates are extremely low regardless of the species. The mechanism underlying the inefficiency is not completely understood; nonetheless studies have revealed complicated anomalies and epigenetic aberrancy associated with somatic cell clones. First, this chapter will overview the epigenetic aberrancy focusing on abnormal DNA methylation in somatic cell clones and the resulting phenotypes. Second, a new nuclear transfer technique will be introduced which has been recently developed to increase the success rates 2-5 fold in mice. This new technique is simply to treat somatic cloned embryos with trichostatin A, a histone deacetylase inhibitor. This success indicates that the ability of oocytes to reprogram the somatic cell nuclei can be artificially improved leading to higher success rates. Taken together, I would finally propose the concept of 'assisted reprogramming technology' for future reproductive biotechnologies.

1. Introduction

1.1. Cloned Animals

Beside scientific demonstration, people had already been familiar with an idea of cloning as seen in the "cloning" of a monkey from Wu-k'ung's hairs in the Chinese classic novel, "Journey to the West" ("Xi You Ji"). In reality, cloned animals have been produced by three distinctive ways: (1) a multiple birth from a single zygote such as identical twins, (2) nuclear transfer of blastomere nuclei, and (3) nuclear transfer of somatic-cell nuclei. Scientifically clones imply no more than animals carrying genetically identical information. In fact, the word clone itself does not guarantee identical epigenetic information, mitochondria or even appearance. Interestingly, even monozygotic twins become epigenetically distinguishable as they get older [1,2]. In this chapter, we focus on one method of cloning, somatic cell nuclear transfer (SCNT).

1.2. Somatic Cell Nuclear Transfer

Nuclear transfer (NT) technique, beginning with Briggs and King's frog experiment in 1952 [3], has come to fruition when the first adult somatic cell cloned animal, a sheep named "Dolly" was reported in 1997 [4]. Subsequently, researchers have succeeded in cloning more than 10 mammals using both embryonic and adult donor somatic cells [5]. The current cloned animals produced by somatic cell nuclear transfer (SCNT) include a wide range of mammals such as cat, cow, and mouse but not any primates [6]. Technically, SCNT consists of four main steps, that is: 1) the transfer of a nucleus from a somatic cell into an enucleated oocyte, followed by 2) oocyte activation, 3) an in vitro culture period, and finally 4) transfer of a preimplantation embryo to a recipient surrogate mother (Figure 1). These successes in somatic cell cloning give promise to applications such as gene manipulation, species preservation, livestock propagation, and cell therapy for medical treatment by nuclear transfer embryonic stem (ES) cells (NT-ES cells) [7-10]. In addition to these applicable aspects, the successes of cloning mammals demonstrate that adult somatic-cell nuclei are reprogrammable to generate an entire organism, and per contra, that the ability to reprogram somatic-cell nuclei to acquire totipotency inheres in mammalian oocytes. Further, it has been demonstrated that even terminally differentiated postmitotic cells can re-enter the cell cycle after NT, and finally give rise to an organism, directly or indirectly via NT-ES cells [11-13]. These cells therefore retain all the genetic information necessary to generate an entire organism, evaluating retention of genetic totipotency during development. Regardless of the striking successes of SCNT, the offspring production rates of cloning animals are extremely low, typically between 0 and 4%, and clones are not perfect copies of the original organism and also carry abnormalities as discussed later. Thus, it is more urgent to improve these techniques as well as to elucidate the mechanisms behind such low efficiencies for their practical applications.

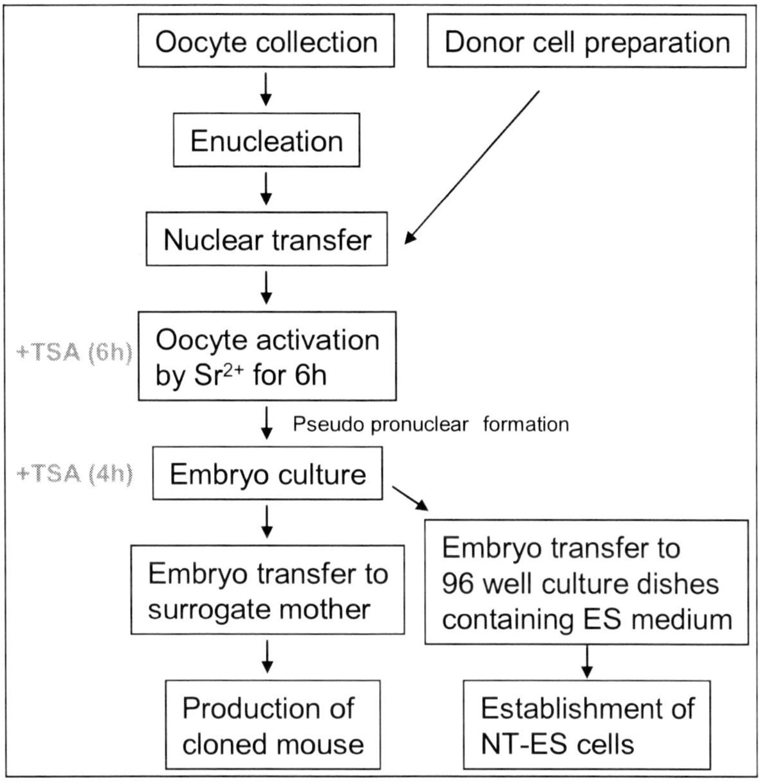

Figure 1. A procedure of somatic cell nuclear transfer in mice including the new method using TSA.

1.3. Summary of Reproductive Biotechnologies

Nowadays in addition to SCNT, the current reproductive biotechnologies enable us to reproduce animals from a variety of cells (Figure 2). For example, intracytoplasmic sperm injection (ICSI), direct injection of a single spermatozoon into an oocyte, has been an indispensable technique for fertility treatment in human as a part of the assisted reproductive techniques (ART) including artificial insemination (AI), embryo transfer, *in vitro* fertilization (IVF). In mice, the offspring can be also obtained by round spermatid injection (ROSI) [14] as well as by ICSI [15]. Further, the successes of the offspring productions using more immature male germ cells, the primary [16] and secondary [17] spermatocyte were reported. All these techniques are expected to be practically used for animal reproduction such as conservation of endangered mammalian species [18]. Regardless of these expectations, SCNT, ROSI and spermatocyte injection techniques do not reach a practical level because of their commonly poor efficiencies. Thus, it is more urgent to improve these techniques as well as to elucidate the mechanisms of such low efficiencies for their practical applications.

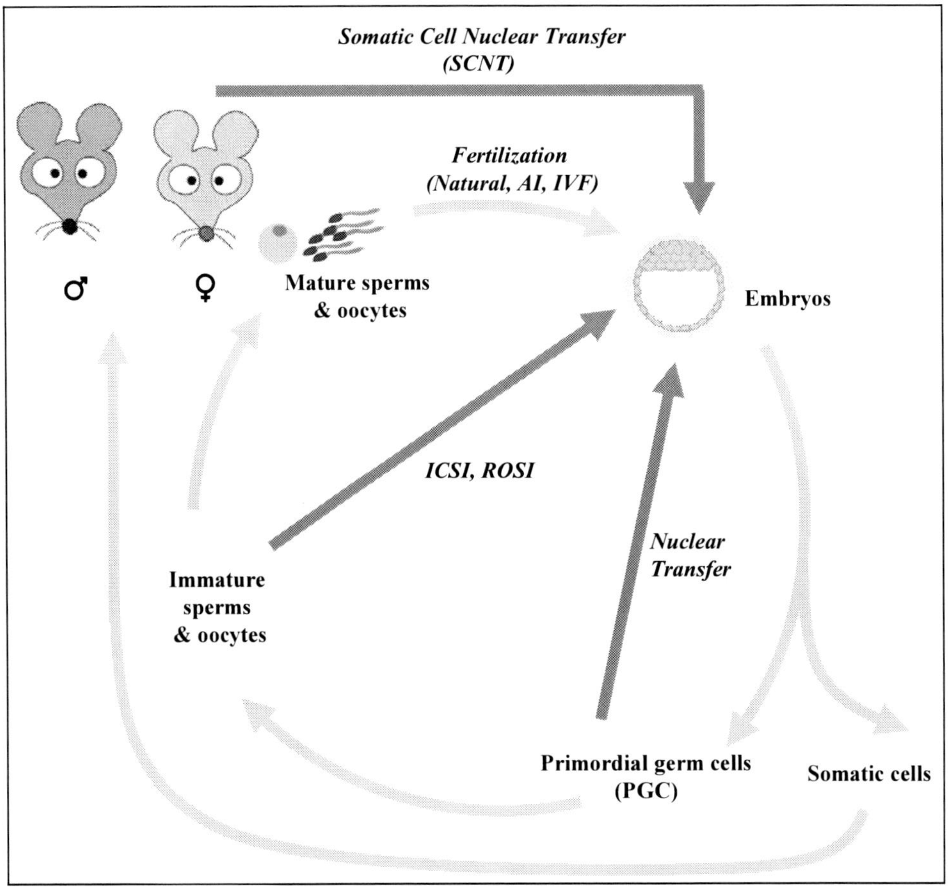

Figure 2. Reproductive Biotechnologies.

2. Developmental Failure and Pathologies after Somatic Cell Nuclear Transfer

2.1. Cumulative Dropout of their Developmental Orbit

Cloning by SCNT is extremely inefficient by the cumulative loss of cloned embryos and fetuses at every stage of development. In mouse, typically about 50% of the reconstructed embryos develop to the morula/blastocyst stage, then about 90% of these cloned embryos showed arrested development between days 5 and 8, shortly after implantation and at the beginning of or just prior to organogenesis. [19,20]. It is noted the success rates of cloned mice is dependent on donor-cell types and strains in cloned mice [13,21-24]. Although the rates of preimplantation death in cattle and sheep clones were not as high as that in mice, more than one-third pregnant females with cloned embryos lose their fetuses during gestation [25,26]. The cumulative loss during development common to somatic cell clones suggests a difficulty to pinpoint a single cause and more likely a reflection of many causes or complex mechanisms.

Another common aberration with high penetrance is that clones have abnormal placentas. For example, Cloned mice at term show placentas two to three times normal size as compared with all cloned mice described to date [23,27-29]. Detailed studies revealed expansion of spongiotrophoblast as well as abnormal gene expression and DNA methylation in the placentas of cloned mouse [30-32]. In cattle, the poor chorioallantoic development in the placentomes, which caused significant loss of clones in the first trimester (35-60 days of pregnancy), and the reduced number of placentomes were observed [25,33,34].

Even after birth, clone survival rate is reduced by neonatal and postnatal death due to a variety of phenotypes. In cattle and sheep, many cloned offspring die within the first 24h of birth. Commonly they show respiratory distress, increased birth weight so called "large offspring syndrome" and cardiovascular abnormalities [33,34]. In mice, clones from a certain strain became obese in adult life [35], while those from Sertoli cells of immature mice died at an unusually early age [36].

Despite these disappointing phenotypes in cloned animals, recent reports on the health profile of cloned animals, especially cows, indicate that the majority of cloned animals were seemingly healthy [37-39]. Specifically, no significant difference has been reported in reproductive development and milk composition between SCNT bovine clones and controls [40-42]. Actually, recently meat and milk products from cloned livestock and offspring are as safe as those from conventional animals for human consumption, according to a draft risk assessment released by U.S. Food and Drug Administration (FDA) [43,44].

Finally, it should be noted that there are reports about no transmission of phenotypes and defects to their progeny as far as they are produced by mating cloned with either cloned or naturally bred animals [45,46] [35] but not by recloning [47].

2.2. Summary

The somatic cell nuclear transfer always show high risk of deaths and abnormalities, which is often cited as a scientific argument against the possible benefits of cloning technology. This may be true as far as the current SCNT technique will not be improved or if we cannot ever understand the mechanisms involved in cloning. But it is worthy of study for the potential to improve human welfare.

3. Molecular Mechanisms under Abnormalities in SCNT Animals

3.1. Abnormal Expression of Non-Imprinted and Imprinted Genes

The unique pattern of mortality during development in clones apparently reflects the inappropriate expression of genes whose effect is exerted at different stages of development. Recent studies have shown that SCNT blastocysts often fail to reactivate many of the genes crucial for normal embryonic development, which is represented by Oct4 [21,48-50]. Interestingly, these reactivation failures by the blastocyst stage seem not to occur unilaterally for all genes required for normal development, but rather selectively and non-stochastically, which directly or indirectly leads to mortality of early cloned embryos [50,51]. Actually, the SCNT bovine embryo's gene expression profiles closely resembled those of the normal embryos [52]. In contrast, DNA microarray analysis in mouse neonatal clones' tissues have demonstrated that normal-appearing clones had a large number of affected genes compared with normal controls, and that many genes showed variation in gene expression among clones, whereas only a few genes were consistently affected in all clones [53]. Thus, abnormal gene expression in clones may more widely occur during redifferentiation for organogenesis probably by magnification of small reprogramming errors downstream in development [52].

It is widely accepted that different marking of imprinted genes is established during gametogenesis and that this pattern is maintained in both parental alleles, so it can be expressed differently in somatic cells. This parent-specific monoallelic expression of imprinted genes is critically important for normal embryo development [54]. Nuclear transfer may affect the imprinting status of the donor genome since the process of reprogramming the genome involves large-scale alterations to epigenetic traits [55]. A recent study that analyzed several imprinted genes in cloned murine blastocysts showed that most of the examined genes displayed aberrant methylation and expression patterns [56]. Term placentas from mouse SCNT exhibit significantly reduced expression of the imprinted genes (*Pegl/Mest*, *Megl/Grb10* and *Meg3/Gtl2*) and the non-imprinted genes (*Igfbp2*, *Igfbp6*, *Vegfr2/Flk1* and *ESX1* [27,31]. An inappropriate expression of both imprinted and non-imprinted genes is also observed in ES clones [27,57]. Thus, expression levels of many of both imprinted and non-imprinted genes are altered in cloned animals.

3.2. Genome-wide Abnormal DNA Methylation

Gene expression is regulated by epigenetic mechanisms such as DNA methylation and chromatin modifications that impose stable but reversible marks on the genome [58-61]. DNA methylation, specifically methylation of the cytosine residue in a CpG dinucleotide sequence, provides heritable information to the DNA as part of the epigenetic code that is not encoded in the nucleotide sequence [62]. DNA methylation has been implicated to participate in a diverse range of cellular functions and pathologies, including tissue-specific gene expression, cell differentiation, genomic imprinting, X-chromosome inactivation, regulation of chromatin structure, carcinogenesis and aging [63]. DNA methylation status on the genomes is not static but rather dynamic through mammalian development. It is also precisely controlled by DNA methylation and demethylation system. Maintenance and establishment of DNA methylation are accomplished by at least three independent catalytically active DNA methyltransferases: Dnmt1, Dnmt3a and Dnmt3b. The essential role of DNA methylation in mammalian development is highlighted by the fact that mutant mice lacking each of the enzymes are not viable and die either during early embryonic development or shortly after birth [63]. On the other hand, the loss of DNA methylation, or DNA demethylation, is accomplished by two distinct ways, active and passive demethylation.

Although the enzymatic basis of active demethylation of 5-methylcytosine *in vivo* has been mysterious, passive demethylation occurs via DNA replication [64].

Genome-wide DNA methylation patterns are known to be dynamic in early mammalian development. Shortly after fertilization, it has been suggested that only the paternal genome is actively demethylated by the end of one-cell stage [65-67]. Subsequently, DNA methylation of maternal genomes also becomes progressively demethylated through a passive DNA replication mechanism as an embryo develops [68] and methylation is found only in some repetitive parts and imprinted genes by the morula stage [68-70]. At the blastocyst stage, genome-wide methylation levels increase rapidly, establishing a differential pattern between the cells of the inner cell mass (ICM) and those of the trophectoderm (TE) [67,71]. Finally those lead to the formation of methylation patterns in the adult [72]. Thus, DNA methylation patterns are dynamic from the early embryogenesis which is required for their proper inheritance for normal development. In view of the pivotal role of DNA methylation in development, recapitulation of the DNA methylation patterns should be required for normal development of SCNT embryos. If not, the cloning-associated abnormal gene expression may be attributed at least partly to the abnormal DNA methylation patterns, involving inadequate or abnormal epigenetic reprogramming of the donor genome.

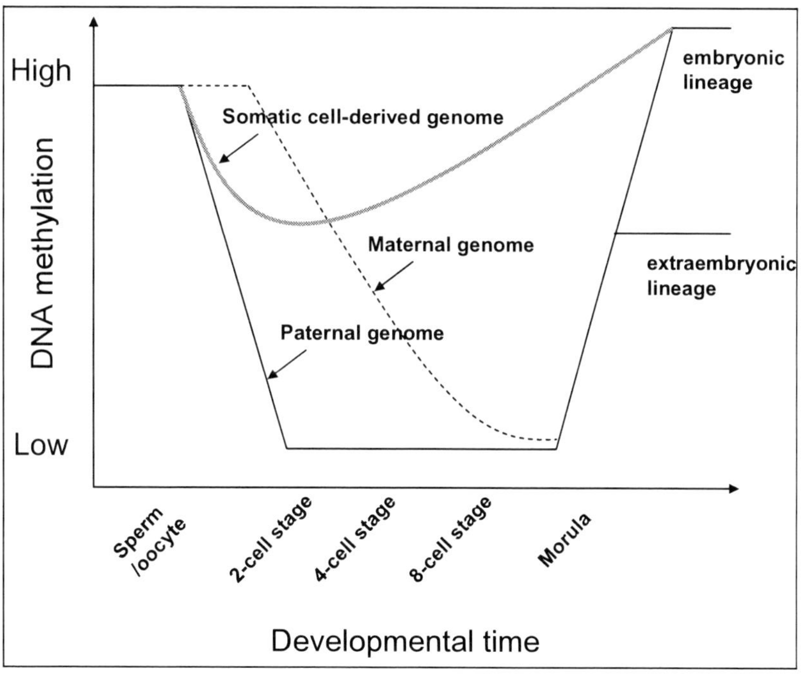

Figure 3. DNA methylation levels throughout pre-implantation development of normal and SCNT bovine embryos.

Unfortunately, studies have revealed that SCNT clones of various species including mouse, cattle and monkey clones have abnormal DNA methylation patterns [56,71,73-82]. These studies show that most SCNT individuals have distinguishable DNA methylation patterns from fertilized controls, that is, abnormally either hyper- or hypo-methylation patterns through their development. In general, the level of methylated DNA in preimplantation embryos of SCNT clones remains much higher than in normal embryos, rather resembling somatic cells (Figure 3). Staining for 5-methyl cytosine elucidated that bovine SCNT embryos fail to reproduce distinguishable parental-chromosome methylation patterns at one-cell stage [76]. Regardless, they develop further with an inefficient passive demethylation process [71,76,77] and even show precocious de novo methylation [71] with euchromatin being abnormally hypomethylated and centromeric heterochromatin being abnormally hypermethylated in the preimplantation development [76], resulting in cloned blastocysts with high DNA methylation on their genomes especially in the trophectoderm cells [71,77]. Thus, different genomic regions might respond differently to demethylation in the oocyte cytoplasm, leading to the distinguishable DNA methylation patterns unique to clones.

Although it is not totally clear yet how these genome-wide aberrant DNA methylation patterns associate with the cause and effect of abnormal development of SCNT clones, it is assumed that the aberrant patterns of DNA methylation contribute to developmental defects of the SCNT embryos. Actually, genome-wide methylation levels of spontaneously aborted SCNT fetuses showed significantly lower than those of surviving SCNT fetuses. Further, seemingly healthy clones had comparable DNA methylation levels with normal cows [83]. Thus, apparently only SCNT clones who have limited alteration in DNA methylation level can survive with normal appearance.

3.3. Abnormal DNA Methylation of Individual Gene Loci

Consistent with the observations of the genome-wide abnormal DNA methylation in SCNT embryos, sequence-level analyses of methylation states after bisulfite treatment have revealed abnormal DNA methylation in SCNT animals. For example, *Bob-B* and satellite I sequences are abnormally hypermethylated in bovine SCNT blastocysts [74]. Regarding single copy genes, the methylation status in the promoter region of *Oct4* which is required for maintenance of the inner cells mass (ICM) as well as pluripotency of ES cells [84,85], has been well-studied since the methylation status does directly affect the *Oct4* expression [86,87]. In somatic cells, the *Oct4* promoter is distinctly methylated, and *Oct4* is not expressed. In contrast, the promoter region is unmethylated in ES cells which highly express *Oct4* [87]. Therefore, the demethylation of this region should be required for reactivation of this gene after SCNT before the 4- to 8-cell stage [88]. Consistent with lower expression of *Oct4* in mouse cloned blastocysts [48-50], the *Oct4* promoter is still significantly methylated even at the 8-cell stage, and then only gradually demethylated by the blastocyst stage [89]. This is quite a contrast with fertilized controls which lose most methylation by the 8-cell stage. Interestingly, the upstream sequences of the other pluripotency regulators such as *Nanog*, *Sox2* and *Foxd3*, are significantly less methylated in somatic cells. These facts seem to highlight the inefficiency of demethylation in the *Oct4* promoter region as one of the rate-limiting factors which determine the low success rates of SCNT animals.

Abnormal DNA methylation patterns as well as abnormal gene expression in imprinted genes are also observed in clones. For example, only 4% of cloned mouse embryos can recapitulate a blastocyst's mode of expression for *H19*, *Snrpn*, *Meg3*, *Igf2*r and *Ascl2*. At the same time, the cloned embryos exhibit parental allele-specific loss of DNA methylation at the differentially methylated regions (DMRs), the imprinting control regions of the *H19* and *Snrpn* [56]. In SCNT sheep, *IGF2R* DMR2 methylation are strongly reduced with almost no demethylation at *H19* [90]. In SCNT Day 17 bovine embryos, *SNRPN* DMR is also strikingly hypomethylated [91] and, also *H19* is found to be biallelically expressed in three of four SCNT-derived calves that died during the perinatal period [92]. Thus, even if not all the imprinted genes are affected by SCNT and not all the clones carry loss of imprinted genes, in any species, imprinted memory can be erased by the demethylation of DMRs through SCNT as an epigenetic error. Further, the susceptibility to error is varied between imprinted genes. On the other hand, aside from gene expression levels, these aberrancies of imprinted genes (i.e., loss of imprinting) are not observed in midgestation cloned murine conceptuses [31,93], suggesting that only the cloned embryos that preserved imprinted expression and methylation can develop to the midgestation stage.

Interestingly, an epigenetic hotspot of aberrant DNA methylation is reported in placentomegaly of cloned mice [32]. The *Spalt-like gene3* (*Sall3*) locus was identified as a hypermethylated region in the placental genome of clones, while the placentas of normal fertilized mice do not show such hypermethylation. This locus has a CpG island containing a tissue-dependent differentially methylated region specific to the trophoblast cell lineage, suggesting an association of the epigenetic error with placental hypertrophy, a common phenotype in cloned mice.

Applying restriction landmark genome scanning (RLGS) in some tissues of healthy cloned mouse also succeeded in identifying four methylation sites which are aberrantly methylated or unmethylated [81]. Interestingly, individual cloned mice have different methylation aberrations.

Thus, in cloned animals, sequence-specific investigations elucidate both hyper- and hypo-methylation can happen depending on gene loci. These aberrant DNA methylation patterns in gene loci may associate directly with abnormal imprinting memory and individual gene expression.

3.4. Abnormal DNA Methylation and X Inactivation

In female mammals, a dosage compensation for X-linked genes is accomplished by inactivation of one of two X chromosomes [94]. This X chromosome inactivation (XCI) proceeds in the early embryo through a complex sequence of events that are not completely understood yet [95]. *Xist*, a noncoding RNA that is transcribed from the silenced X chromosome, coats the chromosome in *cis* and initiates a cascade of epigenetic changes that generate a heritable silent state [96]. In somatic cells, the *Xist* promoter is hypermethylated on the active X and hypomethylated on the inactive X in preimplantation embryos [97,98]. In mice, during the earliest embryogenesis, both X chromosomes are active and subsequently XCI occurs coupled with cellular differentiation. In embryonic cell lineage, the ratio of cells containing inactivated maternal X chromosome to cells containing inactivated paternal X chromosome is approximately 1:1, that is, around 50%. In contrast, 2 of 11 cloned mice kidneys showed skewed X-inactivation ratios (19.2% and 86.8%), suggesting the stochastic distortion of primary choice due to incomplete erasure of previous XCI memory from the donor cell [79]. Further, cloned mouse blastocysts abnormally express both *Xist* alleles and also show abnormal methylation in *Xist* promoter [78]. The similar phenomenon is also observed in cloned cattle. Overexpression of *Xist* in cloned embryos and hypomethylation of *Xist* in the deceased clones are reported as well as patchy patterns of XCI and biallelic repression of X-linked genes [99,100]. Thus, normal XCI following SCNT can occur to some degree in cloned embryos [101]. Neverthless SCNT embryos are unable to regulate XCI consistently probably because of incomplete reprogramming.

3.5. Summary

To date findings from DNA methylation studies on cloned animals suggest genome-wide hypermethylation after SCNT, while sequence-specific investigations show both hyper- and hypo-methylation can happen depending on gene loci. Then, these aberrant DNA methylation patterns affect gene functions including genomic imprinting, X-inactivation and gene expressions. Interestingly, the susceptibility of each locus to epigenetic error by SCNT is not equal but rather biased, suggesting epigenetic hotspots which may differ in species and demonstrate the unique characteristics of clones.

SCNT embryos can have aberrant epigenetic modifications in the beginning which can persist until adulthood. Seemingly healthy clones are likely to survive when they succeed in minimizing these epigenetic errors that are incompatible with embryonic development. It is noted that individual cloned animals have different epigenetic modifications between them. In other words, cloned animals are by no means perfect copies of the original animals and each other in terms of their epigenetic modifications.

4. Improvement of SCNT

4.1. Improvement of Animal Cloning Using Chemicals

The success of somatic cell cloning in animals gives promise to applications such as species preservation, livestock propagation, and cell therapy for medical treatment by NT-ES cells [7-10]. On the other hand, for example, mouse cloning by SCNT has been inefficient since the first cloned mouse, "Cumulina," was born in 1997 [102]. Regardless of many trials to improve the original method, none had exerted a marked influence on the efficiency of cloning. These include the methods of oocyte activation [103], timing of enucleation or injection of nucleus [104] and timing of oocyte activation [105].

Significant successes in improvements of SCNT using chemicals have been reported. Addition of 1% (v/v) DMSO to the activation medium is found to increase the frequency of development to the blastocysts stage, while this improvement did not lead to a higher success rate in the generation of live cloned offspring [105]. DMSO, an amphipathic molecule, is widely used not only as a solvent for water-insoluble substances but also as a cryopreservant for various types of cells. Although the underlying mechanism is not clear yet, a recent study indicates that DMSO has an impact on the epigenetic profile: upregulation of Dnmt3a expression and alteration of genome-wide DNA methylation profiles [106], suggesting DMSO affects reprogramming of somatic cell nuclei, especially their DNA methylation status, after or during oocyte activation. In order to more directly change the epigenetic status before SCNT, trichostatin A (TSA) treated donor cells were used as donor cell nuclei, resulting in increasing bovine cloned blastocyst development compared to controls (35% vs. 25%) [107]. TSA is a histone-deacetylase inhibitor (HDACi), which enhances the pool of acetylated histones [108] and DNA demethylation [87,109]. This result indicates that alteration of preexisting epigenetic marks including DNA methylation of donor cells is one key to improve subsequent development [110].

A more strikingly improvement in mouse cloning has been recently achieved [111,112]. This method is simply to add 5 to 50 nM TSA to activation and culture medium to treat cloned embryos with TSA for 10 h. This TSA treatment led to a more than 5 fold increase in success rate of mouse cloning from cumulus cells without obvious abnormality. Studies indicate that TSA does not only induce not hyperacetylation of histones but suppress remethylation of DNA, leading to the unmethylated status of injected nuclei in oocytes [113]. Again although the underlying mechanism is not completely clear yet, the accumulating evidence illustrates a scheme that TSA reduces DNA methylation levels of transferred somatic nuclei which enhances the reprogramming and subsequent reactivation of genes essential for embryonic development. Further, this TSA cloning technique led to the first success of cloning the adult ICR mouse, an outbred strain which has never been directly cloned before [114]. Actually, only when TSA was used, were both male and female mice cloned from cumulus and fibroblast cells of adult ICR mice with 4-5% success rates. These results demonstrate that this technique not only improves the success rates of cloning from the hybrid strains but also enables mouse cloning from normally "unclonable" strains [114]. And probably more importantly, the TSA treatment of cloned embryos has never caused any additional abnormalities [111,114,115].

4.2. Improvement and Assisted Reprogramming Technology of SCNT

As seen above, during the last decade, the cloning technique has been significantly improved while the effects may not be to a practical level. The successes of those improvements which alter the epigenetic status of donor nuclei strongly suggest that the inefficiency of cloning is to no small extent attributed to the issue of reprogramming which can be artificially assisted by chemicals. In other words, the normal reprogramming ability in oocytes, that is, enzymatic activities involved in reprogramming including HDAC activity may not be optimized for reprogramming of somatic nuclei even though the condition is well-optimized for normal fertilization [112,115]. Therefore, the future study should be focused on identifying enzymes involved in reprogramming and optimizing their activities which may overcome the current difficulties not only of inefficient success rates of cloning but also of cloning phenotypic abnormalities. Thus, to develop the "assisted reprogramming technology" may be the one of ways to go for future assisted reproductive technology.

4.3. Summary

The animal cloning has been inefficient which hinders practical usage of SCNT. The successes of improvement of animal cloning using chemicals strongly support the idea that the inefficiency is not due to the limited number of potentially clonable nuclei existing in donor cells but rather inefficient reprogramming ability of oocytes. In other words, the findings of improved methods lead to an important conclusion that failure or success of cloning may be stochastically determined, which can be artificially controlled and improved in future. Further, these practical improvements of SCNT techniques also provide insight into reprogramming of somatic nuclei.

5. Conclusion

During the last decade after Dolly, SCNT has become a powerful tool for reprogramming studies in a variety of species, specifically in mammals. This technique also holds great promise for future regenerative medicine using patient-tailored ES cells for each person derived from SCNT blastocysts. Different from disappointing abnormalities of cloned animals, it is demonstrated in mice that NT-ES cells are indistinguishable from normal ES cells in terms of gene expression [116,117]. To practically use the gift from SCNT, however, more studies would be required to understand the underlying mechanisms that would improve the current SCNT techniques.

Acknowledgements

I thank Dr. T. Wakayama and my colleagues for useful suggestions. I also acknowledge discussions with Dr. T. Castranio.

References

[1] Petronis, A. (2006) Epigenetics and twins: three variations on the theme, *Trends Genet*, *22*, 347-50.
[2] Fraga, M. F., Ballestar, E., Paz, M. F. et al. (2005) Epigenetic differences arise during the lifetime of monozygotic twins, *Proc Natl Acad Sci U S A*, *102*, 10604-9.
[3] Briggs, R. & King, T. J. (1952) Transplantation of Living Nuclei From Blastula Cells into Enucleated Frogs' Eggs, *Proc Natl Acad Sci U S A*, *38*, 455-63.
[4] Wilmut, I., Schnieke, A. E., McWhir, J., Kind, A. J. & Campbell, K. H. (1997) Viable offspring derived from fetal and adult mammalian cells, *Nature*, *385*, 810-3.
[5] Campbell, K. H., Alberio, R., Choi, I. et al. (2005) Cloning: eight years after Dolly, *Reprod Domest Anim*, *40*, 256-68.

[6] Simerly, C., Dominko, T., Navara, C. et al. (2003) Molecular correlates of primate nuclear transfer failures, *Science*, 300, 297.

[7] Rideout, W. M., 3rd, Hochedlinger, K., Kyba, M., Daley, G. Q. & Jaenisch, R. (2002) Correction of a genetic defect by nuclear transplantation and combined cell and gene therapy, *Cell*, *109*, 17-27.

[8] Wakayama, S., Kishigami, S., Van Thuan, N. et al. (2005) Propagation of an infertile hermaphrodite mouse lacking germ cells by using nuclear transfer and embryonic stem cell technology, *Proc Natl Acad Sci U S A*, *102*, 29-33.

[9] Wakayama, S., Mizutani, E., Kishigami, S. et al. (2005) Mice cloned by nuclear transfer from somatic and ntES cells derived from the same individuals, *J Reprod Dev*, *51*, 765-72.

[10] Wakayama, T., Tabar, V., Rodriguez, I. et al. (2001) Differentiation of embryonic stem cell lines generated from adult somatic cells by nuclear transfer, *Science*, *292*, 740-3.

[11] Eggan, K., Baldwin, K., Tackett, M. et al. (2004) Mice cloned from olfactory sensory neurons, *Nature*, *428*, 44-9.

[12] Li, J., Ishii, T., Feinstein, P. & Mombaerts, P. (2004) Odorant receptor gene choice is reset by nuclear transfer from mouse olfactory sensory neurons, *Nature*, *428*, 393-9.

[13] Sung, L. Y., Gao, S., Shen, H. et al. (2006) Differentiated cells are more efficient than adult stem cells for cloning by somatic cell nuclear transfer, *Nat Genet*.

[14] Ogura, A., Matsuda, J. & Yanagimachi, R. (1994) Birth of normal young after electrofusion of mouse oocytes with round spermatids, *Proc Natl Acad Sci U S A*, *91*, 7460-2.

[15] Kimura, Y. & Yanagimachi, R. (1995) Mouse oocytes injected with testicular spermatozoa or round spermatids can develop into normal offspring, *Development*, *121*, 2397-405.

[16] Ogura, A., Suzuki, O., Tanemura, K. et al. (1998) Development of normal mice from metaphase I oocytes fertilized with primary spermatocytes, *Proc Natl Acad Sci U S A*, *95*, 5611-5.

[17] Kimura, Y. & Yanagimachi, R. (1995) Development of normal mice from oocytes injected with secondary spermatocyte nuclei, *Biol Reprod*, *53*, 855-62.

[18] Andrabi, S. M. & Maxwell, W. M. (2006) A review on reproductive biotechnologies for conservation of endangered mammalian species, *Anim Reprod Sci*.

[19] Ogura, A., Inoue, K., Ogonuki, N. et al. (2002) Phenotypic effects of somatic cell cloning in the mouse, *Cloning Stem Cells*, *4*, 397-405.

[20] Wakayama, T. & Yanagimachi, R. (1999) Cloning the laboratory mouse, *Semin Cell Dev Biol*, *10*, 253-8.

[21] Inoue, K., Ogonuki, N., Miki, H. et al. (2006) Inefficient reprogramming of the hematopoietic stem cell genome following nuclear transfer, *J Cell Sci*, *119*, 1985-91.

[22] Inoue, K., Ogonuki, N., Mochida, K. et al. (2003) Effects of donor cell type and genotype on the efficiency of mouse somatic cell cloning, *Biol Reprod*, *69*, 1394-400.

[23] Wakayama, T. & Yanagimachi, R. (2001) Mouse cloning with nucleus donor cells of different age and type, *Mol Reprod Dev*, *58*, 376-83.

[24] Inoue, K., Wakao, H., Ogonuki, N. et al. (2005) Generation of cloned mice by direct nuclear transfer from natural killer T cells, *Curr Biol*, *15*, 1114-8.

[25] Hill, J. R., Burghardt, R. C., Jones, K. et al. (2000) Evidence for placental abnormality as the major cause of mortality in first-trimester somatic cell cloned bovine fetuses, *Biol Reprod*, *63*, 1787-94.

[26] Wilmut, I., Beaujean, N., de Sousa, P. A. et al. (2002) Somatic cell nuclear transfer, *Nature*, *419*, 583-6.

[27] Humpherys, D., Eggan, K., Akutsu, H. et al. (2001) Epigenetic instability in ES cells and cloned mice, *Science*, *293*, 95-7.

[28] Ono, Y., Shimozawa, N., Ito, M. & Kono, T. (2001) Cloned mice from fetal fibroblast cells arrested at metaphase by a serial nuclear transfer, *Biol Reprod*, *64*, 44-50.

[29] Wakayama, T. & Yanagimachi, R. (1999) Cloning of male mice from adult tail-tip cells, *Nat Genet*, *22*, 127-8.

[30] Tanaka, S., Oda, M., Toyoshima, Y. et al. (2001) Placentomegaly in cloned mouse concepti caused by expansion of the spongiotrophoblast layer, *Biol Reprod*, *65*, 1813-21.

[31] Inoue, K., Kohda, T., Lee, J. et al. (2002) Faithful expression of imprinted genes in cloned mice, *Science*, *295*, 297.

[32] Ohgane, J., Wakayama, T., Senda, S. et al. (2004) The Sall3 locus is an epigenetic hotspot of aberrant DNA methylation associated with placentomegaly of cloned mice, *Genes Cells*, *9*, 253-60.

[33] Chavatte-Palmer, P., Heyman, Y., Richard, C. et al. (2002) Clinical, hormonal, and hematologic characteristics of bovine calves derived from nuclei from somatic cells, *Biol Reprod*, *66*, 1596-603.

[34] Lee, R. S., Peterson, A. J., Donnison, M. J. et al. (2004) Cloned cattle fetuses with the same nuclear genetics are more variable than contemporary half-siblings resulting from artificial insemination and exhibit fetal and placental growth deregulation even in the first trimester, *Biol Reprod*, *70*, 1-11.

[35] Tamashiro, K. L., Wakayama, T., Akutsu, H. et al. (2002) Cloned mice have an obese phenotype not transmitted to their offspring, *Nat Med*, *8*, 262-7.

[36] Ogonuki, N., Inoue, K., Yamamoto, Y. et al. (2002) Early death of mice cloned from somatic cells, *Nat Genet*, *30*, 253-4.

[37] Lanza, R. P., Cibelli, J. B., Faber, D. et al. (2001) Cloned cattle can be healthy and normal, *Science*, *294*, 1893-4.

[38] Cibelli, J. B., Campbell, K. H., Seidel, G. E., West, M. D. & Lanza, R. P. (2002) The health profile of cloned animals, *Nat Biotechnol*, *20*, 13-4.

[39] Chavatte-Palmer, P., Remy, D., Cordonnier, N. et al. (2004) Health status of cloned cattle at different ages, *Cloning Stem Cells*, *6*, 94-100.

[40] Enright, B. P., Taneja, M., Schreiber, D. et al. (2002) Reproductive characteristics of cloned heifers derived from adult somatic cells, *Biol Reprod*, *66*, 291-6.

[41] Walsh, M. K., Lucey, J. A., Govindasamy-Lucey, S., Pace, M. M. & Bishop, M. D. (2003) Comparison of milk produced by cows cloned by nuclear transfer with milk from non-cloned cows, *Cloning Stem Cells*, *5*, 213-9.

[42] Norman, H. D. & Walsh, M. K. (2004) Performance of dairy cattle clones and evaluation of their milk composition, *Cloning Stem Cells*, *6*, 157-64.

[43] Rudenko, L. & Matheson, J. C. (2007) The US FDA and animal cloning: Risk and regulatory approach, *Theriogenology*, *67*, 198-206.
[44] Rudenko, L., Matheson, J. C. & Sundlof, S. F. (2007) Animal cloning and the FDA-the risk assessment paradigm under public scrutiny, *Nat Biotechnol*, *25*, 39-43.
[45] Mir, B., Zaunbrecher, G., Archer, G. S., Friend, T. H. & Piedrahita, J. A. (2005) Progeny of somatic cell nuclear transfer (SCNT) pig clones are phenotypically similar to non-cloned pigs, *Cloning Stem Cells*, *7*, 119-25.
[46] Shimozawa, N., Ono, Y., Kimoto, S. et al. (2002) Abnormalities in cloned mice are not transmitted to the progeny, *Genesis*, *34*, 203-7.
[47] Wakayama, T., Shinkai, Y., Tamashiro, K. L. et al. (2000) Cloning of mice to six generations, *Nature*, *407*, 318-9.
[48] Boiani, M., Eckardt, S., Scholer, H. R. & McLaughlin, K. J. (2002) Oct4 distribution and level in mouse clones: consequences for pluripotency, *Genes Dev*, *16*, 1209-19.
[49] Bortvin, A., Eggan, K., Skaletsky, H. et al. (2003) Incomplete reactivation of Oct4-related genes in mouse embryos cloned from somatic nuclei, *Development*, *130*, 1673-80.
[50] Kishigami, S., Hikichi, T., Van Thuan, N. et al. (2006) Normal specification of the extraembryonic lineage after somatic nuclear transfer, *FEBS Lett*, *580*, 1801-6.
[51] Li, X., Kato, Y. & Tsunoda, Y. (2005) Comparative analysis of development-related gene expression in mouse preimplantation embryos with different developmental potential, *Mol Reprod Dev*, *72*, 152-60.
[52] Smith, S. L., Everts, R. E., Tian, X. C. et al. (2005) Global gene expression profiles reveal significant nuclear reprogramming by the blastocyst stage after cloning, *Proc Natl Acad Sci U S A*, *102*, 17582-7.
[53] Kohda, T., Inoue, K., Ogonuki, N. et al. (2005) Variation in gene expression and aberrantly regulated chromosome regions in cloned mice, *Biol Reprod*, *73*, 1302-11.
[54] Reik, W. & Walter, J. (2001) Genomic imprinting: parental influence on the genome, *Nat Rev Genet*, *2*, 21-32.
[55] Solter, D. (2000) Mammalian cloning: advances and limitations, *Nat Rev Genet*, *1*, 199-207.
[56] Mann, M. R., Chung, Y. G., Nolen, L. D. et al. (2003) Disruption of imprinted gene methylation and expression in cloned preimplantation stage mouse embryos, *Biol Reprod*, *69*, 902-14.
[57] Suemizu, H., Aiba, K., Yoshikawa, T. et al. (2003) Expression profiling of placentomegaly associated with nuclear transplantation of mouse ES cells, *Dev Biol*, *253*, 36-53.
[58] Jaenisch, R. & Bird, A. (2003) Epigenetic regulation of gene expression: how the genome integrates intrinsic and environmental signals, *Nat Genet*, *33 Suppl*, 245-54.
[59] Li, E. (2002) Chromatin modification and epigenetic reprogramming in mammalian development, *Nat Rev Genet*, *3*, 662-73.
[60] Shiota, K. (2004) DNA methylation profiles of CpG islands for cellular differentiation and development in mammals, *Cytogenet Genome Res*, *105*, 325-34.
[61] Caiafa, P. & Zampieri, M. (2005) DNA methylation and chromatin structure: the puzzling CpG islands, *J Cell Biochem*, *94*, 257-65.

[62] Robertson, K. D. & Wolffe, A. P. (2000) DNA methylation in health and disease, *Nat Rev Genet*, *1*, 11-9.

[63] Bird, A. (2002) DNA methylation patterns and epigenetic memory, *Genes Dev*, 16, 6-21.

[64] Wolffe, A. P., Jones, P. L. & Wade, P. A. (1999) DNA demethylation, *Proc Natl Acad Sci U S A*, *96*, 5894-6.

[65] Oswald, J., Engemann, S., Lane, N. et al. (2000) Active demethylation of the paternal genome in the mouse zygote, *Curr Biol*, *10*, 475-8.

[66] Mayer, W., Niveleau, A., Walter, J., Fundele, R. & Haaf, T. (2000) Demethylation of the zygotic paternal genome, *Nature*, *403*, 501-2.

[67] Santos, F., Hendrich, B., Reik, W. & Dean, W. (2002) Dynamic reprogramming of DNA methylation in the early mouse embryo, *Dev Biol*, *241*, 172-82.

[68] Reik, W., Dean, W. & Walter, J. (2001) Epigenetic reprogramming in mammalian development, *Science*, *293*, 1089-93.

[69] Walsh, C. P., Chaillet, J. R. & Bestor, T. H. (1998) Transcription of IAP endogenous retroviruses is constrained by cytosine methylation, *Nat Genet*, *20*, 116-7.

[70] Sanford, J. P., Clark, H. J., Chapman, V. M. & Rossant, J. (1987) Differences in DNA methylation during oogenesis and spermatogenesis and their persistence during early embryogenesis in the mouse, *Genes Dev*, *1*, 1039-46.

[71] Dean, W., Santos, F., Stojkovic, M. et al. (2001) Conservation of methylation reprogramming in mammalian development: aberrant reprogramming in cloned embryos, *Proc Natl Acad Sci U S A*, *98*, 13734-8.

[72] Turker, M. S. (1999) The establishment and maintenance of DNA methylation patterns in mouse somatic cells, *Semin Cancer Biol*, *9*, 329-37.

[73] Dean, W., Bowden, L., Aitchison, A. et al. (1998) Altered imprinted gene methylation and expression in completely ES cell-derived mouse fetuses: association with aberrant phenotypes, *Development*, *125*, 2273-82.

[74] Kang, Y. K., Koo, D. B., Park, J. S. et al. (2001) Aberrant methylation of donor genome in cloned bovine embryos, *Nat Genet*, *28*, 173-7.

[75] Kang, Y. K., Park, J. S., Koo, D. B. et al. (2002) Limited demethylation leaves mosaic-type methylation states in cloned bovine pre-implantation embryos, *Embo J*, *21*, 1092-100.

[76] Bourc'his, D., Le Bourhis, D., Patin, D. et al. (2001) Delayed and incomplete reprogramming of chromosome methylation patterns in bovine cloned embryos, *Curr Biol*, *11*, 1542-6.

[77] Santos, F., Zakhartchenko, V., Stojkovic, M. et al. (2003) Epigenetic marking correlates with developmental potential in cloned bovine preimplantation embryos, *Curr Biol*, *13*, 1116-21.

[78] Nolen, L. D., Gao, S., Han, Z. et al. (2005) X chromosome reactivation and regulation in cloned embryos, *Dev Biol*, *279*, 525-40.

[79] Senda, S., Wakayama, T., Yamazaki, Y. et al. (2004) Skewed X-inactivation in cloned mice, *Biochem Biophys Res Commun*, *321*, 38-44.

[80] Yang, J., Yang, S., Beaujean, N. et al. (2007) Epigenetic marks in cloned rhesus monkey embryos: comparison with counterparts produced in vitro, *Biol Reprod*, *76*, 36-42.

[81] Ohgane, J., Wakayama, T., Kogo, Y. et al. (2001) DNA methylation variation in cloned mice, *Genesis*, *30*, 45-50.

[82] Hiendleder, S., Mund, C., Reichenbach, H. D. et al. (2004) Tissue-specific elevated genomic cytosine methylation levels are associated with an overgrowth phenotype of bovine fetuses derived by in vitro techniques, *Biol Reprod*, 71, 217-23.

[83] Cezar, G. G., Bartolomei, M. S., Forsberg, E. J. et al. (2003) Genome-wide epigenetic alterations in cloned bovine fetuses, *Biol Reprod*, *68*, 1009-14.

[84] Nichols, J., Zevnik, B., Anastassiadis, K. et al. (1998) Formation of pluripotent stem cells in the mammalian embryo depends on the POU transcription factor Oct4, *Cell*, *95*, 379-91.

[85] Niwa, H., Miyazaki, J. & Smith, A. G. (2000) Quantitative expression of Oct-3/4 defines differentiation, dedifferentiation or self-renewal of ES cells, *Nat Genet*, *24*, 372-6.

[86] Gidekel, S. & Bergman, Y. (2002) A unique developmental pattern of Oct-3/4 DNA methylation is controlled by a cis-demodification element, *J Biol Chem*, *277*, 34521-30.

[87] Hattori, N., Nishino, K., Ko, Y. G. et al. (2004) Epigenetic control of mouse Oct-4 gene expression in embryonic stem cells and trophoblast stem cells, *J Biol Chem*, *279*, 17063-9.

[88] Simonsson, S. & Gurdon, J. (2004) DNA demethylation is necessary for the epigenetic reprogramming of somatic cell nuclei, *Nat Cell Biol*, *6*, 984-90.

[89] Yamazaki, Y., Fujita, T. C., Low, E. W. et al. (2006) Gradual DNA demethylation of the Oct4 promoter in cloned mouse embryos, *Mol Reprod Dev*, *73*, 180-8.

[90] Young, L. E., Fernandes, K., McEvoy, T. G. et al. (2001) Epigenetic change in IGF2R is associated with fetal overgrowth after sheep embryo culture, *Nat Genet*, *27*, 153-4.

[91] Lucifero, D., Suzuki, J., Bordignon, V. et al. (2006) Bovine SNRPN methylation imprint in oocytes and day 17 in vitro-produced and somatic cell nuclear transfer embryos, *Biol Reprod*, *75*, 531-8.

[92] Zhang, S., Kubota, C., Yang, L. et al. (2004) Genomic imprinting of H19 in naturally reproduced and cloned cattle, *Biol Reprod*, *71*, 1540-4.

[93] Humpherys, D., Eggan, K., Akutsu, H. et al. (2002) Abnormal gene expression in cloned mice derived from embryonic stem cell and cumulus cell nuclei, *Proc Natl Acad Sci U S A*, *99*, 12889-94.

[94] Lyon, M. F. (1961) Gene action in the X-chromosome of the mouse (Mus musculus L.), *Nature*, *190*, 372-3.

[95] Avner, P. & Heard, E. (2001) X-chromosome inactivation: counting, choice and initiation, *Nat Rev Genet*, *2*, 59-67.

[96] Brown, S. D. (1991) XIST and the mapping of the X chromosome inactivation centre, *Bioessays*, *13*, 607-12.

[97] Norris, D. P., Patel, D., Kay, G. F. et al. (1994) Evidence that random and imprinted Xist expression is controlled by preemptive methylation, *Cell*, *77*, 41-51.

[98] Heard, E., Clerc, P. & Avner, P. (1997) X-chromosome inactivation in mammals, *Annu Rev Genet*, *31*, 571-610.
[99] Xue, F., Tian, X. C., Du, F. et al. (2002) Aberrant patterns of X chromosome inactivation in bovine clones, *Nat Genet*, *31*, 216-20.
[100] Wrenzycki, C., Lucas-Hahn, A., Herrmann, D. et al. (2002) In vitro production and nuclear transfer affect dosage compensation of the X-linked gene transcripts G6PD, PGK, and Xist in preimplantation bovine embryos, *Biol Reprod*, *66*, 127-34.
[101] Eggan, K., Akutsu, H., Hochedlinger, K. et al. (2000) X-Chromosome inactivation in cloned mouse embryos, *Science*, *290*, 1578-81.
[102] Wakayama, T., Perry, A. C., Zuccotti, M., Johnson, K. R. & Yanagimachi, R. (1998) Full-term development of mice from enucleated oocytes injected with cumulus cell nuclei, *Nature*, *394*, 369-74.
[103] Kishikawa, H., Wakayama, T. & Yanagimachi, R. (1999) Comparison of oocyte-activating agents for mouse cloning, *Cloning*, *1*, 153-9.
[104] Wakayama, S., Cibelli, J. B. & Wakayama, T. (2003) Effect of timing of the removal of oocyte chromosomes before or after injection of somatic nucleus on development of NT embryos, *Cloning Stem Cells*, *5*, 181-9.
[105] Wakayama, T. & Yanagimachi, R. (2001) Effect of cytokinesis inhibitors, DMSO and the timing of oocyte activation on mouse cloning using cumulus cell nuclei, *Reproduction*, *122*, 49-60.
[106] Iwatani, M., Ikegami, K., Kremenska, Y. et al. (2006) Dimethyl sulfoxide has an impact on epigenetic profile in mouse embryoid body, *Stem Cells*, *24*, 2549-56.
[107] Enright, B. P., Kubota, C., Yang, X. & Tian, X. C. (2003) Epigenetic characteristics and development of embryos cloned from donor cells treated by trichostatin A or 5-aza-2'-deoxycytidine, *Biol Reprod*, *69*, 896-901.
[108] Yoshida, M., Kijima, M., Akita, M. & Beppu, T. (1990) Potent and specific inhibition of mammalian histone deacetylase both in vivo and in vitro by trichostatin A, *J Biol Chem*, *265*, 17174-9.
[109] Selker, E. U. (1998) Trichostatin A causes selective loss of DNA methylation in Neurospora, *Proc Natl Acad Sci U S A*, *95*, 9430-5.
[110] Blelloch, R., Wang, Z., Meissner, A. et al. (2006) Reprogramming efficiency following somatic cell nuclear transfer is influenced by the differentiation and methylation state of the donor nucleus, *Stem Cells*, *24*, 2007-13.
[111] Kishigami, S., Mizutani, E., Ohta, H. et al. (2006) Significant improvement of mouse cloning technique by treatment with trichostatin A after somatic nuclear transfer, *Biochem Biophys Res Commun*, *340*, 183-9.
[112] Rybouchkin, A., Kato, Y. & Tsunoda, Y. (2006) Role of histone acetylation in reprogramming of somatic nuclei following nuclear transfer, *Biol Reprod*, *74*, 1083-9.
[113] Kishigami, S., Van Thuan, N., Hikichi, T. et al. (2006) Epigenetic abnormalities of the mouse paternal zygotic genome associated with microinsemination of round spermatids, *Dev Biol*, *289*, 195-205.
[114] Kishigami, S., Bui, H. T., Wakayama, S. et al. (2006) Successful Mouse Cloning of an Outbred Strain by Trichostatin A Treatment after Somatic Nuclear Transfer, *J Reprod Dev*.

[115] Kishigami, S., Ohta, H., Mizutani, E. et al. (2006) Harmful or Not: Trichostatin A treatment of embryos generated by ICSI or ROSI, *Cent Eur J Biol*, *1*, 376-85.
[116] Wakayama, S., Jakt, M. L., Suzuki, M. et al. (2006) Equivalency of nuclear transfer-derived embryonic stem cells to those derived from fertilized mouse blastocysts, *Stem Cells*, *24*, 2023-33.
[117] Brambrink, T., Hochedlinger, K., Bell, G. & Jaenisch, R. (2006) ES cells derived from cloned and fertilized blastocysts are transcriptionally and functionally indistinguishable, *Proc Natl Acad Sci U S A*, *103*, 933-8.

In: Progress in DNA Methylation Research
Editor: Hans P. Neumann, pp. 173-188
ISBN: 978-1-60021-722-7

Chapter VI

GENOME-WIDE DNA METHYLATION PROFILING METHODS - FROM RLGS TO MIAMI

Izuho Hatada
Laboratory of Genome Science, Biosignal Genome Resource Center, Institute for Molecular and Cellular Regulation, Gunma University, Maebashi 371-8511 Japan.

ABSTRACT

The extensive genetic analysis of human diseases began more than ten years ago; however, some diseases are explained not by genetic changes but by epigenetic changes. One of the best examples is cancer. It is well recognized that cancer is caused by the transcriptional silencing of tumor suppressor genes via genetic changes; however, one important aspect of cancer genomics is the transcriptional silencing of tumor suppressor genes caused by epigenetic changes such as DNA methylation. Furthermore, epigenetic silencing is a more frequent mechanism of gene inactivation than genetic change in some cancers. Emerging evidence suggests that epigenetic changes are involved in common diseases such as schizophrenia, autism, and diabetes.

Genome-wide DNA methylation profiling has become one of the most exciting frontiers both in epigenetics and in genome sciences. It has the potential to find epigenetic mutations in common diseases. We developed the first genome-wide DNA methylation profiling technology called restriction landmark genomic scanning (RLGS) fifteen years ago. Several thousand loci in a genome can be detected simultaneously by using two-dimensional gel electrophoresis combined with methylation-sensitive restriction enzyme digestion. RLGS has successfully identified imprinted genes and several putative tumor suppressor genes. However, this method cannot be applied to all genes in the genome and requires some skill. Therefore, we developed the microarray-based integrated analysis of methylation by isoschizomers (MIAMI) to solve these problems. This method uses the methylation-sensitive restriction enzyme *Hpa* II for methylation detection and the methylation-insensitive isoschizomer *Msp* I for control to exclude effects other than methylation. Methylation differences between the samples are detected as differences in fluorescence intensities of Cy3 and Cy5 of DNA probes on a microarray. Genome tiling-arrays can be used for this method; therefore, DNA methylation can be scanned throughout the genome.

Applications of RLGS and MIAMI will be described in this chapter. A comparison with other genome-wide profiling will also be discussed.

WHAT IS EPIGNETICS?

Epigenetics is the study of heritable codes other than genetic information encoded by the nucleotides adenine (A), guanine (G), cytosine (C) and thymine (T). Epigenetic codes works as gene switches which regulate gene expression. Therefore, cells or individuals with the same genetic information can differ in phenotype. Cells in the brown and black fur of the tortoiseshell cat are an example of cells with the same genetic information but with different phenotypes. Monozygotic twins and cloned animals are examples of individuals with the same genetic information and different phenotypes.

Epigenetic information is stored using two major modifications to DNA and histone: DNA methylation, which is methylation at the C5 position of cytosine, and histone modifications, such as acetylation, methylation and phosphorylation (Bird, 2002; Jenuwein, 2001). The DNA methylation and the histone H3 methylation at K9 and K27 switch off the expression. On the other hand, DNA histone acetylation and histone H4 methylation switch on the expression. Among these modifications, DNA methylation is the best known and studied. DNA methylation is maintained by DNA methyltransferase 1 (Dnmt1), which can methylate hemimethylated DNA during DNA replication. Therefore, DNA methylation can be maintained and inherited by offspring cells. However, the state of DNA methylation can be also changed by *de novo* methylases called DNA methyltransferase 3a (Dnmt3a) and DNA methyltransferase 3b (Dnmt3b). Most DNA methylation is located at CpG sites although CpG content is relatively low in eukaryotic genomes. This can be explained by the conversion of methylated cytosine to thymine by spontaneous deamination (Fryxell and Zuckerkandl, 2000). Meanwhile, clusters of unmethylated CpG sites, called CpG islands, which are typically located in the 5' regions of genes, are observed in mammals (Klose and Bird, 2006). On the other hand, methylated CpG sites are mostly located in repetitive sequences.

Epigenetics is involved in differentiation, development, genomic imprinting, X inacitivation, chromatin stability and inactivation of the transposon. Epigenetics is also involved in cancer and thought to be involved in other diseases such as schizophrenia and diabetes.

One interesting characteristics of epigenetic is that some epigenetic information can be altered by the environment. For example in rats, offspring nursed longer were found to be more resistant to stress. This can be explained by DNA methylation and suppression of the glucocorticoid receptor gene in the offspring nursed for a short period of time (Weaver et al., 2004).

Another interesting characteristics of epigenetics is that some epigenetic mutations are known to be inherited. For example in plants, an epigenetic mutation of the Lcyc gene, which is responsible for natural variation in floral symmetry, is heritable and co-segregates with the mutant phenotype (Cubas et al., 1999). The epigenetic mutation is DNA hypermethylation which represses the expression of Lcyc. Occasionally the mutant reverts phenotypically during somatic development, correlating with demethylation of Lcyc and restoration of gene expression. In animals, DNA methylation of imprinted genes is heritable, however, it is erased and reestablished in each generation depending on the sex of the individual.

Epigenetics and Diseases

The extensive genetic analysis of human diseases began more than ten years ago; however, some diseases are explained not by genetic changes but by epigenetic changes. One of the best examples is cancer. It is well recognized that cancer is caused by the transcriptional silencing of tumor suppressor genes via genetic changes; however, one of the most exciting frontiers in cancer genomics is transcriptional silencing of tumor suppressor genes through epigenetic changes such as DNA methylation. Furthermore, epigenetic silencing is a more frequent mechanism of gene inactivation than genetic change in some cancers. For example, somatic mutations of *BRCA1* are extremely rare in sporadic breast cancer and ovarian cancers, while this gene is frequently inactivated by DNA methylation (Dobrovic et al., 1997). In addition to the hypermethylation of genes, a genome-wide decrease in methylation is an important characteristic of cancers that is associated with genomic instability and , consequently, causes tumors (Chen et al., 1998; Gaudet et al., 2003; Kanai et al., 1996).

Emerging evidence suggests that epigenetic changes are involved in diseases such as schizophrenia, autism, and diabetes. For example, reduced expression of the REELIN gene in cortical GABAergic interneurons of schizophrenia patients is a consequence of an epigenetic hypermethylation of the gene (Costa et al., 2002). Protracted administration of large doses of methyionine to mice generates some of the behavioral phenotypes observed in schizophrenia (Tremolizzo et al., 2002). This is associated with hypermethylation of the mouse reelin gene promoter. Valproate which reactivates the reelin gene corrects the schizophrenia-like epigenetic behavioral modifications induced by methionine in mice.

Most people believe epigenetic mutations are not inheritable, however, a recent paper suggests an inheritable germline epimutation of *MSH2* in a family with hereditary nonpolyposis colorectal cancer (Chan et al., 2006). From now on, we have to consider the possibility of inheritable epimutation in genetic research.

Genome-wide DNA Methylation Profiling with Restriction Landmark Genomic Scanning (RLGS)

Genome-wide DNA methylation profiling is one of the most exciting frontiers both in epigenetics and in the genome sciences. It has the potential to find epigenetic mutations in common diseases. We developed the first genome-wide DNA methylation profiling technology called restriction landmark genomic scanning (RLGS) fifteen years ago (Hatada et al., 1991). Several thousand loci in a genome can be detected simultaneously by using two-dimensional gel electrophoresis combined with restriction enzyme digestion. This method was originally developed to detect genetic changes, such as deletion in cancer (Hirotsune et al., 1992). However,it can also be applied to the detection of methylation changes when using methylation-sensitive restriction enzymes such as *Not* I (Hatada et al., 1993). RLGS has successfully identified imprinted genes (Hatada et al., 1993) and several putative tumor suppressor genes, such as *SOCS-1, ID4, BMP3B, SLC5A8, TCF-21, SFRP4, TWIST2, FRX1, ALX3, and WIT-1*(Yoshikawa et al., 2001; Yu et al., 2005; Dai et al., 2001: Li et al., 2003; Smith et al., 2006; Liu et al., 2006; Raval et al., 2005; Ohashi et al. 2004; Wimmer et al., 2002; Plass et al. 1999).

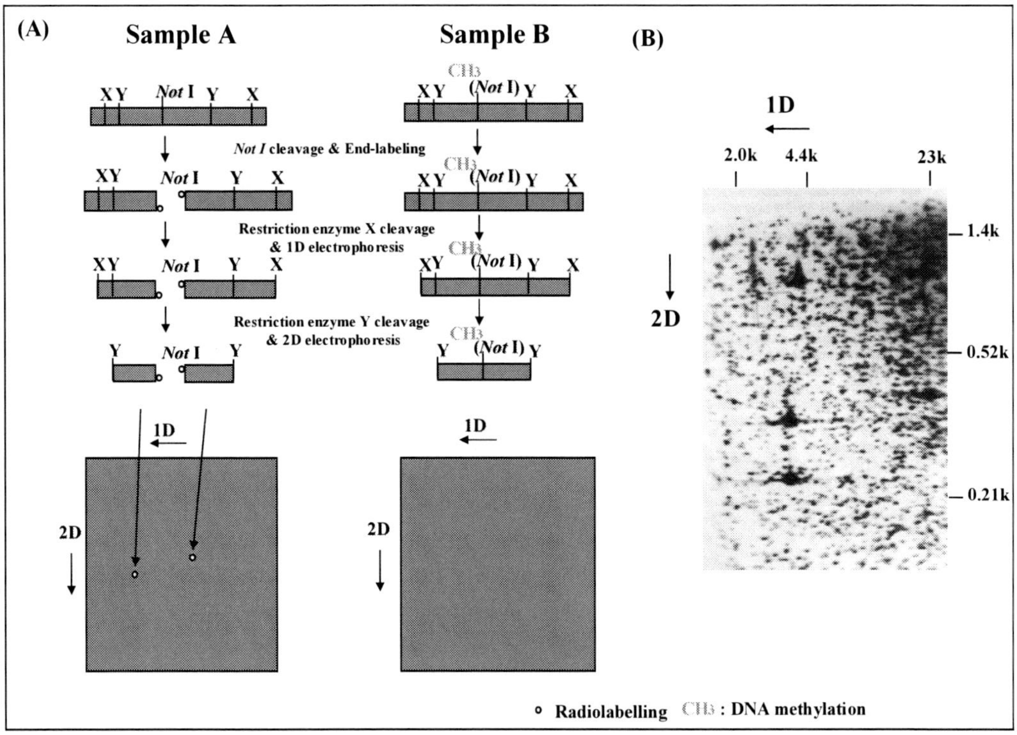

Figure 1. Restriction landmark genomic scanning (RLGS) (A) Schematic flowchart for the RLGS method. Genomic DNA is cleaved with a methylation-sensitive restriction enzyme, *Not* I and radiolabeled. The labeled fragments are subjected to one dimensional electrophoresis after cleavage with restriction enzyme X for size reduction. In-gel cleavage with restriction enzyme Y is followed by a second dimension of electrophoresis. Unmethylated *Not* I sites are detected as spots by radioautography. (B) Application of RLGS to mouse genomic DNA.

The procedure for RLGS is shown in Figure 1A. Genomic DNA is first digested with a rare-cutting methylation-sensitive restriction enzyme such as *Not* I. This enzyme can cleave the GCGGCCGC sequence when it has an unmethylated cytosine at CpG sites; however, it cannot cleave the sequence when it has a methylated cytosine at CpG sites. The cleaved ends are radiolabeled and cleaved with another restriction enzyme (restriction enzyme X) to reduce the fragment to a size appropriate for electrophoresis. One-dimensional agarose gel electrophoresis is followed by in-gel digestion with another restriction enzyme (restriction enzyme Y). After the second dimension of electrophoresis, unmethylated *Not* I sites are detected by autoradiography of the gel as spots. As shown in Figure 1B, several thousand spots can be separately detected at once from mouse genomic DNA, which means the simultaneous detection of several thousand *Not* I sites in the genome. The intensity of the spot is related to the copy number in the genome. This method is quantitative enough to detect the loss of one of two copies of the HRAS gene in a cancer (Hirotsune et al., 1992). This is because RLGS uses the direct detection of end-labeled DNA and does not use any amplification methods. When two RLGS patterns are compared, the lack of a spot or a less intense spot in one sample indicates the presence of a differentially methylated *Not* I site.

Use of RLGS to Identify Imprinted Genes

Mammals inherit two copies of every gene, one from the mother and one from the father. Most genes are expressed from both alleles and functionally equivalent. However, imprinted genes represent an exception to this rule because the expressed allele of the genes is determined by the sex of the parent (Reik et al., 2001). There are two categories of imprinted genes classified by allelic expression patterns; paternally expressed and maternally repressed genes, and maternally expressed and paternally repressed genes. These patterns are a result of epigenetic modifications such as DNA methylation, which are established in the paternal or the maternal germ line and maintained throughout somatic development in the offspring. The parental allele-specific DNA methylation controls the allele-specific expression of imprinted genes. The lethality of embryos with a single parental genome is explained by imprinted genes indispensable for development.

RLGS was used to search for imprinted genes by screening for parental allele-specific methylation. To identify parental allele-specific methylation, profiles of the F1 progeny of reciprocal crosses between two inbred strains, (DBA2 x C57BL/6)F1 and (C57BL/6 x DBA2)F1, were compared. Genes with strain-specific restriction site polymorphisms can be used for screening. If there is no parental allele-specific methylation, both the maternal and the paternal allele can be cleaved with *Not* I and detected as spots, resulting in identical profiles. On the other hand, if one of the parental allele is methylated, only one allele can be cleaved with *Not* I and detected as spots, resulting in a differential profile (Figure 2). The DNA extracted from one of these spots was cloned. This gene, called *U2af1-rs1,* was shown to be a maternally methylated and paternally expressed imprinted gene (Hatada et al., 1993).

Advantages and Disadvantages of the RLGS Method and Technical Advances

Today, there are several genome-wide DNA methylation profiling methods; however, RLGS still offers several advantages. One is its high quantitative power. RLGS is quantitative enough to detect the loss of one of the two alleles of the HRAS gene in a cancer as described (Hirotsune et al., 1992) because it uses the direct detection of end-labeled DNA and does not use an amplification procedure, such as PCR. Another advantage is that it can be applied without any information on an organism's genome such as DNA sequences. For example, RLGS was successfully used to analyze *Citrus Unshiu* (Shimizu et al., 2006). One disadvantage of RLGS method is that it requires spot cloning to obtain the DNA sequence information. However, recent technical advances make it possible to obtain DNA sequence information for each spot using a computer simulated virtual pattern of a two-dimensional gel generated from complete genome information (Matsuyama et al., 2003; Zardo et al., 2002). An integrated genomic and epigenetic analysis can be performed with a combination of high-resolution deletion maps from microarray-based comparative genomic hybridization and RLGS-based methylation analysis using the sequence information for spots (Zardo et al., 2002). Certain subsets of gene-associated CpG islands are preferentially affected by convergent methylation and deletion, including genes that exhibit tumor suppressor activity, such as *SOCS1*. This analysis showed that most aberrant methylation events were focal and independent of deletions, and the rare convergence of these mechanisms can pinpoint biallelic gene inactivation without the use of positional cloning.

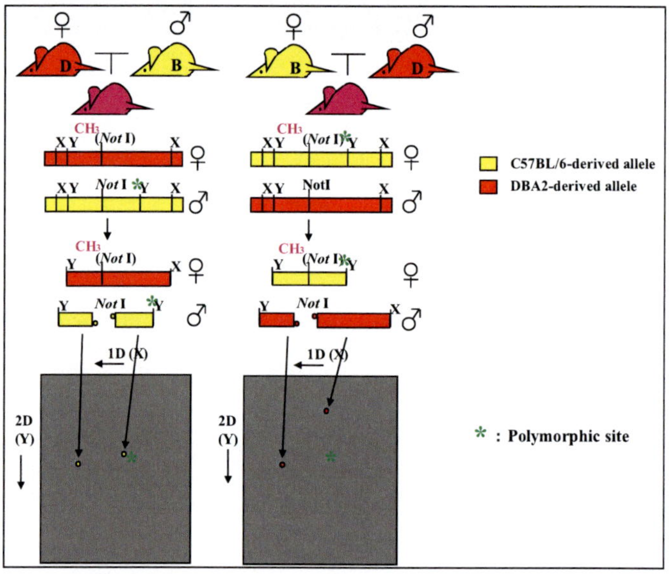

Figure 2. Expected spot profiles derived from a polymorphic gene with maternal methylation. (DBA2 x C57BL/6)F1 and (C57BL/6 x DBA2)F1 were subjected to RLGS. The maternal allele was methylated and only the paternal allele can be detected. Therefore, spot profiles of F1 progeny of reciprocal crosses are different.

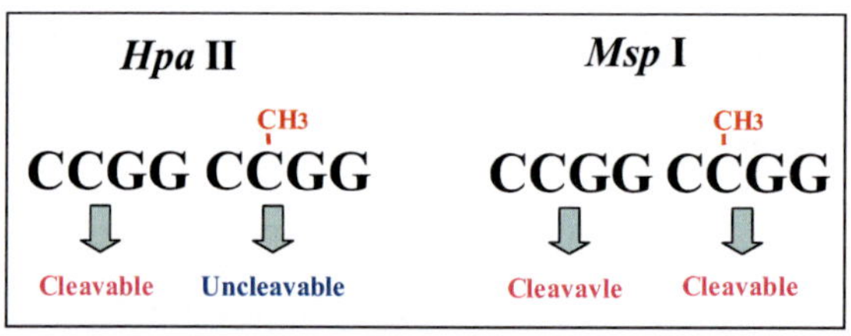

Figure 3. *Hpa* II and *Msp* I. *Hpa* II can cleave unmethylated CCGG sites while it cannot cleave methylated CCGG sites. *Msp* I can cleave both unmethylated and methylated CCGG sites.

MICROARRAY-BASED INTEGRATED ANALYSIS OF METHYLATION BY ISOSCHIZOMERS (MIAMI)

RLGS cannot be applied to the all genes in a genome because only a portion have *Not* I sites and it requires some skill to perform. Therefore, we developed a microarray-based integrated analysis of methylation by isoschizomers (MIAMI) to resolve these problems (Hatada et al., 2006). This method uses a methylation-sensitive restriction enzyme, *Hpa* II, for methylation detection and a methylation-insensitive isoschizomer, *Msp* I, for a control to exclude effects other than methylation. *Hpa* II recognizes and cleaves the unmethylated CCGG sequence; however, it can not cleave the same sequence when the internal C is methylated. On the other hand, an isoschizomer called *Msp* I, cleaves both methylated and unmethylated CCGG sequences (Figure 3). One of the problems in detecting methylation by methylation-sensitive restriction enzymes is the risk of false positives derived from restriction site polymorphism and/or incomplete digestion of one of the samples (Figure 4A). This is because a cleavage difference of the methylation-sensitive enzyme (*Hpa* II) can be derived from both the difference of methylation and a difference other than methylation such as polymorphism and/or imcomplete digestion. If a cleavage difference of the methylation-insensitive isoschizomer (*Msp* I), which is derived only from the difference other than methylation, is also used for analysis, such problems can be solved because *Hpa* II cleavage difference other than methylation can be estimated as *Msp* I cleavage difference. Both the *Hpa* II cleavage difference and *Msp* I cleavage difference are used for MIAMI (Figure 4B). For example, in the case of a restriction site polymorphism at the *Hpa* II site, cleavage by the methylation-insensitive *Msp* I will also differ between samples because both enzymes recognize the same recognition site. Such changes are treated as false positives (Figure 4B). The procedure for MIAMI is presented in Figure 5. The *Hpa* II cleavage difference is detected as follows. *Hpa* II digestion is followed by adaptor ligation, and several cycles of PCR (1st PCR) result in the amplification of regions located between unmethylated *Hpa* II sites. At this stage, only DNA fragments that have methylated *Hpa* II sites before the PCR retain internal *Hpa* II (*Msp* I) sites. Therefore, *Msp* I digestion followed by a 2nd PCR amplification makes it impossible to amplify these methylated fragments. Amplified and enriched unmethylated DNA fragments from two samples are labeled with Cy3 and Cy5, respectively, and co-hybridized to an oligonucleotide microarray made using inkjet technology. The log ratios of fluorescence intensities are used as the *Hpa* II cleavage difference. On the other hand, an identical procedure except for *Msp* I digestion in the first step is performed for the detection of the *Msp* I cleavage difference. MIAMI was successfully used to compare a lung cancer cell line and a normal lung. As a result, 5.7% of the promoters of the genes were found to be hypermethylated in the lung cancer cells compared to normal lung (Hatada et al., 2006). This frequency is higher than most reports, suggesting high sensitivity (Yan et al., 2001). One of the hypermethylated genes, *CIDEB* (Celol death-inducing DFFA-like effector b), was found to be hypermethylated in 71% of primary lung cancers. The *CIDEB* product activates appotosis and is located at 14q11, where LOH frequently occurs in lung cancers (Abujiang et al., 1998). MIAMI also revealed 0.6% of the promoters to be hypomethylated in the same lung cancer cell line. An analysis of these sequences revealed that hypomethylated sequences are significantly low in CpG content, suggesting CpG-poor promoters are sensitive to demethylation. It is intriguing to think that methylated CpG-poor promoters are free from a methyl-CpG-binding protein, which binds to methylated CpG-rich sequences, and is attacked by demethylation activity. MIAMI was successfully applied to the detection of methylation changes in primary lung cancers, neural differentiation and schizophrenia (in preparation).

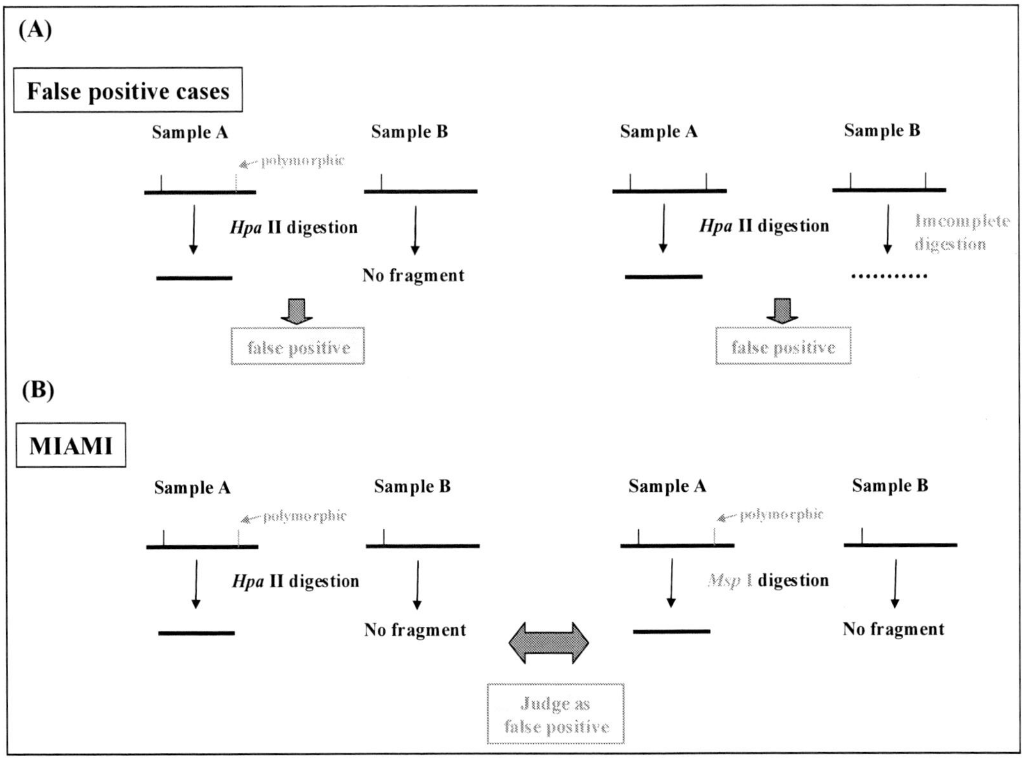

Figure 4. (A) False positives resulting from polymorphic *Hpa* II sites or incomplete digestion of one of the samples. (B) With MIAMI method, cleavage differences other than methylation can be judged as false positves by using the *Msp* I cleavage difference.

Genome Coverage of *Hpa* II Sites in Human Genome

One disadvantage of RLGS is that the *Not* I sites used for detecting methylation do not cover the genome and cover only a portion of the genes. In contrast to *Not* I which recognizes an 8-base sequence, *Hpa* II recognizes a 4-base sequence, suggesting its sites are abundant in genomes. Figure 6A shows the number of *Hpa* II sites in the human genome. There are 2,300,000 *Hpa* II sites in the human genome (Figure 6B) and the average length of *Hpa* II fragments is 1.4 kb. Therefore, *Hpa* II sites are abundant enough to cover the genome.

Whether the distribution of *Hpa* II sites is the same as that of CpG sites, where most methylation occurs, is an important issue. Figure 6B shows that the distribution of *Hpa* II sites is the same as that of CpG sites.

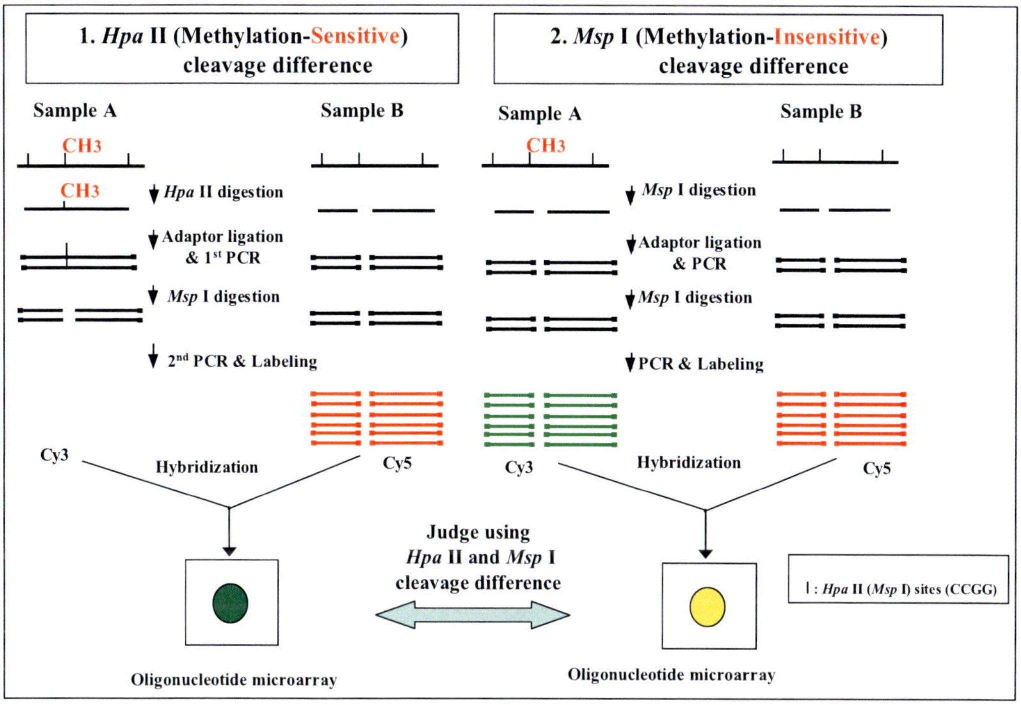

Figure 5. Schematic flowchart for the MIAMI method. Both the *Hpa* II cleavage difference and *Msp* I cleavage difference are used for judging the methylation difference between sample A and sample B. For the *Hpa* II cleavage difference, *Hpa* II-digested genomic DNA is ligated to an adaptor, and subjected to several cycles of PCR (1st PCR). Among the amplified DNA fragments, only the one derived from methylated *Hpa* II sites have internal *Hpa* II. Therefore, digestion with *Msp* I followed by a 2nd main PCR results in amplification of unmethylated DNA. Amplified DNA fragments from two samples are labeled with Cy3 and Cy5, respectively and co-hybridized to an oligonucleotide microarray. For the *Msp* I cleavage difference, the identical procedure is performed except for *Msp* I digestion of genomic DNA in the first step.

Oligonucleotide Microarray Platforms

Recent advances in technology make it possible to commercially supply several microarray platforms, such as photolithographic masked arrays of Affymetrix, photolithographic adaptive optics arrays of NimbleGen, Inkjet arrays of Agilent technology and bead arrays of Illumina. The Agilent Human promoter array covers from 8 kb upstream to 2kb downstream of the transcriptional start sites of the genes. The Affymetrix Human promoter array covers 7.5 kb upstream to 2.4 kb downstream. The custom built array made by an Agilent technology was used for MIAMI. Promoter arrays supplied by Agilent technology are also applicable to MIAMI (unpublished results of I.H.). Agilent technology supplies a useful web-based tool called eArray (http://earray.chem.agilent.com/earray/login.do). This tool enables one to create custom microarrays using oligonucleotide probes tiling several genomes including human, mouse, rat, drosoophila, C elegans, yeast and Arabidopsis. We can choose from these tiling probes and make an original microarray very easily.

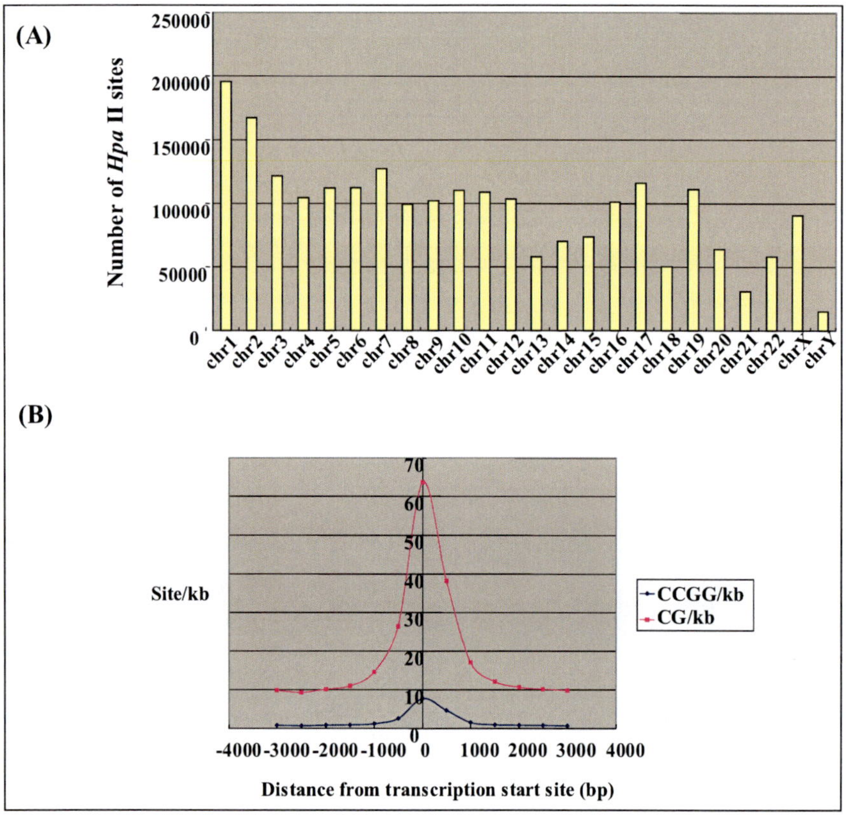

Figure 6. (A) The number of *Hpa* II sites in each chromosome of the human genome. (B) The distribution of *Hpa* II sites and CpG sites from transcriptional start sites.

Methylation Analysis by DNA Immunoprecipitation (MeDIP)

In addition to methylation-sensitive resitriction enzymes, immunoprecipitation with 5-methylacytosine-specific antibody can be used for detection of methylation. Combining of this detection system with a microarray makes it possible to perform genome-wide methylation profiling (Weber et al., 2005). In this method called MeDIP, methylated DNA fragments are enriched by immunoprecipitation with a 5-methylcytosine-specific antibody after sonication and denaturation (Figure 7). Input DNA and immunoprecipitated DNA are labeled with Cy5 and Cy3, respectively, and co-hybridized to a microarray with a CpG island library or a PCR-amplified DNA. The methylation ratio of the genes is calculated from the Input DNA and immunoprecipitaed DNA. MeDIP was applied to a cancer cell line and 0.5% of the genes were found to be hypermethylated compared to those in normal cells. The application of this method to a cancer cell line using a BAC array revealed that the methylation profile of cancer cells was similar to that of normal cells; nevertheless, large genomic segments with hypomethylation in cancer cells resided in gene-poor areas (Weber et al., 2005).

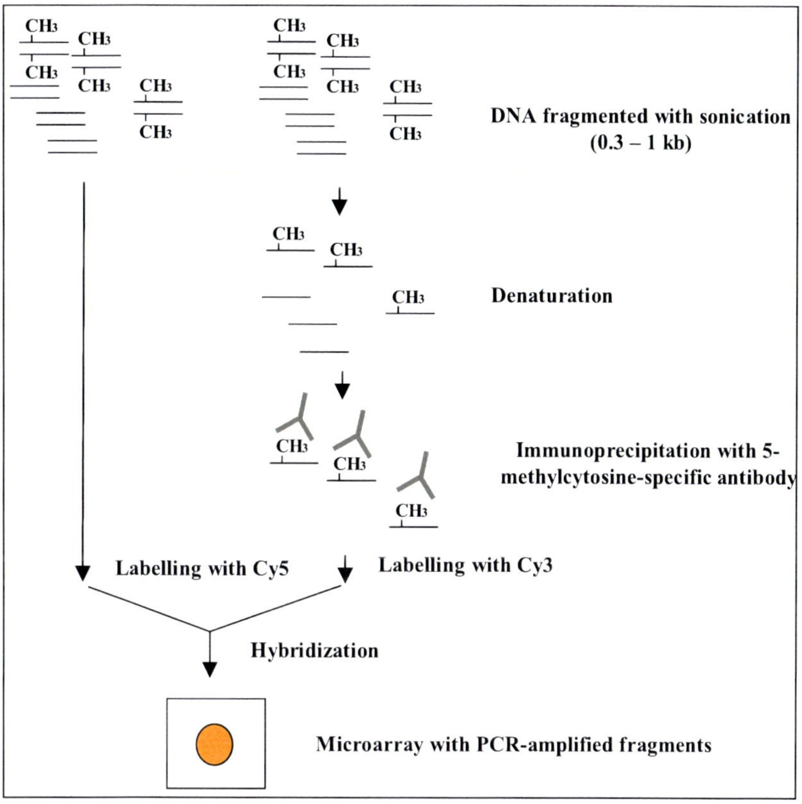

Figure 7. Schematic flowchart for the MeDIP method. Sonicated genomic DNA is denatured and immunoprecipitated with 5-methylcytosine-specific antibody. Input DNA and enriched methylated DNA are labeled with Cy5 and Cy3, respectively, and are cohybridized to a microarray with BAC or CpG island library clones.

Comparison of the MIAMI and MeDIP Methods

Coverage of the genome is an important aspect for selecting a genome-wide profiling method. In this point of view, the methylation detection system and applicability to oligonucleotide probe microarray which can tile the genome were compared between MIAMI and MeDIP. An advantage of the MeDIP method is that, theoretically, all methylated CpG sites can be detected; however, in practical, CpG sites located within 1kb cannot be detected separately because the length of the immunoprecipitation DNA is 0.3-1kb. As for oligonucleotide probe microarrays, MIAMI is applicable, however, it is not clear whether MeDIP can be applied or not. Whether methylated DNA immunoprecipitated with a 5-methylcytosine-specific antibody is applicable to oligonucleotide arrays or not was preliminarily examined (Figure 8, unpublished results of I.H.). Identical 60-mer oligonucleotide microarrays made by Agilent technology using inkjet technology were used for hybridization either with the MIAMI target DNA or with the 5-methylcytosine-specific antibody precipitated target DNA. The methylation of undifferentiated and differentiated P19 embryonic carcinoma cells was compared. Most of the hypomethylated genes detected with the MIAMI method did not show methylation differences with the 5-methylcytosine-specific antibody-precipitated target DNA. Some of the results were confirmed using the combined bisulfite restriction assay (COBRA, Warnecke et al., 1997). These result suggest that 5-methylcytosine-specific antibody-precipitated DNA is not suitable for oligonucleotide probe analysis. Methylated repetitive sequences may interfere with the hybridization results in the MeDIP method. The MeDIP method uses methylated DNA for a target while the MIAMI method uses unmethylated DNA. Most methylated CpG sites are located in repetitive sequences while most unmethylated CpG sites are located near the genes. Therefore, it is easy to analyze the unmethylated CpG sites compared to methylated CpG sites, which are diluted with abundant methylated repetitive sequences. Therefore, the background of MIAMI should be lower than MeDIP. By comparing the amount of unmethylated DNA for each gene between two samples, one can determine the methylation difference between the samples in each gene. Hypermethylation of a gene in one sample and not in the other will produce a loss of unmethylated DNA for the gene in one sample and a presence in the other.

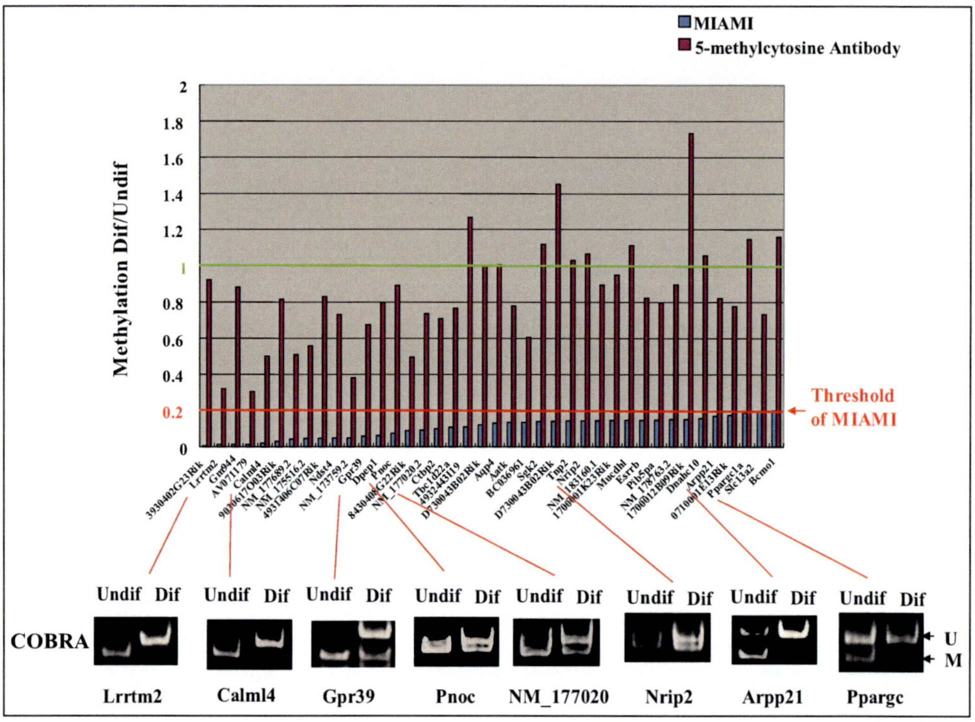

Figure 8. Comparison of MIAMI and the 5-methylcytosine-specific antibody-based method applied to an oligonucleotide microarray made by using Agilent ink-jet technology. Either MIAMI target DNA or 5-methylcytosine-specific antibody-precipitated target DNA were hybridized to identical 60-mer oligonucleotide microarrays. The methylation of undifferentiated (Undif) and differentiated (Dif) P19 embryonic carcinoma cells was compared. The methylation difference of hypomethylated genes in differentiated P19 cells was compared with that of the 5-methylcytosine-specific antibody-based method. Most of the hypomethylated genes detected with the MIAMI method did not show methylation difference in the 5-methylcytosine-specific antibody based method. Some of the results were confirmed by combined bisulfite restriction assay (COBRA).

Other Methylation Detection Methods to Compensate for Genome-wide Profiling

Genome-wide profiling is useful for the screening of genes, however, other methods of detecting methylation applicable to a single gene provide more precise data for that gene. Therefore, it is appropriate to use these methods for detailed analyses of genes selected by genome-wide profiling. For this purpose, bisulfite-based methods are useful because they can detect all methylated cytosines. The bisulfite reaction, which converts unmethylated cytosine to uracil, but does not convert methylated cytosine, was found by Hayatsu thirty years ago (Hayatsu et al., 1970). After PCR, uracil is converted to thymine. Therefore, unmethylated cytosine is read as thymine and methylated cytosine is read as cytosine by sequencing. In a bisulfite-based method called bisulfite sequencing, the PCR products are cloned for sequencing. By sequencing several clones, the methylation of each cytosine is calculated. A problem with this method is the bias derived from cloning. This problem is resolved in other methods called PyroMeth (Uhlmann et al., 2002) and MALDI-TOF mass spectrometry (Schatz et al., 2004) which can determine the methylated cytosine ratio directly from PCR products without cloning. PyroMeth involves a direct DNA sequencing of PCR-amplified bisulfite-treated DNA by pyrosequencing. Pyrosequencing is a sequencing-by-synthesis method in which a cascade of enzymatic reactions yields detectable light, which is proportional to incorporated nucleotides. Essentially, the method allows sequencing of a single strand of DNA by synthesizing the complementary strand along it. Each time a nucleotide, A, C, G or T is incorporated into the growing chain a cascade of enzymatic reactions is triggered which results in a light signal. A disadvantage of pyrosequencing is that it can be only used for sequencing of short stretches of DNA. This situation was improved using MALDI-TOF mass spectrometry (Schatz et al., 2004). In this method, bisulfite-treated DNA is PCR-amplified using a T7 RNA polymerase promoter-tagged primer. Uracil-specific cleaved RNA transcript from the PCR product was analyzed by MALDI-TOF mass spectrometry to give a characteristic mass signal pattern depending on methylation.

References

Abujiang P, Mori TJ, Takahashi T, Tanaka F, Kasyu I, Hitomi S, Hiai H. Loss of heterozygosity (LOH) at 17q and 14q in human lung cancers. *Oncogene*. 1998 Dec 10;17(23):3029-33.

Bird A. DNA methylation patterns and epigenetic memory. *Genes Dev*. 2002; 16(1):6-21.

Chan TL, Yuen ST, Kong CK, Chan YW, Chan AS, Ng WF, Tsui WY, Lo MW, Tam WY, Li VS, Leung SY. Heritable germline epimutation of MSH2 in a family with hereditary nonpolyposis colorectal cancer. *Nat Genet*. 2006; 1178-1183.

Chen RZ, Pettersson U, Beard C, Jackson-Grusby L, Jaenisch R. DNA hypomethylation leads to elevated mutation rates. *Nature*. 1998; 395(6697):89-93.

Costa E, Chen Y, Davis J, Dong E, Noh JS, Tremolizzo L, Veldic M, Grayson DR, Guidotti A. REELIN and Schizophrenia: A Disease at the Interface of the Genome and the Epigenome. *Mol Interv*. 2002; 1:47-57.

Cubas P, Vincent C, Coen E. An epigenetic mutation responsible for natural variation in floral symmetry. *Nature*. 1999; 401:157-161.

Dai Z, Lakshmanan RR, Zhu WG, Smiraglia DJ, Rush LJ, Fruhwald MC, Brena RM, Li B, Wright FA, Ross P, Otterson GA, Plass C. Global methylation profiling of lung cancer identifies novel methylated genes. *Neoplasia*. 2001; 3(4):314-323.

Dobrovic A, Simpfendorfer D. Methylation of the BRCA1 gene in sporadic breast cancer. *Cancer Res*. 1997; 57(16):3347-3350.

Fryxell KJ, Zuckerkandl E. Cytosine deamination plays a primary role in the evolution of mammalian isochores. *Mol Biol Evol*. 2000;17(9):1371-1383.

Gaudet F, Hodgson JG, Eden A, Jackson-Grusby L, Dausman J, Gray JW, Leonhardt H, Jaenisch R. Induction of tumors in mice by genomic hypomethylation. *Science*. 2003; 300(5618):489-492.

Hatada I, Hayashizaki Y, Hirotsune S, Komatsubara H, Mukai T. A genomic scanning method for higher organisms using restriction sites as landmarks. *Proc Natl Acad Sci U S A*. 1991; 88(21):9523-:9527.

Hatada I, Sugama T, Mukai T. A new imprinted gene cloned by a methylation-sensitive genome scanning method. *Nucleic Acids Res*. 1993;21(24):5577-5582.

Hatada I, Fukasawa M, Kimura M, Morita S, Yamada K, Yoshikawa T, Yamanaka S, Endo C, Sakurada A, Sato M, Kondo T, Horii A, Ushijima T, Sasaki H. Genome-wide profiling of promoter methylation in human. *Oncogene*. 2006; 25(21):3059-3064.

Hayatsu H, Wataya Y, Kazushige K. The addition of sodium bisulfite to uracil and to cytosine. *J Am Chem Soc*. 1970; 92(3):724-726.

Hirotsune S, Hatada I, Komatsubara H, Nagai H, Kuma K, Kobayakawa K, Kawara T, Nakagawara A, Fujii K, Mukai T, et al. New approach for detection of amplification in cancer DNA using restriction landmark genomic scanning. *Cancer Res*. 1992 52:3642-3647.

Jenuwein T. Re-SET-ting heterochromatin by histone methyltransferases. *Trends Cell Biol*. 2001; 11(6):266-273.

Kanai Y, Ushijima S, Tsuda H, Sakamoto M, Sugimura T, Hirohashi S. Aberrant DNA methylation on chromosome 16 is an early event in hepatocarcinogenesis. *Jpn J Cancer Res*. 1996; 87(12):1210-1217.

Klose RJ, Bird AP. Genomic DNA methylation: the mark and its mediators. *Trends Biochem Sci*. 2006; 31(2):89-97. in press.

Li H, Myeroff L, Smiraglia D, Romero MF, Pretlow TP, Kasturi L, Lutterbaugh J, Rerko RM, Casey G, Issa JP, Willis J, Willson JK, Plass C, Markowitz SD. SLC5A8, a sodium transporter, is a tumor suppressor gene silenced by methylation in human colon aberrant crypt foci and cancers. *Proc Natl Acad Sci U S A*. 2003; 100(14):8412-8417.

Liu TH, Raval A, Chen SS, Matkovic JJ, Byrd JC, Plass C. CpG island methylation and expression of the secreted frizzled-related protein gene family in chronic lymphocytic leukemia. *Cancer Res*. 2006; 66(2):653-658.

Matsuyama T, Kimura MT, Koike K, Abe T, Nakano T, Asami T, Ebisuzaki T, Held WA, Yoshida S, Nagase H. Global methylation screening in the Arabidopsis thaliana and Mus musculus genome: applications of virtual image restriction landmark genomic scanning (Vi-RLGS). *Nucleic Acids Res*. 2003;31: 4490-4496.

Ohashi Y, Ueda M, Kawase T, Kawakami Y, Toda M. Identification of an epigenetically silenced gene, RFX1, in human glioma cells using restriction landmark genomic scanning. *Oncogene*. 2004; 23(47):7772-7779.

Plass C, Yu F, Yu L, Strout MP, El-Rifai W, Elonen E, Knuutila S, Marcucci G, Young DC, Held WA, Bloomfield CD, Caligiuri MA. Restriction landmark genome scanning for aberrant methylation in primary refractory and relapsed acute myeloid leukemia; involvement of the WIT-1 gene. *Oncogene*. 1999; 18(20):3159-3165.

Raval A, Lucas DM, Matkovic JJ, Bennett KL, Liyanarachchi S, Young DC, Rassenti L, Kipps TJ, Grever MR, Byrd JC, Plass C. TWIST2 demonstrates differential methylation in immunoglobulin variable heavy chain mutated and unmutated chronic lymphocytic leukemia. *J Clin Oncol*. 2005; 23(17):3877-3885.

Reik W, Dean W, Walter J. Epigenetic reprogramming in mammalian development. *Science*. 2001; 293:1089-93.

Schatz P, Dietrich D, Schuster M. Rapid analysis of CpG methylation patterns using RNase T1 cleavage and MALDI-TOF. *Nucleic Acids Res*. 2004; 32(21):e167.

Shimizu T, Fujii H, Shimada T, Endo T, Omura M. Restriction Landmark Genomic Scanning Reveals DNA Diversity And Phylogeny In Satsuma Mandarin Cultivars anuary 14-18, 2006 San Diego, CA

Smith LT, Lin M, Brena RM, Lang JC, Schuller DE, Otterson GA, Morrison CD,Smiraglia DJ, Plass C. Epigenetic regulation of the tumor suppressor gene TCF21 on 6q23-q24 in lung and head and neck cancer. *Proc Natl Acad Sci U S A*. 2006; 103(4):982-987.

Tremolizzo L, Carboni G, Ruzicka WB, Mitchell CP, Sugaya I, Tueting P, Sharma R, Grayson DR, Costa E, Guidotti A. An epigenetic mouse model for molecular and behavioral neuropathologies related to schizophrenia vulnerability. *Proc Natl Acad Sci U S A*. 2002;99:17095-17100.

Uhlmann K, Brinckmann A, Toliat MR, Ritter H, Nurnberg P. Evaluation of a potential epigenetic biomarker by quantitative methyl-single nucleotide polymorphism analysis. *Electrophoresis*. 2002; 23(24):4072-4079.

Warnecke PM, Stirzaker C, Melki JR, Millar DS, Paul CL, Clark SJ. Detection and measurement of PCR bias in quantitative methylation analysis of bisulphite-treated DNA. *Nucleic Acids Res*. 1997;2 5(21):4422-4426.

Weaver IC, Cervoni N, Champagne FA, D'Alessio AC, Sharma S, Seckl JR, Dymov S, Szyf M, Meaney MJ. Epigenetic programming by maternal behavior. *Nat Neurosci*. 2004;7:847-854.

Weber M, Davies JJ, Wittig D, Oakeley EJ, Haase M, Lam WL, Schubeler D. Chromosome-wide and promoter-specific analyses identify sites of differential DNA methylation in normal and transformed human cells. *Nat Genet*. 2005; 37(8):853-862.

Wimmer K, Zhu Xx XX, Rouillard JM, Ambros PF, Lamb BJ, Kuick R, Eckart M, Weinhausl A, Fonatsch C, Hanash SM. Combined restriction landmark genomic scanning and virtual genome scans identify a novel human homeobox gene, ALX3, that is hypermethylated in neuroblastoma. *Genes Chromosomes Cancer*. 2002; 33(3):285-294.

Yan PS, Chen CM, Shi H, Rahmatpanah F, Wei SH, Caldwell CW, Huang TH. Dissecting complex epigenetic alterations in breast cancer using CpG island microarrays. *Cancer Res*. 2001; 61(23):8375-8380.

Yoshikawa H, Matsubara K, Qian GS, Jackson P, Groopman JD, Manning JE, Harris CC, Herman JG. SOCS-1, a negative regulator of the JAK/STAT pathway, is silenced by methylation in human hepatocellular carcinoma and shows growth-suppression activity. *Nat Genet*. 2001; 28(1):29-35.

Yu L, Liu C, Vandeusen J, Becknell B, Dai Z, Wu YZ, Raval A, Liu TH, Ding W, Mao C, Liu S, Smith LT, Lee S, Rassenti L, Marcucci G, Byrd J, Caligiuri MA, Plass C. Global assessment of promoter methylation in a mouse model of cancer identifies ID4 as a putative tumor-suppressor gene in human leukemia. *Nat Genet*. 2005; 37(3):265-274.

Zardo G, Tiirikainen MI, Hong C, Misra A, Feuerstein BG, Volik S, Collins CC, Lamborn KR, Bollen A, Pinkel D, Albertson DG, Costello JF. Integrated genomic and epigenomic analyses pinpoint biallelic gene inactivation in tumors. *Nat Genet*. 2002; 32(3):453-458.

In: Progress in DNA Methylation Research
Editor: Hans P. Neumann, pp. 189-202
ISBN: 978-1-60021-722-7

Chapter VII

EPIGENETIC REGULATION OF THE GENOME DURING NEURODEVELOPMENT AND DISEASE

Paromita Deb-Rinker and P. Roy Walker
Neurobiology Program, Institute for Biological Sciences, National Research Council Canada, Ottawa, Canada K1A 0R6.

ABSTRACT

Neural differentiation is a complex process involving the differential expression of several thousand genes. Most of these genes have been identified in recent years by genome-wide microarray analysis. The challenge now is to understand the mechanisms by which the genome orchestrates this process. Although transcription factor networks play a key role, it is now evident that other levels of regulation are also involved. Of these, epigenetic modifications of DNA and histones is emerging as a major component, not just in terms of gene silencing, but also in the differential expression of developmental genes. To better understand the mechanisms involved in the differentiation of neurons and astrocytes, we initially identified key gene clusters through microarray analysis, such as genes playing a role in RA binding, embryogenesis, regulation of neural fate, differentiation of neurons, etc. Using QPCR to supplement the microarray data we have categorized these genes into stage-specific groups, such as precursor-specific, progenitor cell state-specific, astrocyte-specific and neuron-specific. A common aspect of these genes is that many of them are subject to epigenetic control. In this chapter we will review DNA methylation, particularly in the context of neurodevelopment and also indicate to what extent these marks in the genome become disrupted in disease states.

Introduction

Background

In 1925, Johnson and Coghill were the first to demonstrate the existence of a fifth nucleotide, 5-methyl-deoxycytidine (5-mC), in the nucleic acid of the Tubercle Bacillus. They identified it as a product of hydrolysis when "tuberculinic acid" (the nucleic acid of Tubercle Bacillus), was digested with sulfuric acid. Later, in 1948, Hotchkiss while using paper chromatography to separate purines and pyrimidines from nucleic acids of calf thymus, also observed what he designated 'epicytosine', a minor constituent that migrated at a rate slightly higher than cytosine. Furthermore, he reported that the absorption characteristics of this fraction resembled those of cytosine; however it was distinct from cytosine and was not uracil. Subsequently, Kelly and Smith (1970) discovered that the specificity of some restriction endonucleases depended on whether the DNA was methylated or not and that these methylation-sensitive restriction enzymes could be used to study the methylation status of any given DNA sequence.

From a biological perspective, Scarano (1973) was one of the first scientists to hypothesize that DNA methylation would lead to alterations of gene activity and could be involved in the differentiation process. During the 1970s, the role of DNA methylation in biological phenomena became widely discussed among leading scientists (Holliday and Pugh, 1975; Holliday, 2006; Riggs, 1975). Further elaboration on the concept in the 1980s brought about the description of the CpG island (Bird et al., 1985), its definition (Gardiner-Garden and Frommer, 1987), and its role in modulating gene expression (Toniolo et al., 1988).

The discovery of 5-mC has therefore provoked a challenging search for its functional significance and alerted the scientific community of the importance of DNA methylation in eukaryotic genomes. This search is by no means completed, and investigations on numerous unsolved questions are still continuing. Some of the aspects being investigated are the pattern of DNA methylation in the genome and whether the distribution of 5-mC evolves over time, whether methylation patterns are specific to different cell types and how they are maintained between individuals in a given species, in what way do these patterns modulate chromatin structure and promote or deter protein-DNA interactions etc.

The Methylation of DNA

DNA methylation is the covalent addition of a methyl group to the 5-carbon position of cytosine residues (Figure 1) located in CpG dinucleotides, catalyzed by DNA methyltransferases (DNMTs). The modification of cytosine (C) to 5-mC is the only modification among the nucleotides in mammalian DNA (Lu, 2006). The CpG dinuleotide is not randomly distributed throughout the genome and not all the cytosines within CpGs are modified, giving rise to specific patterns, thereby producing a mechanism for the creation of specific epigenomes. Each epigenome has a unique information content, embodied in the methylation pattern, which in turn, directs a unique expression pattern. One of the most striking features of DNA methylation is the presence of CpG islands. These are defined as regions of DNA of greater than 500 bp with a G+C equal to or greater than 55% and an observed/expected CpG ratio of 0.65 or greater. Typically, these unmethylated GC-rich regions are located at the 5' ends of many human genes (Figure 2). Approximately 60% of human genes are associated with CpG islands, of which the majority appear to be unmethylated at all stages of development and in all tissue types. All housekeeping genes are associated with CpG islands. However, not all islands are associated with housekeeping genes. Because many CpG islands are located at genes that have a tissue-restricted expression pattern, CpG islands can remain methylation-free even when their associated gene is silent. Well-studied examples of tissue-restricted genes which are associated with CpG islands are the human α-globin genes (Antequera and Bird 1993). The covalent modification of cytosines in DNA by methylation, or the removal of this modification is a binary switch which typically leads to either the repression of a gene, often resulting in silencing, or the induction of a previously unexpressed gene.

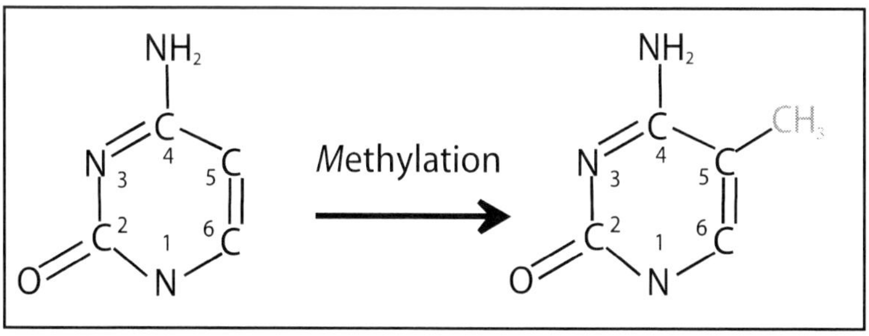

Figure 1. DNA Methylation is the covalent addition of a methyl group, catalyzed by DNA methyltransferase (DNMT), to the 5-carbon of cytosine in a CpG dinucleotide.

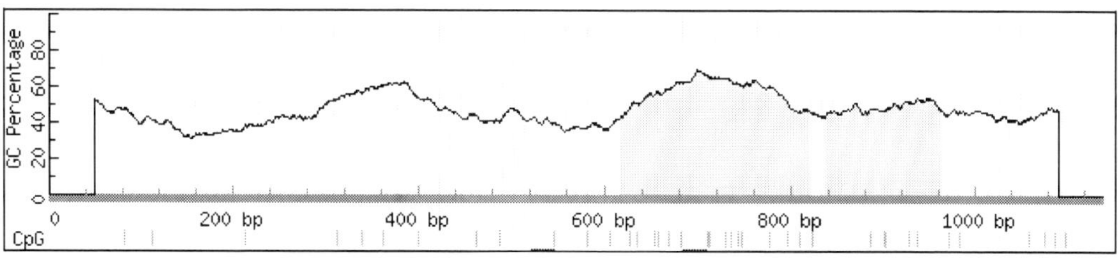

Figure 2. Representative map of CpG islands in the genome. CpG sites are shown as red lines and islands as blue blocks. This figure is an output of MethPrimer (Li and Dahiya, 2002).

Biochemically, there are two different kinds of DNA methylation – maintenance methylation and *de novo* methylation. Maintenance DNA methylation is a post-replication event, since the newly replicated DNA strand is initially unmethylated. DNA methyl transferases recognize sites where the parental strand is methylated and the daughter strand is not (hemi-methylated DNA) and tend to methylate the daughter strand. In this type of methylation, the parental DNA strand with the 5-mC residue still in place, can serve as the template to direct the DNMT to modify the newly synthesized DNA (Doerfler 2005). Subsequently, the information content of the parental genome is transmitted across the cell cycle to both daughter strands.

When a site that was not previously methylated becomes methylated, the process is known as *de novo* methylation. Methyltransferases, DNMT3A and DNMT3B are devoted to *de novo* methylation, whereas DNMT1 acts mainly as a maintenance methyltransferase, as described above. These enzymes produce a mechanism by which information can be inherited at either the level of the organism or across the cell cycle during development. It involves a modification to the genetic material as opposed to an outright mutation. In this way, the information content of the genome is modified without an overt change in DNA base sequence, permitting the same genetic content to generate numerous stable phenotypes (see (Holliday, 2006) for a historical perspective and important concepts and definitions associated with epigenetics and DNA methylation). Not surprisingly, therefore, *de novo* methylation and post-replication DNA methylation play a major role during development and differentiation. DNA methylation is also a mechanism by which the environment of a cell can stably modify its genome. Thus, during development, transient changes in the cellular environment (e.g. exposure to hormones or developmental signals) can commit cells to a particular phenotype. This is now recognized as a fundamental mechanism of development. An initial pattern of DNA methylation is gradually established during early embryogenesis. There is *de novo* methylation in the developing germ cells, but after fertilization, when the zygote starts to divide; extensive demethylation of both the paternal and maternal genomes takes place. DNMT1 is excluded from the nucleus at this stage. In the developing pre-implantation embryo, *de novo* methyltransferases re-establish DNA methylation during implantation and germ layer and subsequent cell-type specific differentiation (Wu and Sun, 2006). Thus DNA methylation dynamically changes during development. However, it is stably inherited through cell divisions and not changed by short term changes of environmental signals. DNA methylation changes during the course of aging - it is known that DNA methylation decreases with age, particularly in the brain (Klein et al., 1991).

Functional Significance of DNA Methylation

Fundamentally, DNA methylation modulates the expression of the genome. Initially, this was considered to be a mechanism to irreversibly block the ability of specific regions to be expressed. Examples of such repression are silencing and imprinting. More recently, it has become apparent that DNA methylation plays a much more dynamic role in the control of gene expression, particularly during development.

It is long established that in mammals, DNA methylation is one of the main players of long-term gene silencing involved in important biological phenomena, such as X chromosome inactivation and genomic imprinting (Li, 2002). Mammalian X-chromosome inactivation is an example of epigenetic gene regulation, where the majority of genes on one of the X chromosomes in the female are inactivated. DNA hypermethylation is observed at the 5' region of genes on the inactive X chromosome, however, methylation is assumed to be a late event in the inactivaton process. It is thought to be an important factor in maintaining the inactive state rather than initiating it (Chang et al., 2006). Alleles on the inactive X chromosome are often methylated in promoter regions, while alleles on the active X chromosome are normally unmethylated (Grant and Chapman, 1988).

Genomic imprinting is an epigenetic modification that inactivates one allele of a gene in a parent-of-origin dependent manner. It occurs primarily by allelic-specific methylation of cytosines in areas of DNA that are rich in CpG dinucleotides, known as imprinting control regions (ICRs). These imprints are established in the germ line and are propagated stably through many rounds of DNA replication during development. They affect transcription directly, by influencing the binding of methyl-sensitive DNA binding proteins, and/or indirectly, by altering regional chromatin structure. Imprinted genes are therefore, expressed selectively from either the maternal or paternal allele and are functionally haploid as a consequence.

In addition to long-term and normally irreversible changes in the genome's ability to be expressed, there is now reason to believe that DNA methylation also plays a more dynamic role in gene expression. Highly specific patterns of 5-mC residues exist all over the genome. These patterns appear to be different in each cell type and in each region of the genome. Tissue-specific methylation allows gene expression levels to change in specific cells at specific times (Kress et al., 2001). This capability also permits different cell types to respond differently to the same set of signals. The distribution of methylated cytosine residues in DNA undergoes characteristic changes during differentiation within a cell lineage, and changes in methylation status (usually the loss of methylated cytosines from promoter regions) are correlated with the activation of many tissue-specific genes during differentiation in vivo (Cedar, 1988). More recent evidence supports a role for DNA methylation in controlling tissue-specific expression in adult somatic tissue, such as the expression of the Maspin gene (SERPINB5), thus indicating a role for the establishment and/or the maintenance of cell-type restricted gene expression (Futscher et al., 2002). Dynamic changes in DNA methylation regulate tissue-specific gene expression of a set of primordial germ cell-specific genes that are coordinately activated upon germ cell entry into the gonads (Maatouk et al., 2006). These results provide *in vivo* evidence that postmigratory germ cell-specific genes are silenced by DNA methylation in both premigratory and somatic cells. In this way, epigenomes created during development give rise to stable tissue-specific gene expression patterns. Moreover, within a given tissue or cell type, DNA methylation can dynamically regulate the expression of individual genes. This also provides a mechanism by which the environment to which a given cell-type is exposed, can directly and inheritably alter the activity of the genome.

Regulation of Neural Development

During embryonic and early postnatal development of the mammalian nervous system, neural stem cells proliferate, then stop dividing and differentiate into neurons, glia and other cell types. This diversity in the phenotypic properties of the cells involve complex regulation of gene expression and protein function and more genes are expressed in the nervous system than in any other tissue (Mattson, 2003). DNA methylation is believed to play a key role in regulating the proliferation and differentiation of neural stem cells as well as the nature of the final differentiated state. When neural progenitor cells first undergo neuronal differentiation, neuronal lineage genes are activated and the genes required for glial fates are repressed, since neurogenesis precedes gliogenesis. DNA methylation may also control the expression of genes critical for differentiation of astrocytes, but much less is known about their development. Astrocyte differentiation is dependent in part, upon activation of the transcription factor STAT3. It has been shown that a CpG dinucleotide within a STAT3 binding element in the glial fibrillary acidic protein (GFAP) promoter is methylated in progenitor cells and becomes demethylated immediately prior to differentiation of astrocytes; methylation of the promoter prevents gene activation by STAT3 (Takizawa et al., 2001). This is an excellent example of a key role for methylation of a cell type-specific promoter in cell lineage specification in the developing brain.

To illustrate the critical and diverse roles that DNA methylation plays in the process of neural development, here we will briefly discuss changes in DNA methylation of promoters and/or enhancers of four genes - Oct4 (POU5F1), Nanog, Nitric oxide Synthase 3 (NOS3) and Sox2, in human NT2 neural precursors cells induced to differentiate with retinoic acid (Deb-Rinker et al., 2005; Bani-Yaghoub et al., 2006). Figure 3 shows the distinctive expression patterns of these genes. Oct4 and Nanog are two genes essential for the maintenance of the stem cell state and must be down-regulated in order for the cells to exit this state and establish a differentiated function. The Sox2 gene is also expressed at high levels in stem cells, but shows a more gradual decline and is believed to control cell cycle exit as well as differentiation into either neurons or astrocytes. NOS3 in contrast, is neither expressed in stem cells nor in differentiated neurons and astrocytes, but is transiently expressed during development to also control cell cycle exit and the onset of differentiation.

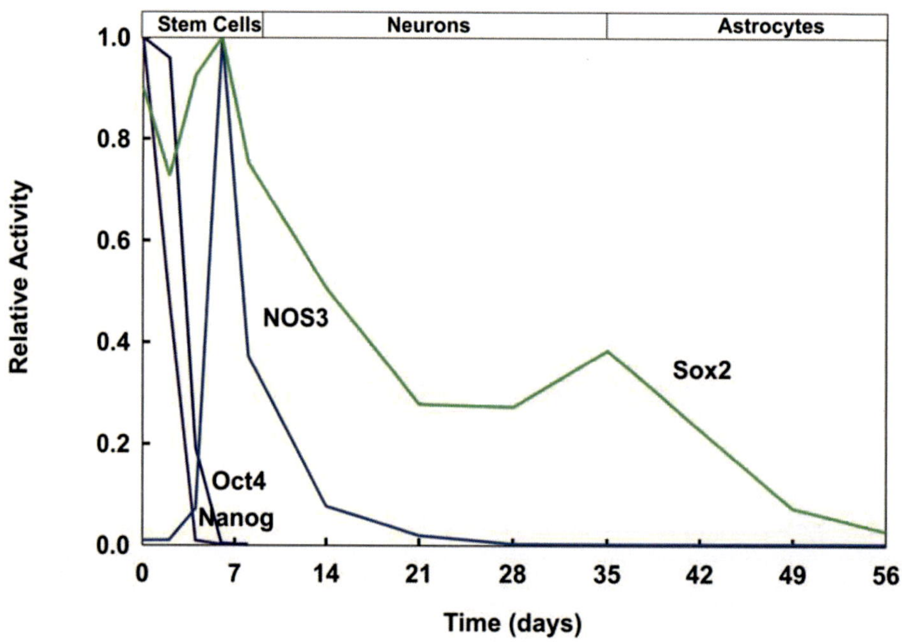

Figure 3. Expression patterns of Nanog, Oct4, NOS3 and Sox2 in human NT2 cells induced to differentiate into neurons and astrocytes with retinoic acid.

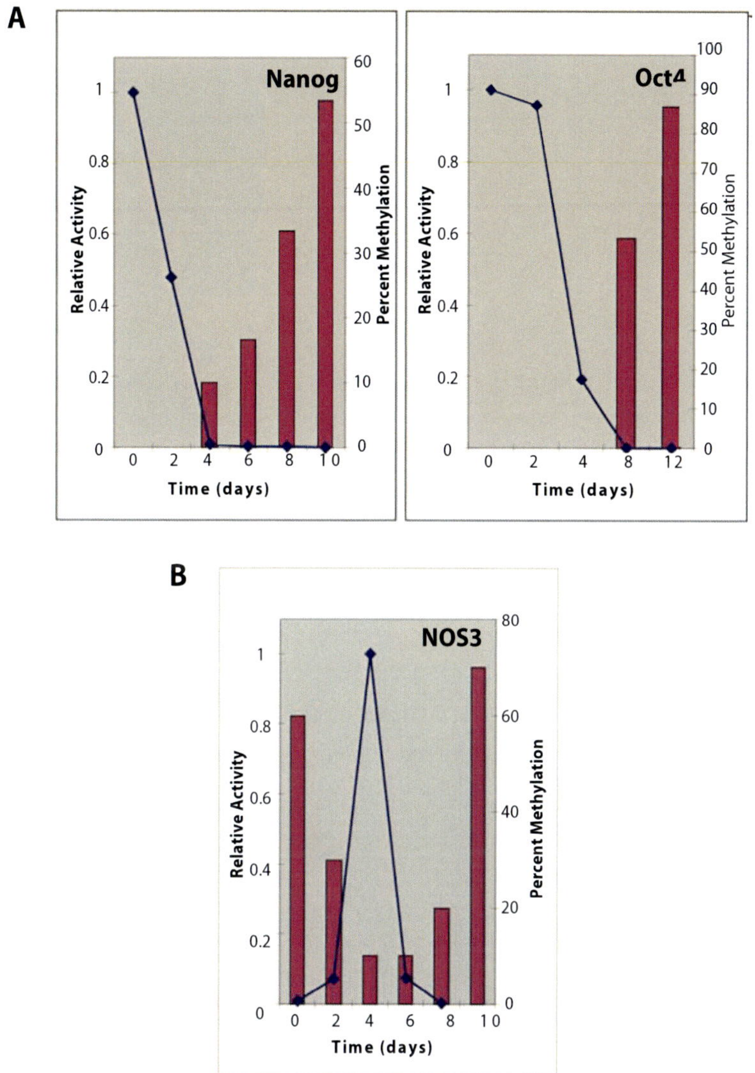

Figure 4. Graphical comparison of expression (lines) and methylation (bars) levels of A. Nanog and Oct4 and B. Nos3 genes.

The activities of both Nanog and Oct4 decline almost immediately after the addition of morphogens to permit the cells to commence the differentiation process (Figure 4). CpGs in the promoter regions of both genes become extensively methylated as the genes are irreversibly silenced. Interestingly, in both cases, there is a significant delay between decreasing levels of transcript and increased methylation, suggesting that methylation *per se* is not the primary event in switching off gene transcription, but is involved in a secondary step of gene silencing. In the case of NOS3, which is only transiently expressed during the developmental process, there is a dynamic change in the methylation status of certain CpGs in the upstream region of the gene (Figure 4). In the stem cell state, these sites are methylated and the gene is repressed. During development, the sites are dynamically demethylated to permit a narrow window of expression between days 4 and 12 and then the sites are remethylated and the gene is silenced in the differentiated neuronal cells. This is one of the few examples which demonstrate that methylation controls transient gene expression during development as opposed to simply altering the end-state expression of a gene.

The changes in the methylation status of the regulatory regions of the Sox2 gene illustrate other facets of the role of methylation in directing cell type-specific gene expression (i.e. the creation of epigenomes). Sox2 is an intronless gene situated entirely within a CpG island (Figure 5). Since CpG islands are characteristically unmethylated, there are no changes in the methylation status of CpGs in the promoter regions of the gene at any stage of development or in somatic cells. The expression of the Sox2 gene is however, controlled by 2 enhancers, one located approximately 4 Kb upstream of the gene and a second enhancer located approximately 4 Kb downstream of the gene. In the undifferentiated cells, the CpGs in both of these regions are unmethylated and the gene is highly expressed. As the cells differentiate into either neurons or astrocytes, the Sox2 gene is turned off. Co-incident with the repression of the gene is an increase in the methylation of the upstream enhancer in both neurons and astrocytes. However, there is differential methylation of the CpGs in the downstream enhancer. In neurons these sites are methylated, whereas in astrocytes they are not, thereby producing a different epigenome in each cell type. Thus, differential methylation during development creates a difference in the ability of the Sox2 gene to be expressed at a later time. Consequently, in neurons, where both the upstream and downstream enhancers are methylated, the Sox2 gene can never be re-expressed. In astrocytes on the other hand, the lack of methylation of the downstream enhancer permits the Sox2 gene to be re-expressed when cells undergo proliferation.

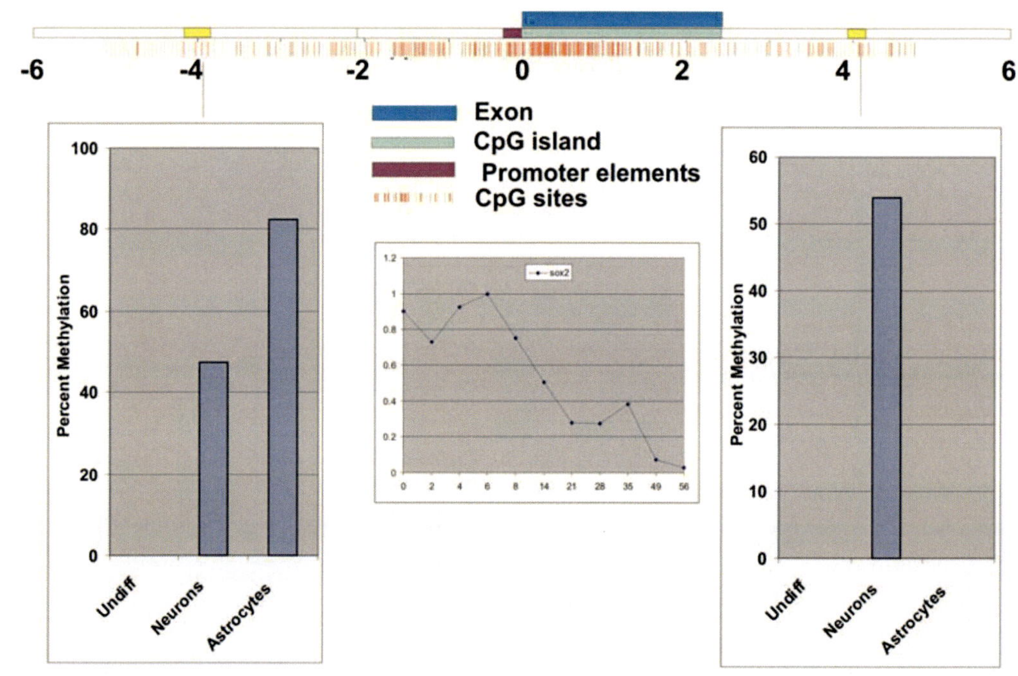

Figure 5. Kinetics of change in Sox2 gene expression and methylation status of the upstream (at -4 kb) and downstream (at 4 kb) Sox2 enhancers.

Mechanisms for the Modulation of the Activity of the Genome

The methylation of DNA is only one component of a much more complex apparatus that controls the expression of the genome in an epigenetic fashion. Genomic DNA is packaged along with histones into chromatin. DNA wrapped tightly with histones produces highly condensed chromatin fibers (heterochromatin) which is resistant to transcriptional activation, whereas DNA that is loosely associated with histones and present within unfolded chromatin structures (euchromatin) is permissive to transcriptional activation (Gan et al., 2007). Covalent modification of these DNA-associated proteins can also modulate the expression of the genome, thereby expanding the epigenetic paradigm. Histone tails can be modified by acetylation, methylation, phosphorylation, ubiquitylation, ADP-ribosylation and sumoylation (Fish and Marsden, 2006) creating an additional layer of epigenetic control.

Acetylation of lysine residues located on the N-terminal tails of H3 and H4 has been the most comprehensively studied post-translational histone modification. Gene silencing is associated with histone deacetylation, which is catalyzed in human cells by at least three classes of histone deacetylases (HDACs, class I, II, III). In general, acetylated histones are located at the promoters of transcriptionally active genes. Lysine methylation has also been extensively studied and has been shown to have divergent effects on transcription. Methylation of lysine 4 of histone H3 is a signature of transcriptionally active chromatin, particularly at the promoter and 5'coding region of genes, however, lysine 9 methylation of histone H3 is a silencing signal that facilitates the recruitment of heterochromatin protein 1. The addition of a methyl group to cytosine changes the major groove of the DNA helix and can directly influence its interaction with proteins. DNA methylation can suppress gene transcription by binding methyl-CpG-binding proteins, such as MeCP2 and MBD2, which attract and tether chromatin inactivation complexes containing HDACs and histone methyltransferases (HMTs) to the methylated sequences, promoting local condensation of the chromatin into a transcriptionally inactive configuration (Chang et al., 2006).

In one of the few studies on the kinetics of DNA methylation, Mutskov and Felsenfeld (2004) recently studied the transcriptional silencing of stably integrated transgenes and specifically excluded DNA methylation as the primary causative event. They propose the loss of histone acetylation and H3 K4 di- and trimethylation as early steps in a sequence of events leading up to the process of silencing. Initially, the gene is repressed reversibly since transcription can be partially reactivated by inhibition of histone deacetylases. Methylation of histone H3 K9 and of DNA cytosines in the promoters are later events that could 'lock' this stable silenced chromatin state. Thus, changes in gene expression may in some cases precede changes in DNA methylation (Gautsch and Wilson, 1983; Deb-Rinker et al., 2005). Methylation has a maintenance role in controlling gene expression but additional mechanisms are necessary to initiate gene expression in the early embryo. Hypermethylation may thus be a consequence, rather than a cause of gene inactivation during embryogenesis. Whether these rules hold true for endogenous genes as well, has not been determined. However, it has indeed been suggested in the case of mouse Oct-4 (Gidekel and Bergman, 2002).

Methylation and Disease

There is now substantial evidence showing that abnormalities in DNA methylation are the basis for a growing number of diseases (Robertson, 2005). These diseases may be caused by an alteration in specific patterns of CpG methylation or in the enzymes and proteins normally responsible for producing such patterns. In cancer, alterations in DNA methylation manifesting itself as a global hypomethylation, is among the earliest and most predominant molecular change. DNA methylation helps stabilize chromatin, and hypomethylation can lead to genomic instability by predisposing DNA to strand breakage and recombination. Furthermore, promoter CpG hypomethylation can result in gene re-activation, also contributing to abnormal gene expression. In contrast to hypomethylation, aberrant hypermethylation of CpG islands can lead to inappropriate gene silencing and disease states. For instance, methylation of CpG islands of tumor suppressor genes leads to their silencing, and predisposes to carcinogenesis (Attwood et al., 2002).

The importance of epigenetics in disease development is further evidenced by reports that several inherited diseases, such as Prader–Willi syndrome (PWS), Angelman syndrome (AS), Beckwith–Wiedemann syndrome (BWS), Rett's syndrome (RTT), and ICF syndrome, are due to aberrant epigenetic alterations. Studies of cultured neurons have shown that variations in methylation and acetylation can trigger neuronal cell death. For example, exposure of neurons to conditions that deplete the methionine pool necessary for methlyation induces apoptosis and histone deacetylase inhibitors can induce apoptosis of cultured neurons (Mattson, 2003). Dietary factors that influence DNA methylation may affect the risk of neurodegenerative disorders. For example, individuals with low dietary folate intake are at increased risk of Alzheimer's and Parkinson's diseases. Other agents that induce DNA methylation changes include nitric oxide, estradiol, estrogen, prolactin and steroid hormones (Abdolmaleky et al., 2004). A significant correlation has been found between risk of Alzheimer's disease and high plasma levels of homocysteine, as well as low levels of folic acid, and vitamins B6 and B12, vitamins essential for the transformation of homocysteine to methionine (Scarpa et al., 2003). These disease associations are thought to be interrelated to increased homocysteine and decreased concentration of *S*-adenosylmethionine (SAM), a main methyl donor with subsequent hypomethylation of DNA, that make vascular structures and neurons more susceptible to damage and apoptosis. When folate levels are low, SAM is depleted, resulting in the reduction of DNA methylation, thereby enhancing gene transcription and DNA strand breakage. This could also lead to impaired DNA repair resulting in genetic mutations or triggering apoptosis. It has been shown that DNA methylation is involved in amyloid-beta-protein precursor (APP) processing and amyloid-beta (Aβ) production through the regulation of Presenilin 1 (PS1) gene expression, and that exogenous SAM can silence the gene reducing Aβ, suggesting the use of this drug for the treatment of the disease (Scarpa et al., 2006).

Given that changes in DNA methylation are quite stable, it is likely that alterations in the patterns of DNA methylation could give rise to the development of new, highly specific disease diagnostics.

CONCLUSION

Epigenetic modifications of DNA and histones is emerging as a major component, not just in terms of gene silencing, but also in the differential expression of developmental genes. These modifications which include DNA methylation and histone acetylation are covalent, heritable marks that stably alter the function of the genome without changing its sequence. In this way, DNA methylation and the histone code can act as high-level controls which mediate the determination of cell fate by altering localized chromatin structure and activating or silencing genes. Many developmental genes appear to be subject to epigenetic control. DNA methylation is used to silence genes of pluripotency (eg Oct-4, Nanog), permitting cells to differentiate. DNA methylation also controls genes such as Sox2 and NOS3 that are transiently expressed in intermediate cell types, but not in differentiated phenotypes. Preliminary data has shown changes in the degree of methylation in Alzheimer brains. Thus epigenetics also appears to play a role in the etiology of neurodegenerative diseases and modulation of these modifications could be a key to brain repair.

ACKNOWLEDGEMENTS

The authors would like to thank Dao Ly and Anna Jezierski for technical assistance.

REFERENCES

Abdolmaleky, HM; Smith, CL; Faraone, SV; Shafa, R; Stone, W; Glatt, SJ; Tsuang, MT. Methylomics in psychiatry: Modulation of gene-environment interactions may be through DNA methylation. *Am. J. Med. Genet. B Neuropsychiatr Genet.*, 2004, 127, 51-59.

Antequera, F; Bird, A. Number of CpG islands and genes in human and mouse. *Proc. Natl. Acad. Sci.,* 1993, 90, 11995-11999.

Attwood, JT; Yung, RL; Richardson, BC. DNA methylation and the regulation of gene transcription. Cell. *Mol. Life Sci.*, 2002, 59, 241–257.

Bani-Yaghoub, M; Tremblay, RG; Lei, JX; Zhang, D; Zurakowski, B; Sandhu, JK; Smith, B; Ribecco-Lutkiewicz, M; Kennedy, J; Walker, PR; Sikorska, M. Role of Sox2 in the development of the mouse neocortex. *Dev. Biol.*, 2006, 295, 52-66.

Bird, A; Taggart, M; Frommer, M; Miller, OJ; Macleod, D. A fraction of the mouse genome that is derived from islands of nonmethylated, CpG-rich DNA. *Cell,* 1985, 40, 91–99.

Cedar, H. DNA methylation and gene activity. *Cell,* 1988, 53, 3-4.

Chang, SC; Tucker, T; Thorogood, NP; Brown, CJ. Mechanisms of X-chromosome inactivation. *Front. Biosci.*, 2006, 11, 852-866.

Deb-Rinker, P; Ly, D; Jezierski, A; Sikorska, M; Walker, PR. Sequential DNA methylation of the Nanog and Oct-4 upstream regions in human NT2 cells during neuronal differentiation. *J. Biol. Chem.*, 2005, 280, 6257-6260.

Doerfler, W. On the biological significance of DNA methylation. *Biochemistry (Mosc),* 2005, 70, 505-524.

Fish, JE; Marsden, PA. Endothelial nitric oxide synthase: insight into cell-specific gene regulation in the vascular endothelium. *Cell. Mol. Life Sci.,* 2006, 3, 144–162.

Futscher, BW; Oshiro, MM; Wozniak, RJ; Holtan, N; Hanigan, CL; Duan, H; Domann, FE. Role for DNA methylation in the control of cell type specific maspin expression. *Nat. Genet.,* 2002, 31, 175-179.

Gan, Q; Yoshida, T; McDonald, OG; Owens, GK. Concise review: Epigenetic Mechanisms Contribute to Pluripotency and Cell Lineage Determination of Embryonic Stem Cells. *Stem Cells*, 2007, 25, 2-9.

Gardiner-Garden, M; Frommer, M. CpG islands in vertebrate genomes. *J. Mol. Biol.*, 1987, 196, 261–282.

Gautsch, JW; Wilson, MC. Delayed de novo methylation in teratocarcinoma suggests additional tissue-specific mechanisms for controlling gene expression. *Nature*, 1983, 301, 32-37.

Gidekel, S; Bergman, Y. A unique developmental pattern of Oct-3/4 DNA methylation is controlled by a cis-demodification element. *J. Biol. Chem.,* 2002, 277, 34521-34530.

Grant, SG; Chapman, VM. Mechanisms of X-chromosome regulation. *Annu. Rev. Genet.,* 1988, 22, 199-233.

Holliday, R; Pugh, JE. DNA modification mechanisms and gene activity during development. *Science,* 1975, 187, 226–232.

Holliday, R. Epigenetics A Historical Overview. *Epigenetics*, 2006, 1, 76-80.

Hotchkiss, RD. The Quantitative Separation of Purines, Pyrimidines, and Nucleosides by Paper Chromatography. *J. Biol. Chem.,* 1948, 175, 315-332.

Johnson, TB; Coghill, RD. Researches on Pyrimidines. The Discovery of 5-Methyl-Cytosine in Tuberculinic Acis, the nucleic acid of the Tubercle Bacillus. *J. Am. Chem. Soc.,* 1925, 47, 2838-2844.

Kelly, TJ; Smith, HO. A restriction enzyme from Hemophilus influenzae. II. *J. Mol. Biol.,* 1970, 28, 393-409.

Klein, CB; Conway, K; Wang, XW; Bhamra, RK; Lin, XH; Cohen, MD; Annab, L; Barrett, JC; Costa, M. Senescence of nickel-transformed cells by an X chromosome: possible epigenetic control. *Science,* 1991, 251, 796-799.

Kress, C; Thomassin, H; Grange, T. Local DNA demethylation in vertebrates: How could it be performed and targeted? *FEBS Lett.,* 2001, 494, 135-140.

Li, LC; Dahiya, R. MethPrimer: designing primers for methylation PCRs. *Bioinformatics*, 2002, 18, 1427-1431.

Lu, Q; Qiu, X; Hu, N; Wen, H; Su, Y; Richardson, BC. Epigenetics, disease, and therapeutic interventions. *Ageing Res. Rev.,* 2006, 5, 449-467.

Maatouk, DM; Kellam, LD; Mann, MR; Lei, H; Li, E; Bartolomei, MS; Resnick, JL. DNA methylation is a primary mechanism for silencing postmigratory primordial germ cell genes in both germ cell and somatic cell lineages. *Development,* 2006, 133, 3411-3418.

Mattson, MP. Methylation and acetylation in nervous system development and neurodegenerative disorders. *Ageing Res. Rev.*, 2003, 2, 329-342.

Mutskov, V; Felsenfeld, G. Silencing of transgene transcription precedes methylation of promoter DNA and histone H3 lysine *EMBO J.,* 2004, 923, 138-149.

Riggs, AD. X inactivation, differentiation, and DNA methylation. *Cytogenet. Cell Genet.,* 1975,14, 9–25.

Robertson, KD. DNA methylation and human disease. *Nat. Rev. Genet.,* 2005, 6, 597-610.

Scarano, E. Letter: DNA methylation. *Nature,* 1973, 246, 539.

Scarpa, S; Fuso, A; D'Anselmi, F; Cavallaro, RA. Presenilin 1 gene silencing by S-adenosylmethionine: a treatment for Alzheimer disease? *FEBS Lett.,* 2003, 24; 145-148.

Scarpa, S; Cavallaro, RA; D'Anselmi, F; Fuso, A. Gene silencing through methylation: an epigenetic intervention on Alzheimer disease. *J. Alzheimers Dis*., 2006, 9, 407-414.

Takizawa; T; Nakashima, K; Namihira, M; Ochiai, W; Uemura, A; Yanagisawa, M; Fujita, N; Nakao, M; Taga, T. DNA methylation is a critical cell-intrinsic determinant of astrocyte differentiation in the fetal brain. *Dev. Cell,* 2001, 1, 749–758.

Toniolo, D; Martini, G; Migeon, BR; Dono, R. Expression of the G6PD locus on the human X chromosome is associated with demethylation of three CpG islands within 100 kb of DNA. *EMBO J.,* 1988, 7, 401–406.

Wu, H; Sun, Y.E. Epigenetic Regulation of Stem Cell Differentiation. *Pediatr. Res,.* 2006, 59, 21R-25R.

In: Progress in DNA Methylation Research
Editor: Hans P. Neumann, pp. 203-224
ISBN: 978-1-60021-722-7

Chapter VIII

POST-SYNTHETIC METHYLATION OF MACROMOLECULES IN SYNCHRONIZED MAMMALIAN CELLS

*Pietro Volpe**
Department of Biology of the University of Rome "Tor Vergata",
Scientific Research Street 1, 00133 Rome, Italy.

ABSTRACT

The analysis performed in *HeLa* cells demonstrates that post-synthetic methylation of almost the half population of mitochondrial DNA (mtDNA) is cell-cycle (CC) dependent, in harmony with the timing of post-synthetic methylation of macromolecules functioning in gene expression. At variance with nuclear DNA (nDNA), which is synthesized and methylated in the S phase, ribosomal, transfer and messenger RNAs as well as ribosomal and nuclear SDS-soluble proteins are synthesized in G_1 and G_2 but methylated in S. The replicating mtDNA exhibits two fractions: the first is detectable during S in synchrony with nDNA replication; the second is detectable during G_2 when replication of nDNA is switched-off. The methylation process involves the mtDNA revealed in S, while the mtDNA revealed in G_2 remains unmethylated. This diversification unequivocally accounts for heterogeneity of mitochondria, in a cell world with which they interact maintaining a large margin of autonomy.

Keywords: nDNA and mtDNA; rRNA, tRNA and polyA-containing pre-mRNA; nascent polypeptide chains, ribosomal proteins, saline- and SDS-soluble nuclear proteins.

* Correspondence concerning this article should be addressed to Prof. Pietro Volpe, Department of Biology of the University of Rome "Tor Vergata", Via della Ricerca Scientifica 1, Rome, Italy. E-mail: volpe@bio.uniroma2.it.

Abbreviations

(CC) Cell cycle; (mtDNA) mitochondrial DNA; (nDNA) nuclear DNA; (hTGc) human tissue transglutaminase gene; (MEM) minimum essential medium; (TCA) trichloroacetic acid; (SDS) sodium dodecyl sulphate; (Act) Actinomycin D; (Pur) Puromycin; (m^5C) 5-methylcytosine; (AdoMet) S-adenosyl-L-methionine; ([^{3}H]Met) [^{3}H]methyl-L-methionine; ([^{14}C]Met) [^{14}C]methyl-L-methionine; ([^{14}C]1-Met) [^{14}C]1-L-methionine; ([^{3}H]AdoMet) [^{3}H]methyl-S-adenosyl-L-methionine.

Introduction

The biochemical knowledge so far achieved indicated that, differently from mammalian DNA where 5-methylcytosine (m^5C) is the unique modified base [1], the targets for methyl groups increase to a notable extent along RNAs and proteins: the former are methylated in all four bases [2] and in ribose [3]; the latter are methylated at numerous amino acid sites [4] (Figure 1).

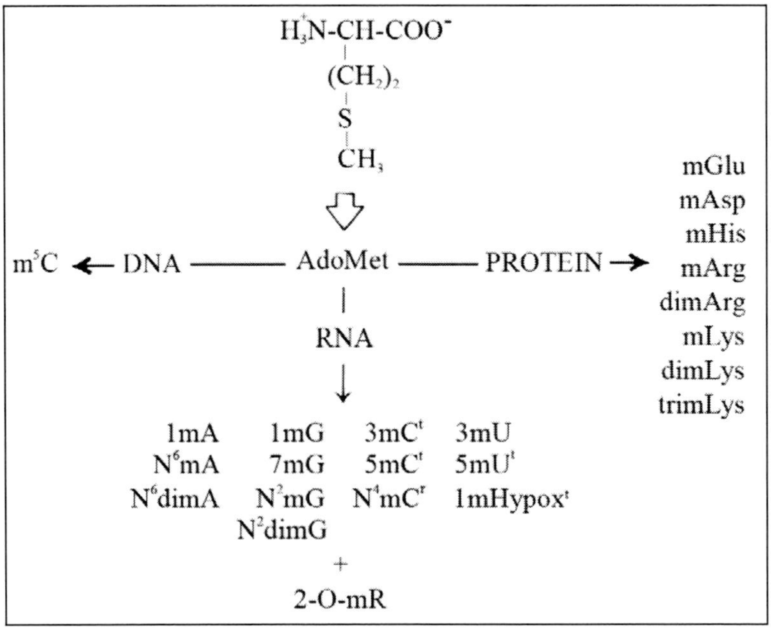

Figure 1. *Methyl groups released from methionine via AdoMet are attached to a different number of targets on DNA, RNA and protein.* In mammalian DNA, only C may be methylated. In RNA, all four bases plus ribose may be methylated (*t*, transfer; *r*, ribosomal). In protein, the carboxylic group of glutamic and aspartic acids may be monomethylated, the imidazole nitrogen of histidine may be monomethylated, the guanidinium group of arginine may be mono- or dimethylated, the epsilon group of lysine may be mono-, di- or trimethylated.

On such a basis, by using a new method developed to measure the lengths of the whole CC and its single phases [5,6], it was discovered that methylation takes place at the level of the replication fork along semiconservatively newly replicating nDNA chains to guarantee, in the double helix, the genetic maintenance of dimethylated palindromic 5'-m^{5}CpG-3'/3'-Gpm^{5}C-5' words [7]. This fact, confirmed by reactions with restriction endonucleases [8] and direct separation of new and old nDNA strands [9], represented an important platform not only for studies on time-correlation between synthesis and modification of a number of nDNA sequences [10,11] but also for studies on time-correlation between post-transcriptional RNA [12] and post-translational protein [13] methylation. A quantitative analysis, reported earlier for 45S pre-rRNA, facilitated these studies. In this precursor, methylation was found to be confined to the sequences ultimately integrated into ribosomes, namely into 18S and 28S rRNAs [14]. In mammalian cells, they contain 116 methyl groups: 45 in 18S and 71 in 28S [15]. In ribose, 28S was found to contain about 50% of its methyl groups and 18S was found to contain about 40% of its methyl groups [3,14]. Despite these calculations, the RNA methylation timing was still not defined, although it became evident that the 45S pre-rRNA transcription and processing occur as alternate phenomena early and late during CC [16]. On the other hand, the rate of peptide elongation on polyribosomes, as measured throgh a 3-min pulse labelling with radioactive amino acids, presented maximal values in G_1 and G_2 [17]. These peaks are independent of the amino acid pool [18] or uptake [19]. They take place 1 hr after the appearance of similar peaks of newly synthesized pre-mRNAs [20] and correspond to an increase in polyribosome content which is higher in G_2 than in G_1 [17]. The rate of peptide synthesis on polyribosomes is *vice versa* higher in G_1 than in G_2 even if these are smaller in G_1 than in G_2 [17].

The present contribution proved that synthesis of new mtDNA rings is characterized by two peaks, in S and G_2, when methylation selectively occurs on those which are constructed in S. The mtDNA population, built in G_2, remains unmethylated *vs.* an intriguing CC-dependent chain of events. While the new nDNA strands are replicating and methylated in S, the various RNA and protein species are essentially synthesized in G_1 and G_2 but methylated in S: a model shedding light on nucleo-cytoplasmic interactions.

Materials and Methods

Cell Synchronization

HeLa cells were grown and synchronized in suspension using a reversible double thymidine treatment [5]. The entire duration of their CC and the lengths of their CC-phases were measured as described before [5,6]. The siliconized spinners were areated with 5% CO_2. The growth in suspension and the induced synchrony served to yield large amounts of cells and, in turn, large amounts of purified macromolecules.

DNA, RNA and Protein Labelling

Synthesis and methylation of mtDNA were estimated by exploiting [^{14}C]methyl-L-methionine ([^{14}C]Met) as common labelled precursor (the analysis was made with minor modification of the method described for the detection of synthesis and methylation of nDNA [7]). RNA synthesis [22,23] was detected with [^{14}C]uridine, [^{3}H]uridine or [^{14}C]Met. When this ^{14}C-labelled amino acid was used, the measurement of synthesis of RNA (varying from the measurement of nDNA synthesis given by labelling of its A, G and T [7,9]) was based on labelling of its A and G aromatic rings only, since the radioactive carbon atom released from the methyl group of methionine enters the purine ring but not the pyrimidine ring (the C and U rings are unlabelled in RNA). Methylation of RNA was measured by subtracting, from its total [^{14}C]Met-labelling, the residual radioactivity which remains after inhibition of transcription with Actinomycin D (Act). Protein synthesis was studied with [^{3}H]leucine [17], [^{3}H]methyl-L-methionine ([^{3}H]Met) or [^{14}C]1-L-methionine ([^{14}C]1-Met) [13]. Protein methylation was carried out, in intact cells, according to Paik and Kim [4,13] (through labelling with both [^{3}H]Met and [^{14}C]1-Met) and, in isolated nuclei [7], with [^{3}H]methyl-S-adenosyl-L-methionine ([^{3}H]AdoMet).

Purification of Labelled Macromolecules

Mitochondrial and nuclear DNAs were sedimented through ultracentrifugation in CsCl gradients [21]. PolyA-containing pre-mRNA was divided on oligo-dT cellulose [20] from previously extracted total RNA [16,23]. Transfer and ribosomal RNAs were analyzed on 15-30% sucrose gradients [16]. Nitrogen bases of hydrolyzed nucleic acids were separated by High Performance Liquid Chromatography [9,24]. Ribosomal subunits, ribosomes, polyribosomes and peptide chains nascent on polyribosomes were analyzed on 7-47% sucrose gradients [17]. Proteins of pooled 40S, 60S and 80S ribosomal complex [17] were purified according to Schindler and Davies [25]. Saline- and SDS-soluble nuclear proteins were extracted as described elsewhere [13]. After the dansylation reaction [17,18], the concentration and the radioactivity of amino acids in the protein hydrolysates [26] were established on the basis of their fluorescence on micropolyamide sheet chromatograms [17,18]. Transcription and translation were inhibited by Act [27] and Pur [28], respectively.

Media and Chemicals

The Joklik-modified minimum essential medium (MEM) and the methionine-free MEM, both without Ca^{2+}, were supplied by GIBCO. The calf serum, from GIBCO, when added to the methionine-free MEM in given conditions of cell labelling, was depleted of free amino acids on Sephadex-G50 and purified through KSII filters [28]. The [^{3}H]uridine (28 Ci/mmol), [^{14}C]uridine (56.7 mCi/mmol), [^{3}H]leucine (35.5 Ci/mmol), [^{14}C]Met (15.2 mCi/mmol), [^{14}C]1-Met (57 Ci/mmol) and [^{3}H]AdoMet (77 Ci/mmol) precursors were furnished by NEN. The [^{3}H]Met (6 Ci/mmol) precursor was provided by the Radiochemical Center.

RESULTS

Replication of mtDNA occurs in S and G_2 but its Methylation Takes Place in S

MtDNA Synthesized in S Manifests Physico-chemical Properties which are different from those of mtDNA Synthesized in G_2

In the past, chronology of replicative synthesis of mtDNA was a subject of considerable uncertainity. On the one hand, it was reported that in *Tetrahymena* and *Physarum* mtDNA is produced at a rather constant rate throughout the whole interphase [29,30]; on the other, it was claimed that while its periodic synthesis can be observed in *yeast* [31] a wave of its replication can be seen in *Chang's* cells between phases S and M [32]. To ascertain this timing, new experiments were performed in *HeLa* cells [5,6]. They showed that mtDNA is synthesized in two different CC-stages: a first peak of incorporation of [^{3}H]thymidine into mtDNA occurs in S, almost in synchrony with nDNA replication; a second peak of incorporation of [^{3}H]thymidine into mtDNA occurs in G_2 [21]. In CsCl, these two S-phase and G_2-phase mtDNAs were found to be characterized by buoyant densities of 1.701 g/cm^3 and 1.688 g/cm^3, respectively [21]. This work confirms that, when mtDNA incorporates the radioactive carbon atom coming from the methyl group of [^{14}C]Met into proper A, G and T [7,9], two well expressed peaks of its synthesis can be observed: the first, in S, representing approximately 65% of the total replicated mtDNA; the second, in G_2, representing approximately 35% of the total replicated mtDNA (Figure 2a). These different percentages assumed particular interest when judged from the amount of labelled bases contained in each of them: the content of labelled T is always the lowest both in the replicating S-phase and G_2-phase mtDNAs; in the S-phase mtDNA the labelling of G is higher than that of A and in the G_2-phase mtDNA the labelling of A is higher than that of G. This marked inversion of base composition further accounts for a clearcut mitochondrial polymorphism [21].

Methylation defines the Specificity of S-phase mtDNA

It was supposed that mtDNA should contain m^{6}A, since mitochondria are thought to have a prokaryotic origin [24]. However, this methylated base was not found in mtDNA [33]. So, the possibilities were that mtDNA should either be not methylated at all or it should contain m^{5}C according to the fashion of nDNA methylation [7,9]. In the literature there was just one report claiming that m^{5}C is absent from mtDNA [34], but there were many data suggesting that, in addition to the presence of a DNA methyltransferase activity in isolated mitochondria [35,36], mtDNAs of *slime mold* [37] and *Vertebrates* [33,38] do contain m^{5}C. The present results, achieved in the same experimental conditions chosen for detection of mtDNA synthesis, are in harmony with these observations: methylation of mtDNA is shown by incorporation of the labelled carbon atom coming from the methyl group of methionine into its m^{5}C only (Figure 2b). Thus, a sharp wave of m^{5}C labelling exclusively occurs on the population of mtDNA which replicates in S and is characterized in CsCl by a buoyant density of 1.701 g/cm^3. The mtDNA population, replicating in G_2 and characterized in CsCl by a buoyant density of 1.688 g/cm^3, does not show any m^{5}C peak.

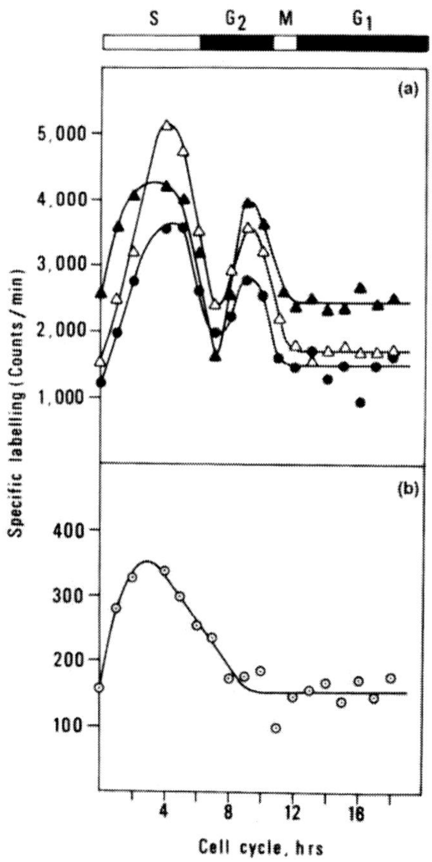

Figure 2. *Synthesis of mtDNA shows peaks in G_1 and G_2 but its methylation occurs in S.* At 1 hr-intervals during the CC, 100 10^6 cells were suspended in 25 ml of methionine-lacking Joklik-modified MEM and treated with 225 μCi [^{14}C]Met (the MEM was supplemented with 10% calf serum depleted of free amino acids on Sephadex G50 [28]). The labelling lasted 1 hr at 37 °C under stirring and 5% CO_2 in air [7]. After addition of 2-mg carrier calf thymus DNA, mtDNA was purified [21], hydrolyzed to bases and chromatographed [7,9,24]: synthesis (a) was shown by incorporation of the labelled carbon atom coming from the methyl group of methionine into its A (▲-▲), G (Δ-Δ) and T (●-●) bases; methylation (b) was shown by incorporation of the labelled carbon atom coming from the methyl group of methionine into its m^5C modified base (o-o).

Replication and Methylation of nDNA take Place in S

The timing of *in vivo* nDNA methylation in animal cells was poorly understood: it was thought that nDNA methylation follows nDNA synthesis by 2 min [39]; it was claimed that nDNA methylation would continue for several hrs after the end of nDNA synthesis [40]; isolated nuclei, which in the absence of dATP, dGTP, dCTP and dTTP did not synthesize nDNA [41], were able to methylate the nDNA chains produced before their isolation [42]. For all these reasons, it seemed appropriate to investigate a correlation between synthesis and methylation of nDNA in synchronized cells [5,6] and in the nuclei isolated from them [43].

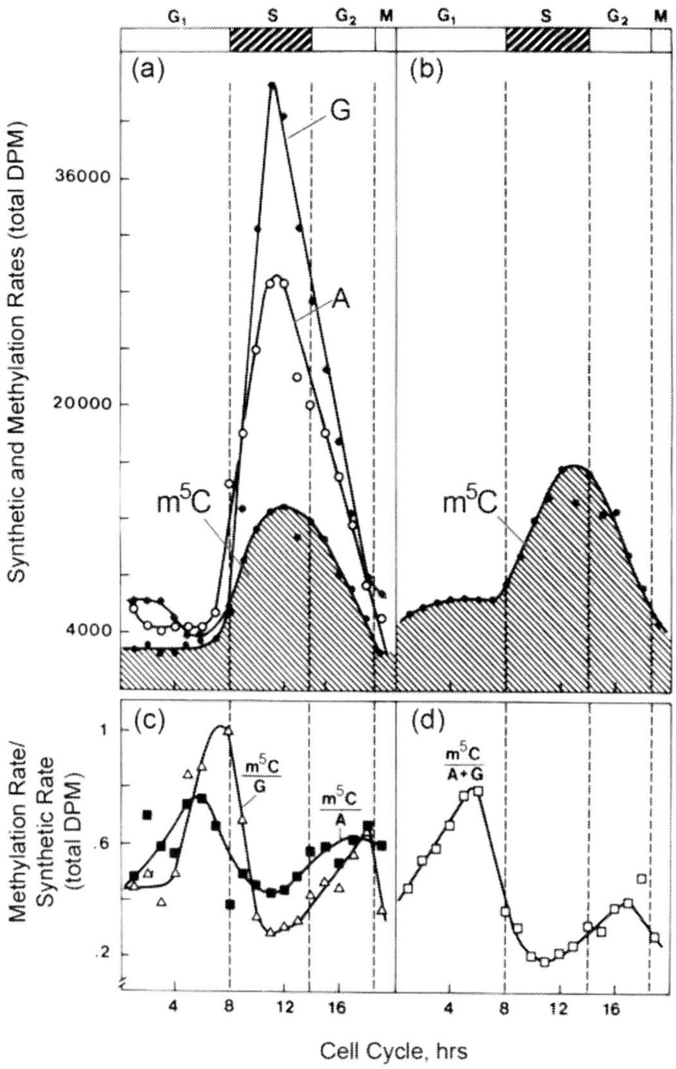

Figure 3. *Synthesis and methylation of nDNA occur in S.* (a) At 1-hr intervals during CC the intact cells were treated with [^{14}C]Met and their nDNA was analyzed as described for figure 2. In this case, the radioactivity detected in bases A (o-o), G (✱-✱) and T (not shown) accounts for synthesis and the radioactivity seen in m^{5}C (shaded area) accounts for methylation [7,9,24]. (b) At 1-hr intervals during CC nuclei were constantly isolated from 100 10^6 cells [7], washed in 0.003 M Tris-HCl at pH 7.4 and incubated in 0.3 ml Tris buffer with 25 μCi [^{3}H]AdoMet [7]. The incubation lasted 30 min at 37 °C under stirring. Two mg-carrier calf thymus DNA was added while nDNA extraction [7], hydrolysis and chromatography were performed as for figure 2 [7,9,24]. In this case, the methylation shown by m^{5}C (shaded area) occurred in the absence of A, G or T labelling. The rate of methylation *vs.* that of synthesis was determined in intact cells (c) or isolated nuclei (d) on the basis of the radioactivity carried by the corresponding nDNA m^{5}C, A and G bases.

Methylation Continues by half an hr on newly Synthesized nDNA Strands

At intervals of 1 hr during the whole CC, the cells were treated for 30 min with [^{14}C]Met: one expected that radioactivity found in nDNA A and G, similarly to that found in nDNA T (not shown), should account for synthesis of these bases going through the oxidation C_1-

chain, while the radioactivity found in nDNA m^{5}C should account for C methylation [43]. In agreement with this, the results clarified two processes: in intact cells, nDNA methylation does follow nDNA synthesis in S (Figure 3a); in isolated nuclei, in the absence of synthesis [41], nDNA methylation still goes on in correspondence of the S phase (Figure 3b). In other words, methylation continues for at least 30 min when nDNA synthesis does not occur.

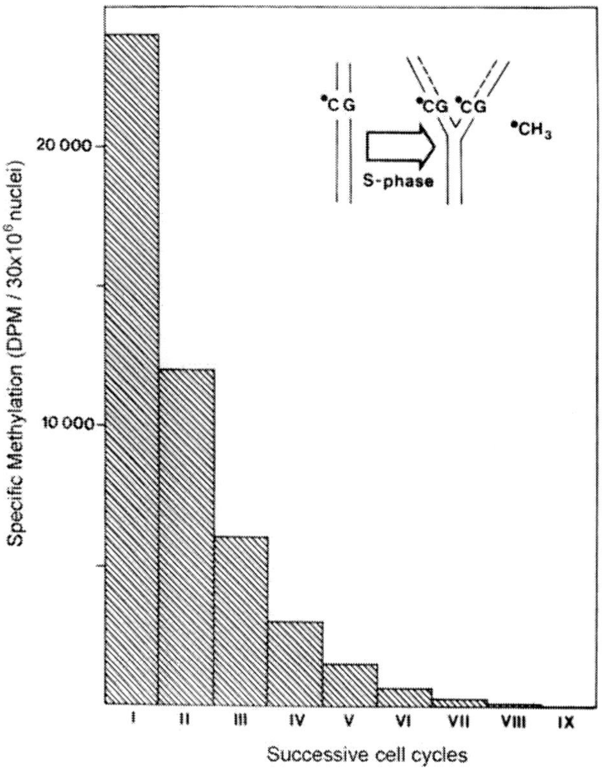

Figure 4. *The methylated nDNA strands are semiconservatively transmitted from one cell generation to the next.* Two billion cells were grown and synchronized in suspension [6]. After removal of the nucleoside from the culture, they were adjusted to a density of 0.5 10^6/ml and treated with 10 μCi/ml [^{14}C]Met (the labelling was performed in a Spinner under stirring and 5% CO_2 in air). At the end of the first S-phase, the cells were washed and resuspended in a fresh medium [6], without any additional radioactive precursor. After 14 hrs (i. e. a complete CC), 30 10^6 cells were drawn to purify nDNA [21]. This was hydrolyzed to bases and chromatographed [7]. The radioactivity found in A, G and T (normalized to the nDNA ODs at 256 nm) measures synthesis (not shown); the radioactivity found in m^{5}C measures methylation (shaded columns) [7,9,24]. The nDNA analysis was repeated under the same conditions for 10 CCs while, after each re-cycle, the cells were washed and resuspended in a fresh medium constantly maintaining their density at 0.5 10^6/ml.

The Pattern of nDNA Methylation is Semiconservatively Transmitted from a Cell Generation to the Next

The nDNA was extensively methylated with [^{14}C]Met during CC [44]: in figure 4 the first shaded column shows the amount of its methylation reached. After treatment with [^{14}C]Met, the cells were washed and replaced in a fresh medium without any addition of labelled precursor and the washings were repeated for many cell generations: in figure 4 one can observe that, during the successive CCs, the nDNA methylation constantly decreases by 50%. This result corresponds to what one expected from a semiconservative mechanism of transmission of m^5C. Particularly it signifies that this methylated base, once formed on the double helix, is not easily removed. Despite some previous information suggested that parental strands are slightly methylated [45], one of the interpretations of figure 4 was that, in agreement with [7-9], new nDNA strands are methylated in order to exactly copy the old ones.

The nDNA Methylation Rate in S differs from that in G_1 and G_2

An interesting result emerged when the rate of nDNA methylation was compared to that of nDNA synthesis, either in intact cells (Figure 3c) or in isolated nuclei (Figure 3d). There appeared no doubt that methylation follows synthesis during S and that the extra-S phase methylation is minimal (Figure 3a,b). Nevertheless, the rate of this minimal extra-S phase methylation, when normalized to the rate of the corresponding synthesis, is much higher than that observed during S. This accounts for variation in the rates of *maintenance* and *de novo* methylations: the semiconservative methylation pattern on bulk nDNA (Figure 4) would be maintained, in S, at a relatively lower rate; any *de novo* methylation would involve given nDNA sequences [10,11], in G_1 and G_2, at a relatively higher rate. Notwithstanding the fact that such an interpretation looks extremely attractive if related to studies on CC-dependent regulation of gene activity [11], one should not exclude, however, that an aliquot of "fast" methylation, during the extra-S phase time, might deal with nDNA-repair [9].

Most RNAs are Transcribed in G_1 and G_2 but Methylated in S

Synthesis and Methylation of tRNAs and rRNAs are Significantly CC-Dependent

The labelling of these two RNA classes, representing the largest populations of ribonucleic acid, was performed using the [^{14}C]Met precursor [7,9] as specified in Methods. The 4S tRNAs and 18-20S, 28-32S and 45S rRNAs were analyzed by adding Act to the medium. It turned out that, if a 2-µg/ml Act concentration is sufficient to suppress for 60-80% the labelling of transcripts, a 2 to 10-µg/ml Act concentration saves 20-40% of their residual labelling which accounts for net RNA methylation (Figure 5). In another experiment, the various RNA species were treated with [^{14}C]Met both in the absence and in the presence of 2-µg/ml Act: without inhibitor, three peaks of labelling in S, G_2 and G_1 appear (the 45S rRNA is the only one not to show a peak in G_2); with inhibitor, the two peaks of labelling in G_1 and G_2 disappear and the peak of labelling in S remains (Figure 6). The two peaks of labelling in G_1 and G_2 were interpreted in terms of transcription [16], in agreement with the fact that the direct incorporation of [^{14}C]uridine into RNAs occurs during these two phases (Figure 7). In any case, either [^{14}C]Met (Figure 6) or [^{14}C]uridine (Figure 7) were used, the rate of RNA labelling is always higher in G_1 than in G_2, without considering that the rate of total pre-rRNA synthesis dominates over that of total pre-rRNA processing, in G_1, and the rate of total

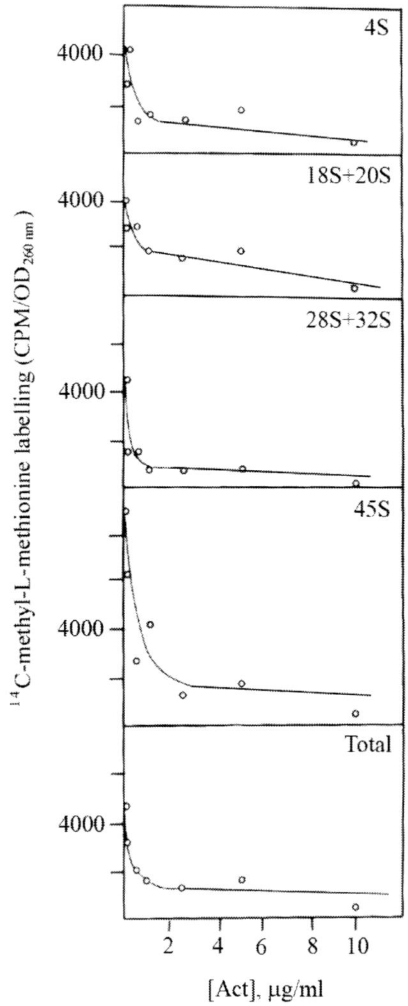

Figure 5. *RNA methylation is revealed by inhibition of transcription with Act.* 30 10^6 unsynchronized cells, suspended in 3 ml of MEM with 2% calf serum (containing 0 to 10 μg/ml Act and 15 μCi [^{14}C]Met) were incubated under shaking for 50 min at 37 °C. The treated cells were washed twice with Hanks and disrupted to analyze their macromolecules in sucrose gradients [16] fractioned with an ISCO instrument [17]. The fractions were treated with 10% TCA. Their precipitated material was sedimented on Millipore filters, washed with 5% TCA and tested for radioactivity in a Packard 460 CD radiospectrometer (using the Insta-gel as scintillation liquid). Each point, along the curves, is the sum of the radioactivity detected in the recovered RNA peaks. All values are the mean of three experiments.

pre-rRNA processing dominates over that of total pre-rRNA synthesis, in G_2 (Figure 7a,b). In sum, the Act-resistant S-peak of labelling accounts in all cases for net methylation, even if with quantitative variations: for tRNA, the level of methylation in S is higher compared to that of transcription in both G_1 and G_2; for rRNAs, the level of methylation in S is lower compared to that of transcription in G_1 (Figure 6).

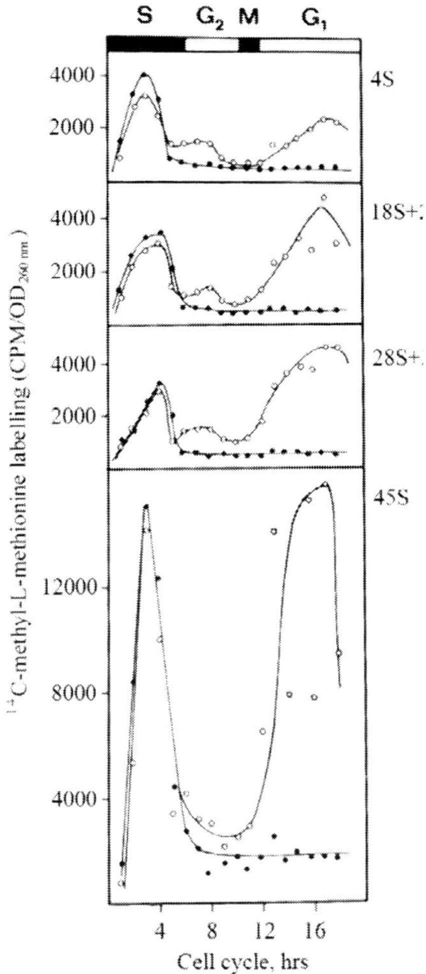

Figure 6. *Ribosomal and transfer RNAs are transcribed in G_1 and G_2 and methylated in S.* At 1-hr intervals during CC, while 50 10^6 control cells, suspended in 5 ml of MEM supplemented with 2% calf serum, were treated with 25 μCi [^{14}C]Met, 50 10^6 experimental cells, suspended in 5 ml of MEM supplemented with 2% calf serum, were treated with 25 μCi [^{14}C]Met and 2 μg/ml Act. All samples were incubated under shaking and 5% CO_2 in air for 50 min at 37 °C. Then they were washed twice with 15 ml Hanks and centrifuged for 5 min at 1,500 rev./min. RNA was extracted, tested for the absence of DNA residues [16] and analyzed on sucrose gradients as for figure 5. For each sample 12 ODs at 256 nm were run. The points along the four pairs of curves summarize the specific radioactivity of the fractions collected. Samples without Act (o-o); samples with Act (●-●). The values are the mean of two experiments.

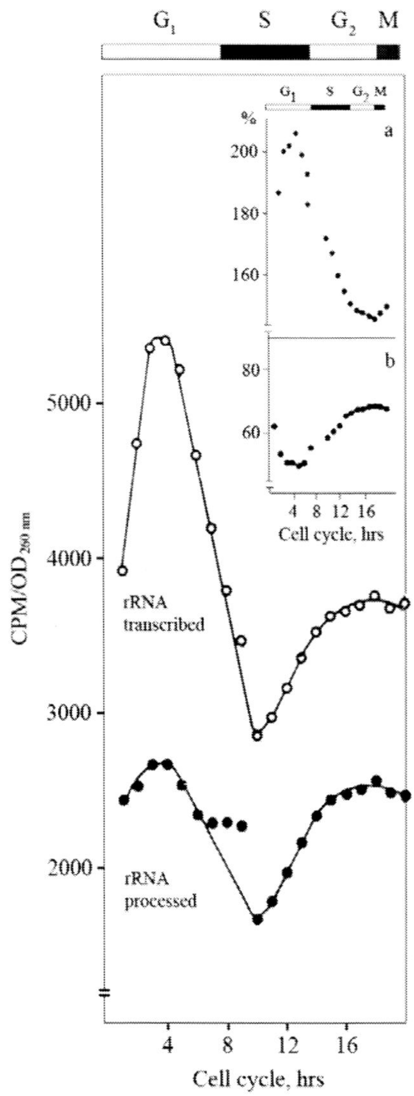

Figure 7. *Synthesis and processing of total rRNA are CC-dependent.* The experimental conditions were those of figure 3, except that samples ($50 \cdot 10^6$ cells in 5 ml of MEM containing 2% calf serum) were labelled for 50 min with 5 μCi [^{14}C]uridine, without Act. The insert shows the ratios between the rates of rRNA transcribed and rRNA processed (a) and between the rates of rRNA processed and rRNA transcribed (b). The curves were drawn by the UNIVAC Computer following a polynomial regression analysis with the method of least square for n = 4.

PolyA-Containing pre-mRNA is Transcribed in G_1 and G_2 and Methylated in S

The experiments showed that, when compared to the amount of transcription of total RNA, if the amount of transcription of polyA-containing pre-mRNA (2,043 CPM) corresponds to 1.3%, the amount of transcription of main RNA (151,000 CPM) corresponds to 98.7% (Figure 8A). Figure 8B illustrates, instead, the CC-dependent development of the entire polyA-containing pre-mRNA transcription: as expected [20], its rate reveals a significant increment in G_1 and a rather modest increment in G_2. A similar picture is reproduced when the rate of transcription of polyA-containing pre-mRNA was measured in the absence or in the presence of Act: in the first case, the labelling due to [^{3}H]uridine increases in G_1 and G_2; in the second, these two peaks disappear (Table I). To detect synthesis and methylation, the cells were treated with [^{14}C]Met at 1-hr intervals during the whole CC

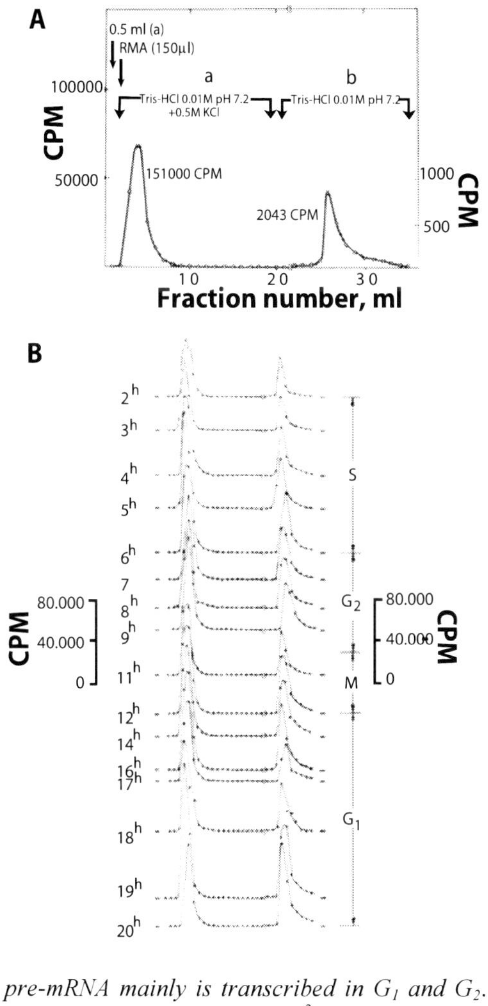

Figure 8. *PolyA-containing pre-mRNA mainly is transcribed in G_1 and G_2.* 100 10^6 cells in 10 ml of MEM containing 2% calf serum were treated with 10 µCi [^{3}H]uridine for 20 min at 37 °C under shaking and 5% CO_2 in air. RNA was extracted from isolated nuclei [20]. (A) A sample (with 5 ODs at 256 nm in 150 µl of 0.01 M Tris-HCl at pH 7.2 with 0.5 M KCl) was percolated through oligo-dT cellulose [20]: most radioactivity (a), carried by rRNA, tRNA and pre-mRNA without polyA, was eluted with higher ionic strength (0.01 M Tris-HCl at pH 7.2 containing 0.5 M KCl); the remaining part of radioactivity (b), carried by polyA-containing pre-mRNA, was eluted with lower ionic strength (0.01 M Tris-HCl at pH 7.2). (B) The sequence of peaks on the left shows the CC-development of total RNA labelling eluted at higher ionic strength; the sequence of peaks on the right shows the CC-development of polyA-containing pre-mRNA labelling eluted at lower ionic strength.

Table I. ***The bulk polyA-containing pre-mRNA is synthesized in G_1 and G_2 and methylated in S.*** **Using the experimental conditions of figure 8, pulses of 20 min with 10 μC [^{14}C]Met were given to 100·10^6 cells suspended in 10 ml of MEM containing 2% calf serum. The bulk fraction of polyA-containing pre-mRNA was separated from total RNA on oligo-dT cellulose. A first set of cell samples was incubated at 37 °C, under shaking and 5% CO_2 in air, without Act. A second set of cell samples was incubated in these same conditions in the presence of 5 μg/ml Act. Another two sets of cell samples were labelled for 20 min with [^{3}H]uridine (lμCi/ml), with and without 5 μg/ml Act.**

CC phases* (in hrs)		CPM/OD$_{256nm}$			
		[^{14}C]Met-labelling		[^{3}H]uridine-labelling**	
		- Act	**+ Act**	**- Act**	**+ Act**
G_1	1	189	38	495	182
	2	260	41	2,128	115
	3	347	52	3,051	216
	4	621	101	6,321	188
	5	831	89	4,181	214
	6	445	67	2,431	126
	7	212	40	4,351	84
	8	210	28	4,181	15
S	9	181	121	3,005	28
	10	241	185	1,241	15
	11	324	310	684	41
	12	412	384	631	16
	13	216	208	920	38
	14	141	128	821	24
G_2	15	384	29	2,121	110
	16	621	21	4,561	98
	17	411	54	5,324	114
	18	318	61	4,124	90
	19	98	38	1,112	116
M	20	105	25	381	45

*In the experiments the CC-timing started from the S phase. **The measurements were performed as for figure 8B. The values represent the mean of three experiments.

and the labelled polyA-containing pre-mRNA was recovered from oligo-dT cellulose as for figure 8A. It was found that, if in the absence of Act three peaks of labelling appear in G_1, S and G_2, in the presence of the inhibitor the two peaks in G_1 and G_2 are suppressed and the peak in S remains (Table I). To mention that the Act-sensitive labellings in G_1 and G_2 correspond to the main waves of transcription of polyA-containing pre-mRNA observed when [^{3}H]uridine was used as labelled precursor, whereas the Act-insensitive labelling in S confirms its net methylation as for tRNA and rRNAs. A minor transcription obviously would involve in S those pre-mRNAs which codify for histones and other proteins [20].

Most Proteins are Translated in G_1 and G_2 but Methylated in S

A 3-min Pulse Labelling with Radioactive Precursors is not Sufficient to Reveal Methylation of Peptide Chains Nascent on Polyribosomes

A preliminary experiment showed that a 3-min treatment with [^{3}H]leucine is sufficient to reveal the elongation of peptide chains on polyribosomes. In agreement with [17], this synthetic process occurs with maximal rates in G_1 and G_2 (Table IID). Analogously with the experiments showing that Act suppresses the RNA synthesis in G_1 and G_2 but does not modify the RNA methylation in S (Table I), an attempt was made to verify whether or not Pur, known as an inhibitor of translation [28], would suppress the protein synthesis in G_1 and G_2 without affecting the protein methylation in S. It turned out that, in the presence of Pur, no wave of labelling, accounting for peptide methylation, appears (Table IID), as confirmed when a double labelling with [^{3}H]Met and [^{14}C]1-Met was used. In this case, with a pulse labelling lasting 3 min [17], an occurrence of methylation would have been detected on the basis of the ^{3}H/^{14}C ratio, because both methionines enter the peptide backbone and because the ^{3}H-labelled methyl groups link its side chains [4]. However, the results showed that the ^{3}H/^{14}C ratio of the investigated amino acids remains constant throughout the interphase (Table IIA). For this reason it was assumed that, considering that the nascent peptides are able to reach primary, secondary or tertiary structures directly during their building on polyribosomes, methylation would involve them after their release from the translational machinery and not during their elongation. The next experiments served to establish the duration of the minimal pulse labelling required to methylate the proteins which enter as constituent parts of a biostructure.

A 30-min Pulse Labelling with Radioactive Precursors shows that Methylation of RIbosomal Proteins Occurs in S

A heavy methylation of proteins, constituting the 40S, 60S and 80S ribosomal complex [17], is detected when the synchronized cells were treated for 30 min with [^{3}H]Met and [^{14}C]1-Met. This methylation is well evidenced by the ^{3}H/^{14}C ratio of the considered labelled amino acids: it shows a clearcut increase in S (Table IIB).

A 60-min Pulse Labelling with [^{3}H]AdoMet Reveals that the SDS-Soluble Nuclear Proteins are Methylated in S

The CC-dependence of post-translational methylation is evidenced lastly by *in vitro* experiments with isolated nuclei [13,28]. At 1-hr intervals throughout the entire interphase, these were treated for 60 min with [^{3}H]AdoMet. Then, their saline- and SDS-soluble protein fractions were extracted, purified and tested for radioactivity [13]: the saline-soluble nuclear proteins show no methylation, at any stage of the CC; the SDS-soluble nuclear proteins show a methylation in correspondence of the S phase (Table IIC).

Table II. ***The bulk protein is synthesized in G_1 and G_2 and methylated in S.*** **At 1-hr intervals during CC, 100·10^6 cells suspended in 10 ml of methionine-lacking MEM containing 2% calf serum (depleted of free amino acids on Sephadex G50 [28]) were incubated with 50 μCi [^{3}H]Met and 25 μCi [^{14}C]1-Met and kept at 37 °C for 3 min (A) or 30 min (B) under shaking and 5% CO_2 in air. The peptides nascent on polyribosomes (A) and those extracted from the 40S+60S+80S ribosomal complex (B) were hydrolyzed, while their amino acids were dansylated and chromatographed [17,18] to detect total radioactivity of the methylated arginine, lysine and histidine. Nuclei were isolated [7] from 100·10^6 cells and treated for 60 min at 37 °C in 1-ml buffer with 10 μCi [^{3}H]AdoMet: their saline- and SDS-soluble protein fractions were extracted [13] and tested for radioactivity in 10% TCA-precipitates (C). In a control experiment (D), 100·10^6 cells were labelled for 3 min in the conditions of (A) with 1 μCi/ml [^{3}H]leucine, with and without 5 μg/ml Pur [28]. In this case, the labelling of the peptide chains nascent on polyribosomes was also measured in 10% TCA-precipitates [17].**

		(A)	**(B)**	**(C)**		**(D)**	
CC phases* (in hrs)		**Peptides nascent on polysomes**	**Ribosomal proteins** (40S+60S+80S)	**Nuclear proteins**		**Peptides nascent on polysomes**	
				Saline-soluble	**SDS-soluble**	**-Pur**	**+Pur**
		(^{3}H/^{14}C)**		(DPM/mg protein)		(DPM/OD_{260nm})	
G_1	1	2.15	1.76	530	1,100	1,510	98
	2	2.18	1.64	538	1,148	1,890	84
	3	2.36	1.85	581	1,235	3,400	47
	4	2.61	2.31	510	1,250	3,800	58
	5	1.18	2.05	460	1,128	4,300	65
	6	1.96	2.14	413	1,074	5,100	34
	7	2.16	3.18	388	1,051	2,150	95
	8	2.36	3.64	318	1,038	880	18
S	9	2.18	5.04	412	1,565	755	25
	10	2.10	7.81	421	2,188	1,100	34
	11	2.18	8.94	476	3,754	800	41
	12	2.36	9.11	510	4,520	735	51
	13	1.94	9.87	438	6,156	900	28
	14	2.38	11.04	490	7,540	1,250	18
G_2	15	2.18	3.84	470	3,211	2,350	36
	16	2.18	0.24	485	1,042	2,624	27
	17	2.14	0.81	510	921	3,050	31
	18	2.09	0.66	481	840	1,200	95
	19	2.14	1.05	344	1,050	800	76
M	20	2.14	0.48	381	1,038	635	18

*In the experiments the CC-timing started from the S phase. **The ^{3}H- and ^{14}C-incorporated DPMs varied in the ranges of 150-1,500 and 40-400, respectively, providing an average ^{3}H/^{14}C ratio of 3.75. The values represent the mean of three experiments.

CONCLUSION

Methylated sites in biological macromolecules are not to be considered errors of Nature. For instance, mtDNA (as biogenetic bridge between pro- and eukaryotic cells) could have lost the bacterial DNA m^6A [24], but why should it conserve the sole m^5C for about two billions of years, in harmony with the maintenance of this modified base in nDNA [46]? Whatever the answer to this intriguing question, it will come from Molecular Evolution (Figure 9). The aim of this investigation was to confront the time-correlation between synthesis and post-synthetic methylation of mtDNA, occurring in the cytosol, with the time-correlation between synthesis and post-synthetic methylation of nDNA, RNAs and proteins, occurring partly in the nucleus and partly in the cytosol.

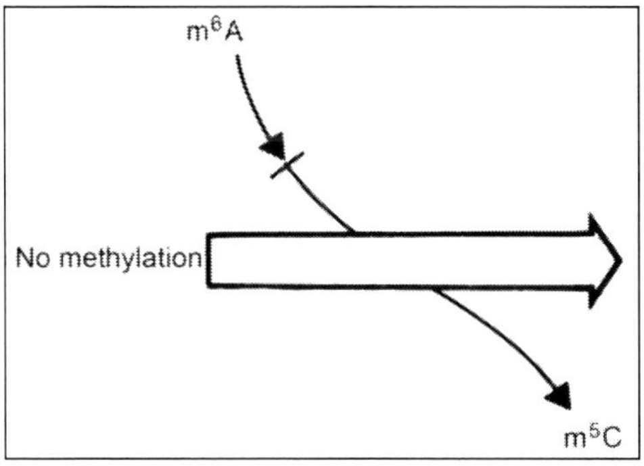

Figure 9. *MtDNA in its evolution acquires the methylation language of nDNA.* In agreement with the hypothesis about the prokaryotic biogenesis of mitochondria, mtDNA should carry N^6-methyladenine (m^6A). Alternatively, mtDNA should either be unmethylated or it should carry m^5C as host cell nDNA does.

Let us first emphasize the studies providing a background information on nDNA. It was suggested that a major *maintenance* nDNA methylation [7,9] follows the nDNA replication in the course of S [6], while a minor *de novo* nDNA methylation takes place during the extra-S-phase time [7,47]. The nature of the maintenance nDNA methylation in S was revealed by the presence of m^5C on the semiconservatively newly replicating nDNA chains made heavier in alkaline CsCl through previous substitution of T with bromodeoxyuridine [9]. Then, the discovery that methylation could be a tool to switch-off gene activity [43] (as reviewed in [48]) conformed with the finding that in S the m^5C modified base is preferentially located along non-coding *promoter* and *intron* sequences [43,46] (as reviewed in [49,50]).

The pre-existing knowledge on RNA methylation, mainly dealing with a variety of possible functions when it occurs on different RNA species, paved the way to a detailed research on its timing [16]. It was shown that, during CC, there is a chronology of 45S rRNA transcription in G_1, 45S rRNA methylation in S and 45S rRNA processing in G_2 [44]. It was also evidenced that the methylated 18S and 28S rRNAs are incorporated, in S, into 40S and 60S ribosomal subunits [14], while the inhibition of rRNA methylation with cycloleucine, which suppresses the AdoMet synthesis, was not found to prevent ribosome assembly: it only delays this assembly through the accumulation of intermediate rRNA sequences in the nucleus [51]. In this case, the function of methylation seemed nothing but a modulation of the whole 45S rRNA processing [16,44]. A modulating role of methylation was also suggested for the processing of pre-mRNA [52]. Two kinds of methylated structures were found in it: the first is the 5'-terminal methyl cap [53] (which enhances the mRNA affinity for the initiation factor eIF-4B [54] and protects the mRNA against the 5'-exonuclease degradation [55]); the second is the internal set of N^6-methyladenine residues, occurring in the Gpm^{6}ApC and Apm^{6}ApC sequences [56] (believed to constitute the recognition sites for processing enzymes [52]). Anyway, the fact that cycloleucine inhibition of 45S rRNA methylation was found to lead to an accumulation of ribosomal intermediates in the nucleus favoured the idea that methylation would facilitate the RNA stabilization *tout-court* by modulating the RNA processing [51]. This idea particularly agreed with the belief that tRNA methylation would prevent the unnecessary pairing of given sequences beneath its chains [2]. On this basis, it was thought that methylation would contribute to the progress of any RNA macromolecule from the nucleus to the cytosol [2,16].

As for proteins, it was proposed that methylation is responsible for the definition of their specific functions. For instance, histones, once methylated, accelerate chromatinization [4,57,58]; methylation facilitates the adaptation of transducer proteins in chemotaxis and modulates the enzyme activity in several instances [4]. Other proteins, after their methylation in S, were found to orient the clockwise progression of the CC phases [4].

The experiments described herewith enrich the information on the complex mosaic world of all these macromolecular events, since an effort was made to understand the basic rules of their time-order. It was established that, if mtDNA is replicated in the S and G_2 phases but methylated in S (Figure 2), nDNA is replicated and methylated in the same S (Figure 3). After this, it was suggested that tRNA, rRNA and polyA-containing pre-mRNA are mainly transcribed in G_1 and G_2 and methylated in S (Figure 6 and Table I). In the case of proteins, the situation is circumstantiated: even if the peptides nascent on polyribosomes do not appear to be methylated in G_1 and G_2 when a 3-min pulse labelling was used, the ribosomal and the SDS-soluble nuclear proteins are heavily methylated in S with a longer pulse labelling (Table II). This suggested that methylation does not immediately follow elongation, because it involves peptides which have already entered a given biostructure. Therefore, whatever the type of proteins methylated in S, the delay between their synthesis and their methylation lasts a few hours for peptides produced in G_1 (early methylation) and a much longer lapse of time for peptides produced in G_2 (late methylation). The proteins methylated late go through mitosis and undergo an almost full CC, before being methylated. The literature in the field supported the idea of an S-phase dependence of protein methylation but did not exclude that a wider range of interphase, mainly after S, would allow some protein methylation [59,60]. To summarize, if half the population of mtDNA (the one duplicating in G_2) remains unmethylated, the post-synthetic methylation of the other half (the one duplicating in S) develops during this same S in harmony with methylation of nDNA, RNAs and proteins. By virtue of the fact that the pool size of methionine, universal donor of methyl groups, does not appreciably change in S, G_1 and G_2 [18,19], the S-phase methylation of macromolecules might deal with a simultaneous genetic control of all their methyltransferases. A clear demonstration in this regard was provided for nDNA methyltransferas [61,62]. A point which deserves attention.

ACKNOWLEDGEMENTS

The collaboration of T. Menna, C. Buono, D. Geraci, A. Granieri, F. Morelli, L. Trionfetti and T. Eremenko is warmly appreciated for experiments performed at the International Institute of Genetics and Biophysics, CNR, Naples, and the former Institute of Experimental Medicine, CNR, Rome. The investigation was supported by the National Research Council, the Ministry of the University and Scientific and Technological Research and the University of Rome "Tor Vergata".

REFERENCES

[1] Vanyushin BF, Mazin AL, Vasiliev VK & Belozersky AN. The content of 5-methyl-cytosine in animal DNA: the species and tissue specificity. *Biochim Bioph Acta* (1973) 19: 138-152.

[2] Iwanami Y & Brown GH. Methylated bases of transfer ribonucleic acid from *HeLa* and *L* cells. *Arch Biochem Biophys* (1968) 124: 472-482.

[3] Brown GM & Attardi G. Methylation of nucleic acids in *HeLa* cells. *Biochem Biophys Res Commun* (1965) 20: 228-302.

[4] Paik WK & Kim S. Protein methylation. *Science* (1971) 174: 114-119.

[5] Volpe P & Eremenko T. A method to measure the length of each phase of the cell cycle in spinner cultures. *Exp Cell Res* (1970) 60: 456-458.

[6] Volpe P & Eremenko T. A method for measuring cell cycle phases in suspension cultures. *Meth Cell Biol* (1973) 6: 113-126.

[7] Geraci D, Eremenko T, Cocchiara R, Granieri A, Scarano E & Volpe P. Correlation between synthesis and methylation of DNA in *HeLa* cells. *Biochem Biophys Res Commun* (1974) 57: 353-361.

[8] Bird AP. Use of restriction enzymes to study eukaryotic DNA methylation: II. The simmetry of methylated sites support semiconservative copying of the methylation pattern. *J Mol Biol* (1978) 118: 49-60.

[9] Eremenko T, Palitti F, Morelli F, Whitehead EP & Volpe P. Hypomethylation of repair patches in *HeLa* cells. *Mol Riol Rep* (1985) 10: 177-182.

[10] Eremenko T, Timofeeva MY & Volpe P. Organization, modification and replication of the human genome: III. Temporal order of replication and methylation of palindromic, repeated and unique *HeLa* nDNA sequences. *Mol Biol Rep* (1980) 6: 131-136.

[11] Cacciamani T, Virgili S, Centurelli M, Bertoli E, Eremenko T & Volpe P. Specific methylation of the CpG-rich domains in the promoter of the human tissue transglutaminase gene. *Gene* (2002) 297: 103-112.

[12] Volpe P, Menna T & Eremenko T. Differential gene expression during the cell life cycle. *Macromol Funct Cell* (1979) 1: 79-108.

[13] Eremenko T, Cimarra P, Giuditta A & Volpe P. Synthesis and methylation of proteins during the *HeLa* cell cycle. *Proc Natl Acad Lincei* (1979) 66: 153-163.

[14] Wagner EK, Penman S & Ingram VM. Methylation patterns of *HeLa* cell ribosomal RNA and its nuclear precursors. *J Mol Biol* (1967) 29: 371-389.

[15] Salim M & Maden BE. Early and late methylations in *HeLa* cell ribosome maturation. *Nature* (1973) 244: 334-336.

[16] Volpe P, Menna T & Eremenko T. RNA transcription and processing as a functlon of the cell cycle. *Bull Mol Biol Med* (1976) 1: 18-28.

[17] Eremenko T & Volpe P. Polysome translational state during the cell cycle. *Eur J Biochem* (1975) 52: 203-210.

[18] Eremenko T, Menna T & Volpe P. Transport, pool and incorporation of precursors: II. Timing of the amino acid pool. *Microbiologica* (1979) 2: 1-11.

[19] Eremenko T, Menna T & Volpe P. Transport, pool and incorporation of precursors: III. Timing of the amino acid incorporation. *Microbiologica* (1979) 2: 137-145.

[20] Volpe P & Eremenko T. Resumption of transcription following inhibition of DNA methylation. *Macromol Funct Cell* (1984) 3: 59-69.

[21] Volpe P & Eremenko T. Nuclear and cytoplasmic DNA synthesis during the mitotic cycle of *HeLa* cells. *Eur J Biochem* (1973) 32: 227-232.

[22] Volpe P & Giuditta A. Kinetics of RNA labelling in fractions enriched with neuroglia and neurons. *Nature* (1967) 216: 154-155.

[23] Volpe P & Giuditta A. Biosynthesis of RNA in neuron and glia enriched fractions. *Brain Res* (1967) 6: 228-240.

[24] Duranti T, La Teana A, Cacciamani T & Volpe P. The prokaryotic origin of the pathways for synthesis and post-synthetic modification of DNA. *RNA Biol* (2006) 3: 49-53.

[25] Schindler D & Davies J. Methylation of proteins in 60S ribosomal subunits from *S. cerevisiae*. *FEBS-Lett* (1977) 75: 187-191.

[26] Shepherd GR, Hardin JM & Noland BJ. Methylation of lysine residues of histone fractions in synchronized mammalian cells. *Arch Biochem Biophys* (1971) 143: 1-5.

[27] Eremenko T, Benedetto A & Volpe P. Poliovirus replication during the *HeLa* cell life cycle. *Nature New Biol* (1972) 237: 114-116.

[28] Volpe P & Eremenko T. Quantitative studies on cell proteins in suspension cultures. *Eur J Biochem* (1970) 12: 195-200.

[29] Parsons JA & Rustad RC. The distribution of DNA among dividing mitochondria of *Tetrahymena pyriformis*. *J Cell Biol* (1968) 37: 683-693.

[30] Broun R & Evans T. Replication of nuclear satellite and mitochondrial DNA in mitotic cycle of *Physarum*. *Biochim Biophys Acta* (1969) 182: 511-522.

[31] Smith D, Tauro P, Schweizer E & Halvorson HO. The replication of mitochondrial DNA during the cell cycle in *Saccaromyces lactis*. *Proc Natl Acad Sci USA* (1968) 60: 936-941.

[32] Koch J and Stokstad ELR. Incorporation of [^{3}H]thymidine into nuclear and mitochondrial DNA in synchronized mammalian cells. *Eur J Biochem* (1967) 3: 1-6.

[33] Vanyushin BF & Kirnos MD. Structure of animal mitochondrial DNA. *Biochim Bioph Acta* (1977) 575: 323-336.

[34] Dawid IB. 5-methylcytidylic acid: Absence from mitochondrial DNA of frogs and *HeLa* cells. *Science* (1974) 184: 80-81.

[35] Sheid B, Srinivasan PR & Borek E. Deoxyribonucleic acid methylase of mammalian tissues. *Biochemistry* (1968) 7: 280-285.

[36] Nass MMK. Differential methylation of mitochondrial and nuclear DNA in cultured mouse, hamster and virus-transformed cells: *in vivo* and *in vitro* methylation. *J Mol Biol* (1973) 80: 155-175.

[37] Evans HH & Evans TE. Methylation of deoxyribonucleic acid of *Physarum polycephalum* at various periods during the mitotic cycle. *J Mol Biol* (1970) 245: 6436-6441.

[38] Vanyushin BF & Kirnos MD. The nucleotide composition and pyrimidine clusters in DNA from beef heart mitochondria. *FEBS-Lett* (1974) 39: 195-199.

[39] Kappler JA. The kinetics of DNA methylation in cultures of mouse adrenal cell line. *J Cell Physiol* (1970) 75: 21-32.

[40] Burdon RH & Adams RLP. The *in vivo* methylation of DNA in mouse fibroblasts. *Biochim Biophys Acta* (1969) 174: 322-329.

[41] Lynch WE, Brown RF, Umeda T, Langreth SG & Lieberman I. Synthesis of deoxyribonucleic acid by isolated liver nuclei. *J Biol Chem* (1970) 245: 3911-3916.

[42] Burdon RH. Enzymatic modification of chromosomal macromolecules: I. DNA and protein methylation in mouse tumour cell chromatin. *Biochim Biophys Acta* (1971) 232: 359-370.

[43] Volpe P & Eremenko T. Preferential methylation of regulatory gene sequences in *HeLa* cells. *FEBS-Lett* (1974) 44: 121-126.

[44] Volpe P & Eremenko T. Cell and virus macromolecular modifications during the cell life. *Macromol Funct Cell* (1982) 2: 179-195.

[45] Evans HH, Evans TE & Littman S. Methylation of parental and progeny DNA strands in *Physarum polycephalum*. *J Mol Biol* (1973) 74: 563-570.

[46] Volpe P. The Language of Methylation in Genomics of Eukaryotes. *Biochemistry (Mosc)* (2005) 70: 584-595.

[47] Volpe P. The gene expression during the cell life cycle. In Quagliariello E, Palmieri F & Singer TP, Editors. *Horizons in Biochemistry and Biophysics.* Reading: Addison-Wesley Publishers (1976), Vol. 2, pp. 285-340.

[48] Liau MG, Chang CF, Saunders GF & Tsai YH. S-Adenosylhomocysteine hydrolases as the primary target enzymes in androgen regulation of methylated complexes. *Arch Biochem Biophys* (1981) 208: 261-272.

[49] Maclean N & Hilder VA. Mechanisms of chromatin activation and repression. *Internat Rev Cytol* (1977) 48: 1-54.

[50] Tentravahi U, Guntaka RV, Erlanger BF & Miller OJ. Amplified ribosomal RNA genes in a rat hepatoma cell line are enriched in 5-methylcytosine. *Proc Natl Acad Sci USA* (1981) 78: 489-493.

[51] Caboche M & Bachellerie JP. RNA methylation and control of eukaryotic RNA biosynthesis. *Eur J Biochem* (1977) 74: 19-29.

[52] Finkel D & Groner Y. Methylation of adenosine residues (m^6A) in pre-mRNA are important for formation of late *Simian Virus 40* mRNAs. *Virology* (1983) 131: 409-425.

[53] Furuichi Y, Morgan M, Shatkin AJ, Jelinek W, Salditt-Georgieff M & Darnell JE. Methylated, blocked 5' termini in *HeLa* cell mRNA. *Proc Natl Acad Sci USA* (1975) 72: 1904-1908.

[54] Banerjee AK. 5'-terminal cap structure in eukaryotic messenger ribonucleic acids. *Microbiol Rev* (1980) 44: 175-205.

[55] Furuichi Y, La Fiandra A & Shatkin AJ. 5'-terminal structure and mRNA stability. *Nature* (1977) 266: 235-239.

[56] Chang FN, Navickas IJ, Clement AU & Budzilowicz C. Identification of the methylated ribosomal proteins in *HeLa* cells and the fluctuation of methylation during the cell cycle. *Biochim Biophys Acta* (1978) 518: 89-94.

[57] Borun TW, Pearson D & Paik WK. Studies of histone methylation during the *HeLa* S3 cell cycle. *J Biol Chem* (1972) 247: 4288-4298.

[58] Thomas G, Lange HW & Hempel K. Kinetics of histone methylation *in vivo* and its relation to the cell cycle in *Ehrlich Ascites* tumor cells. *Eur J Biochem* (1975) 51: 609-615.

[59] Chang FN, Navickas IJ, Chang CN & Dancis BM. Methylation of ribosomal proteins in *HeLa* cells. *Arch Biochem Biophys* (1976) 172: 627-633.

[60] Goldenberg CJ & Eliceri GL. Methylation of ribosomal proteins in *HeLa* cells. *Biochim Biophy Acta* (1977) 479: 220-234.

[61] Delfini C, Alfani E, De Venezia V, Oberholtzer G, Tomasello C, Eremenko T & Volpe P. Cell-cycle dependence and properties of the *HeLa* cell DNA polymerase system. *Proc Natl Acad Sci USA* (1985) 82: 2220-2224.

[62] Eremenko T, Delfini C, Crema AL, Alfani E & Volpe P. Uncoupling of the DNA polymerase and methylase systems leads to hypomethylation of repair patches. *Macromol Funct Cell* (1988) 5: 37-41.

In: Progress in DNA Methylation Research
Editor: Hans P. Neumann, pp. 225-242
ISBN: 978-1-60021-722-7

Chapter IX

IN SEARCH OF ASYMMETRIC CYTOSINE METHYLATION IN EXONS 5-8 OF *P53*

Sofia Kouidou[1,*], Elisavet Georgiou[1], Andigoni Malousi[2], Athena Kyrkou[1], Anastasios Kyventidis[1], Aikaterini Fragou[1], Theodora Katopodi[1], Nicos Maglaveras[2] and Georgios Tzimagiorgis[1]

[1]Department of Biological Chemistry, School of Medicine, Aristotle University of Thessaloniki, Thessaloniki, Greece;

[2]Lab. of Medical Informatics, School of Medicine, Aristotle University of Thessaloniki, Thessaloniki, Greece.

ABSTRACT

The presence of symmetric methylation in the promoter sequences and its effect on transcriptional inactivation in cancer has been well documented. Asymmetric cytosine methylation, very frequent in plants, is associated with transient or permanent gene inactivation. Evidence on this type of methylation in mammalian organisms is scarce and mostly concerns epigenetic modifications in embryonic stem cells or inactivation of viral sequences by the mammalian host. In two reports however, asymmetric methylation has been also shown in cancerous cells, in *p53,* either in promoter (<u>C</u>CG) or oligoC sequences in exon 5.

The IARC's *p53* somatic mutation database is an excellent source of genetic information reflecting the positive association between cytosine methylation and genetic instability (C>T and G>A mutations) in carcinogenesis. Analysis of the distribution and the frequency of silent and nonsense mutations could be of particular importance in the elucidation of the possible role of asymmetric non-CpG cytosine methylation in genetic instability. Recently, we have shown that in certain cancers, the existence of high rates of

* Correspondence concerning this article should be addressed to Sofia Kouidou, PhD, Associate Professor, Lab. of Biological Chemistry, School of Medicine Aristotle University 54124 Thessaloniki, GREECE. Tel.: +30 2310 999088; Fax.: +30 2310 999004; email: kouidou@auth.gr.

silent transitions observed in specific exonic non-CpG cytosines and the frequently observed G:C>A:T transitions in codons of low functional impact could be indicative for the presence of asymmetric cytosine methylation.

The data presented here reveal that using the same tools of mutation analysis, i.e. distribution and frequency of silent and nonsense mutations and comparison with the rates of transitions at CpG sequences (which are always methylated), it can be concluded that the presence of extensive asymmetric methylation in several other cancers, oral, skin, esophagus stomach and in sarcomas besides lung and breast cancer is also highly probably involved in carcinogenesis. The frequency of G:C>A:T mutations at CpG and non-CpG G:C base pairs increases later in life compared to all other silent and nonsense mutations. Asymmetric cytosine methylation in some sporadic cancers (i.e. adrenal, ovarian) is probably focused in specific sites while in colon/colorectal and hemopoietic cancers this modification is probably absent. Extensive asymmetric methylation is also experimentally shown by direct sequencing of $NaHSO_3$-modified DNA in oral and lung cancer specimens. Furthermore, we provide experimental evidence for the expression, in the same lung cancer specimens, of aberrant transcripts.

In conclusion, detailed analysis of *p53* somatic mutation database and experimental evidence are in support of the frequent presence of spread, or focused exonic asymmetric cytosine methylation in certain cancers which might be related to the frequent presence of aberrant transcripts, experimentally detected in lung cancer.

INTRODUCTION

Several studies have shown that expression of a large number of genes is subject to epigenetic regulation by promoter methylation [1-3]. The epigenetic modification patterns of tumor suppressor genes, which are regulated through promoter methylation, have been extensively investigated in association with carcinogenesis [4]. Another alternative epigenetic regulatory mechanism involves the expression of homologous RNAs which interfere with the expression of full length transcripts [5]. However, the regulatory mechanisms and the tissue-specific expression of specific isoforms as well as the possibility of production of mRNA isoforms resulting either by alternative splicing or by premature nonsense codons occasionally participate in autoregulatory gene expression [6]. It is possible that these regulatory mechanisms of gene expression have not yet been resolved even in the case of well characterized genes. An interesting example is the *p53* gene which is probably one of the best characterized genes very frequently mutated in most cancers. Although the expression of different TP53 isoforms is tissue-specific [7], the underlying regulatory mechanisms still remain largely unknown. Moreover, the presence of several TP53 isoforms probably contributes to the complexity of the results obtained by immunoprecipitation analysis [8,9]. Furthermore, the importance of a large number of its mutations has been the subject of debate and extensive investigation [9-19].

The reduced stability of methylated vs. unmethylated cytosine is well documented and the rate of C>T mutations is considered to be strongly enhanced by the presence of cytosine methylation [20]. The relative increase of methylated cytosine instability is addressed by several investigators [10,17]. High rates of C>T transitions are frequently attributed to the presence of methylation in extended sequences which are not expressed in embryonic stem cells [21]. Conversely, it has been proposed that the considerable increase of instability at CC and CCC sequences might be due to the presence of asymmetric methylation [22,23].

Methyltransferase I (DNMT1) is considered to be responsible for the maintenance of symmetric CpG methylation, and Dnmt3a and b are responsible for *de novo* methylation of CpG sequences [24,25]. Besides silencing of gene promoters, it has been suggested, based on the very high frequency of CpGs in consensus sequences related to splicing, that symmetric methylation could be involved in splicing. The process of asymmetric cytosine methylation at non-CpG G:C base pairs associated with the expression of Dnmt2 in Drosophila melanogaster [26], or other enzymes in plants [27], still remains largely uncharacterized in mammals. All these enzymes, however, appear to be functionally highly specialized. There is presently growing evidence that asymmetric cytosine methylation, which plays an important role in X chromosome silencing, is also involved in the inactivation of foreign DNA [28] and embryonic stem cell regulation [29]. Asymmetric methylation could also be involved in transient epigenetic gene inactivation and regulation of tissue-specific or transiently expressed sequences involved in the regulation of complex genes [30].

Recently, we provided experimental evidence for the presence of site-specific asymmetric methylation in *p53* in lung cancer and correlated it to the instability of this gene in the same region of exon 5 in the form of G:C>A:T silent mutations [31]. In another study we showed that analysis of silent mutations in *p53* reveals a very high tissue specific increase of C>T transitions at CC and CCC sequences which are tentative targets of epigenetic modification [23]. Finally, we provided evidence that a considerable percentage of the codons, which are tentatively asymmetrically methylated in breast cancer, is adjacent to codon 133, the initiation site of several TP53 splice isoforms suppressed in the breast [32]. These data, and the corroborating evidence for the presence of tissue- and site-specific increase of cytosine instability in different cancers, obtained by the mutation analysis shown below, provide sound evidence for the involvement of asymmetric methylation in *p53* in several cancers. This evidence is presently supported by corroborating experimental data on the presence of epigenetic modification in *p53* exon 5 in more than one cancer (oral and lung) and is also supplemented by experimental data for the expression of irregular transcripts in the same lung carcinoma specimens tested, which might play a significant regulatory role in overall TP53 activity. Finally, we discuss the existing evidence on the tentative association between asymmetric methylation and transcriptional regulation.

Computational Analysis Data in Support of Asymmetric Methylation in *p53*

Distribution of G:C>A:T Silent Transitions in Different Cancers

The search for asymmetric methylation has been based on the analysis of the IARC *p53* mutation database (release R11). The best criterion for analyzing genetic instability is probably to estimate the elevated rate of silent mutations of *p53* in specific sequences. In the following section we present the analysis of the reported silent mutations in different cancers. In Figure 1 the distribution of these mutations clearly reveals considerable differences in the frequencies of G:C>A:T silent transitions among different cancers. Depending on the frequency of these mutations (shown in Table 1), we are able to identify three different categories of cancers. In the first category shown in this Figure 1A, (breast, bronchus/lung, skin, stomach and connective tissue cancers) C>T transitions are more frequent than in other cancers (Table 1), clustered at specific regions of exons 5 and 7 (Figure 1A) and particularly in sequences 126-161 which contain the initiation codon of Δ133 TP53 isoforms and an Sp1/3 consensus binding site. For cancers in category B (prostate, brain, ovary, oral and adrenal gland), these transitions are significantly less frequent, although significant clustering of both C>T and G>A is observed at specific codons shown with arrows, such as codon 177 (10.35% and 19.05% of total silent mutations in prostate and brain cancers respectively, and 0.45% of the total mutations in these cancers) indicating the tentative presence of an underlying, destabilizing, DNA-related factor in this codon (Figure 1B). The presence of transitions in codons 126-161 is significantly lower in these cancers compared to category A. In category C (colon, colorectal, and hemopoietic cancers) these mutations are scarce and randomly distributed (Figure 1C). This site-specific instability in these cancers is mostly observed as C>T rather than G>A mutations particularly at CC sequences (Table 1) and is significantly lower compared to other cancers. Mutations in codons 126-161 and CC sequences are particularly rare. Thus, the results obtained from analysis of silent mutations are in support of the presence of extensive asymmetric cytosine methylation not only in lung cancer which has been shown experimentally [31], but also in other cancers included in category A. In cancers of the first category, this methylation appears to be spreading in different cytosine residues while it is more focused at CCs in cancers of the second category. On the contrary, no indications for the presence of asymmetric methylation could be deduced based on the analysis of silent mutations for cancers of the lower digestive (colon, colorectal) and the hemopoietic systems.

The mutations included in Figure 1 clearly reveal the distinctions of the three cancer categories and represent approximately 2/3 of the total mutations presently reported for *p53*. It should be noted however, that several cancers which were not included in the discussion presented above (e.g. esophagus, liver) also clearly exhibit the spreading of non-CpG cytosine methylation.

Table 1. Relative frequency and sequence context of silent C>T mutations of *p53* in different cancer categories compared with total mutations. C>T transitions at CC are related to asymmetric methylation. Codons 126-161 include the Δ133p53 initiation codon and an SP1/3 consensus binding site.

Cancer category	Total mutations	Non-CpG	Non-CpG 126-161	Non-CpG at CC	p Non-CpG	p exons 5,6	p CpG
A	7343	147	50	68	0.006[1]	0.025[1]	0.335
B	3135	38	10	23	0.002[2]	0.132[2]	0.029
C	3504	18	5	12	0[3]	0[3]	0.001

[1] χ^2 analysis of mutation frequencies between categories A and B;
[2] χ^2 analysis of mutation frequencies between categories B and C;
[3] χ^2 analysis of mutation frequencies between categories A and C.

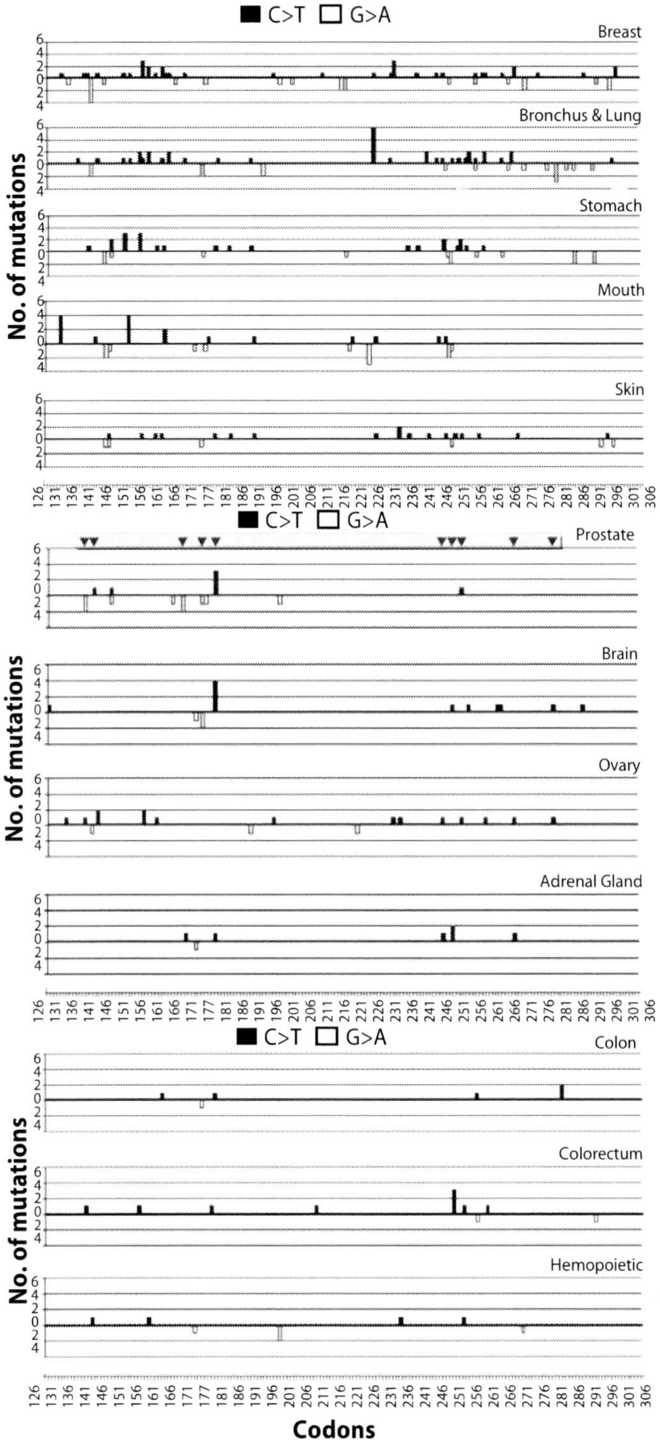

Figure 1. Distribution of silent C>T and G>A mutations in *p53* exons 5-8 in different cancers. The arrows indicate frequently mutated codons. Cancer categories A, B and C. Frequently mutated codons in cancers of the second category.

Frequencies of Clustered Nonsense G:C>A:T Transitions in Exons 5 and 6

Adjacent nonsense mutations in exons 5-6 giving rise to truncated *p53* transcripts which encode for part of the TP53 DNA-binding region and lack the DNA tetramerization domain, necessary for TP53 activity, are probably subject to the same selection requirements, if expressed. Significant frequency deviations for specific C>T transitions in this region, might probably reflect site-specific instability of cytosine as a result of epigenetic modification events. The site-specific cytosine instability in exons 5 and 6 observed as nonsense mutations was investigated by comparing the relative frequencies of clustered nonsense mutations resulting from G:C>A:T transitions in methylated (CpG) and non-CpG G:C-containing codons of exons 5 and 6. This type of analysis (Table 2) reveals that similarly to silent mutations, the total frequency of C>T transitions leading to nonsense mutations in these exons in cancers of category A (Figure 1A) is significantly higher than in categories B and C and becomes comparable to the total rate of the same transitions at CpG sequences. On the contrary, the frequency of C>T transitions at CpG cytosines which provide nonsense mutations is not considerably higher in the same cancers of category A compared to B, while cancers in category C have a high frequency of CpG transitions, probably due to environmentally induced instability, or other selection parameters. Furthermore, in certain cancers, we do not observe any nonsense mutations in certain codons (144, 146, 192) while in others (136, 165 and 167) there are frequent mutations in all cancers. These observations on the distribution and frequency of nonsense mutations are in agreement with the findings from silent mutations further supporting the presence of asymmetric methylation, rather than tissue-specific instability related to exogenous carcinogens, or should otherwise influence the rates CpG and non-CpG cytosines similarly.

Table 2. Relative frequencies of nonsense mutations in *p53* resulting from C>T transitions at CpG and non-CpG G:C base pairs in cancers from categories A-C in exons 5-6 or 5-8

Cancer category	Total mutations	Non-CpG	Non-CpG exons 5,6	CpG	p* Non-CpG	p* exons 5,6	p* CpG	p** Non-CpG
A	7343	96	92	177	0.001[1]	0.001[1]	0.661[1]	0.005[1]
B	3135	17	17	71	0.256[2]	0.256[2]	0[2]	0.513[2]
C	3504	27	27	141	0.015[3]	0.015[3]	0[3]	0[3]

[1] x^2 analysis of mutation frequencies between categories A and B;

[2] x^2 analysis of mutation frequencies between categories B and C;

[3] x^2 analysis of mutation frequencies between categories A and C;

* Analyzed vs. total mutations;

** Analyzed vs. CpG mutations.

Comparison of the total Frequencies of Nonsense G:C>A:T Transitions at CpG and non-CpG G:C Base Pairs in Exons 5-8 in Specific Cancers

In order to further investigate the tentative presence of methylation in exons 5-8 of *p53* in individual cancers, we also analyzed the relative stabilities of CpG and non-CpG cytosines leading to nonsense mutations in these exons in all three categories of cancers described above. This analysis provides evidence on cytosine methylation stability and CpG repair rates in different organs, relative to instability at non-CpG sites and thus, allows us to draw conclusions with respect to the tentative presence of other cytosine destabilizing factors (asymmetric methylation) in these cancers. The reported frequencies for each cancer are compared to lung cancer, which has been shown to exhibit asymmetric methylation and is highly susceptible to environmentally induced genetic instability at specific CpG sequences (hotspots).

It is evident from Table 3 that the ratio of nonsense mutations resulting from non-CpG and CpG G:C>A:T transitions in cancers of category A is high in bladder, breast, lung, esophagus, mouth and prostate cancers, and is considerably lower in colon and colorectal cancers. The nonsense mutation frequencies differences among cancers in the first and second categories are not significantly different from those of lung cancer, but considerable differences are observed compared to cancers in category C. In tissues where methylated cytosines are more labile due to specific environmental conditions [33], (skin and stomach, both in the category A), instability of all cytosines is high and the differences between CpG and non-CpG cytosines are smaller, though again higher compared to cancers in category C.

Table 3. Relative instability of CpG and non-CpG cytosines which lead to nonsense mutations resulting from G:C>A:T transitions in exons 5-8 of *p53*

Cancer category	Topography	Total mutations	G:C>A:T at CpG transitions	G:C>A:T at non-CpG	p	CpG vs. non-CpG
A	Bronchus & Lung	172	26	34		1,31
A	Bladder	85	20	20	0.512	1,00
A	Oral	90	18	16	0.370	0,89
A	Esophagus	157	48	38	0.138	0,79
A	Breast	155	62	42	0.044	0,68
C	Colorectum	124	65	19	0	0,29
C	Colon	78	55	8	0	0,15
A	Skin	74	31	15	0.014	0,48
A	Stomach	60	33	11	0.001	0,33

* Oral: Floor of mouth, Head and Neck, NOS, Other unspecified parts of mouth.

Age Dependence of Silent, Nonsense and Missense Transitions

The average age at which all missense and silent mutations are observed shows little variation (56, 52 and 56, 71 years respectively). However, careful analysis of the age distribution of different types of mutations reveals distinct differences. Thus, the rate of silent and nonsense G:C>A:T transitions in the *p53* mutation database at any G:C base pair (CpG or non-CpG), relative to that of total nonsense and silent mutations reported at a specific age-interval (Figure 2) increases with age and is maximal between the ages of 50 and 70 years. On the contrary, the relative frequency of non-G:C>A:T silent and nonsense mutations reaches a maximum value at a younger age (41-50 years) and shows a definite decrease by the age of 80 years (Figure 3). Although the available data are limited, this difference is considerable between the age of 51-60 years (scatter plots, not shown). Taken together, while the genetic instability observed as silent and nonsense mutations at G:C base pairs, is considerable even at early ages in sporadic cancer (15.25% until 30 years of age), it assumes a substantial frequency of the total genetic instability later in life (21.38% at 51-60 years of age).

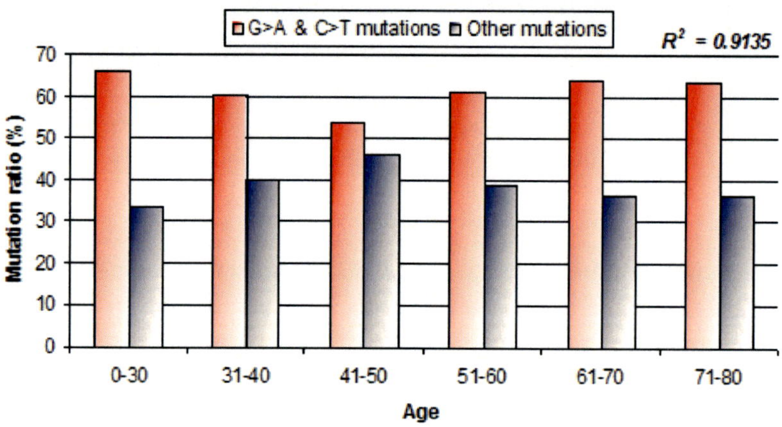

Figure 2. Age-dependence of all GC>AT (CpG or non-CpG) transitions compared to all other silent and nonsense mutations. The relative frequency in both cases was reported vs. total silent and nonsense mutations at each age interval. R^2: Linear Square Regression value.

EXPERIMENTAL EVIDENCE ON ASYMMETRIC METHYLATION AND TRANSCRIPTIONAL ABERRATIONS IN *P53* EXON 5

p53 Exon 5 Methylation Status in Oral and Lung Cancer Specimens

The presence of cytosine methylation, which could be associated with the observed increase in cytosine instability, has also been verified by experimental data in oral, as well as in lung carcinoma specimens. For this purpose we analyzed the methylation pattern of exon 5 of *p53* gene in $NaHSO_3$-modified DNA from oral and lung cancerous and non-cancerous specimens. Genomic DNA isolated from cancerous and non-cancerous tissue specimens was modified and amplified using primers designed to recognize modified *p53* exon 5 sequences. Representative results of modified-DNA sequencing data for codons (159-188) from two oral cancers and one lung specimens are shown in Figure 3. As expected, in all DNA specimens all CpG sequences were methylated. However, in all oral cancer specimens similarly to the lung specimens, extensive methylation was also observed in non-CpG sequences; in the non-cancerous specimen only very limited cytosine methylation was observed. Sequencing of the amplified modified-DNA revealed frequent site-specific cytosine methylation in all cancerous specimens tested.

Two areas of extensive methylation in codons 159-188 of exon 5 are evident in a. the oligoC region (codons 175-179 and b. in the sequence preceding the splice site, codons 181-183 (Figure 3). The cytosines which are methylated in most cancerous specimens are indicated with an asterisk.

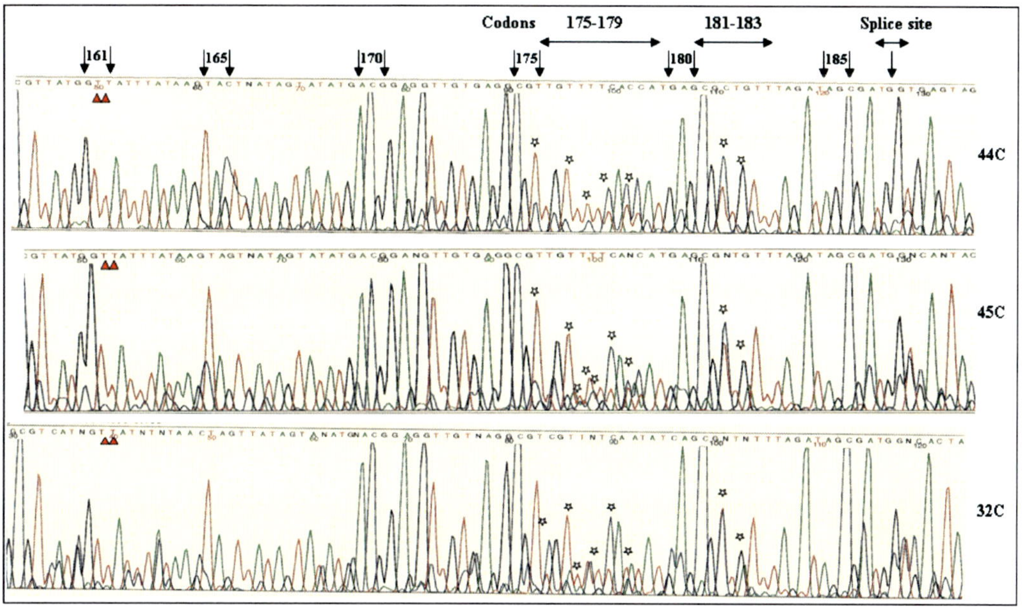

Figure 3. Direct methylation sequencing of *p53* exon 5, corresponding to codons 158—188. Modified DNA (200ng) from oral (44C and 45C) and lung (32C) cancer specimens was amplified using 1.25 U of Platinum Taq DNA polymerase (Invitrogen) in a total volume of 50μl (1.5 mM $MgCl_2$, 10mM Tris-HCl pH 8.3, 0.2 mM dNTPs) containing 100 pmol of the following primers: 5S0023: 5′ GTTTTGTGTAGTTGTGGGTTG 3′ and 5A4063: 5′ CTCCAACCCCAACTACTCACC 3′. Cycling conditions were: 94 °C x 2 min; 94 °C x 45s; 60 °C x 45s; 72 °C x 1 min, 40 cycles, followed by a step at 72 °C for 10 min. PCR products were gel extracted, purified and sequenced on an ABI 373A DNA sequencer (Applied Biosystems). The sites of non-CpG cytosine methylation are shown with an asterisk. Codon numbers are shown above. Red arrowheads show sites of complete modification of cytosines in codon 161. C, cancerous specimens.

p53 mRNA Transcription

In order to explore the possibility that besides *p53* exon 5 hypermethylation, the lung cancer specimens tested might exhibit abnormal transcription profiles, we performed RT-

PCR analysis of *p53* transcripts in the region spanning exons 4-6 and 6-9. As it is evident from the Figure 5, the RT-PCR products from exons 4-6 contrary to the products derived from exons 6-9 were not amplified in some of the specimens tested. In all RT-PCR reactions, β-actin was used as an internal control. These data revealed that abnormal transcription of *p53* exons 4-6 is a frequent characteristic among lung carcinoma specimens (Figure 3, and Kouidou et al., 2005 [31]). Using these primers, we observed complete absence of *p53* transcript in the corresponding specimens (Figure 4, lane 8C) and reduced transcription in others (Figure 4, lanes 7C, 8C and 20C). At the same time, all specimens yield cDNA amplification products for exons 6-9 though the relative abundance of the transcript is lower in some of the specimens (Figure 4, lanes 8C and 8NC). In all cases, to exclude the possibility that the reduced expression in exons 4-6 results from a small deletion in exon 4 and the neighbouring area, exons 4-9 of the *p53* gene have been amplified and analyzed by PCR and SSCP analysis. None of these specimens showed any deletions in the coding region of the gene (data not shown).

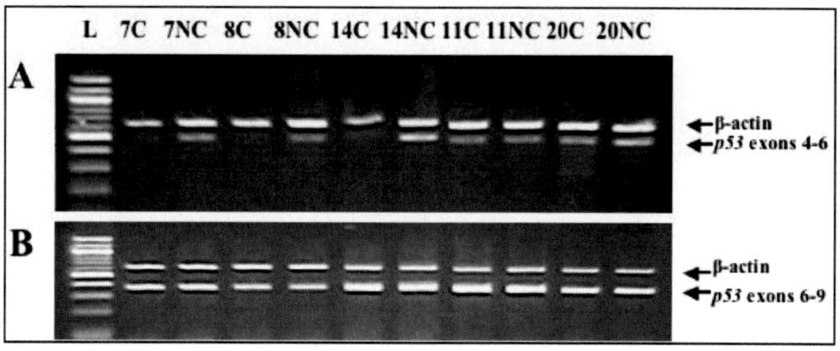

Figure 4. Semi-quantitative RT-PCR analysis of exons 4-6 and 6-9 of *p53* gene. Agarose gel electrophoresis of RT-PCR amplification products of exons 4-6 (A) and exons 6-9 (B) of *p53* gene using total RNA from cancerous (C) and non-cancerous (NC) lung specimens. In all cases, amplification of β-actin cDNA was used as an internal control. L, 100bp DNA ladder. In each case, cDNA was synthesized using 2 μg of total RNA, in the presence of 0.5 μg oligo dT primer and 200U of Superscript II Reverse Transcriptase (Invitrogen) according to the manufacturer. Then, 200 ng cDNA was amplified using of 1.25 U Taq DNA polymerase (Recombinant, Invitrogen) in a total volume of 50μl (1.5 mM $MgCl_2$, 10mM Tris-HCl pH 8.3, 0.2 mM dNTPs) containing 100 pmol of each primer. The primers used in exons 4-6 amplification of the *p53* gene were the following: Ex 4/6 S : 5′ - AAGCAATGGATGATTTGATGC - 3′ (sense) and Ex 4/6A: 5′-CCTTCCACTCGGATAAGATGC–3′ (antisense) and for exons 6-9 were: Ex 6/9S: 5′-GCATCTTATCCGAGTGGAAGG–3 (sense) and Ex 6/9 A: 5′-GGAGAGGAGCTGGTGTTGTTGG- 3′(antisense). Cycling conditions were: 94 °C x 2 min; 94 °C x 45s; 57 °C x 45s; 72 °C x 1.5 min, 40 cycles, followed by a step at 72 °C for 10 min.

To further verify the part of the *p53* coding sequence, i.e. part of exons 4 and 5, might not be expressed, we then amplified the gene region from exons 4-9 (Figure 5A and 5B). Indeed, in the specimens exhibiting abnormal amplification, the full length transcript (816 bp) was not detected (Figure 5B). Instead, we detected specific mRNAs of lower molecular weights (Figure 5A and B).

Transcripts A and B were further analyzed by sequencing. In A, exons 7 and 8 were expressed, as well as exon 6 though splicing anomalies were observed; exon 5 was not detected. Product B corresponded to a *p53* transcript containing residues between codon 38 to

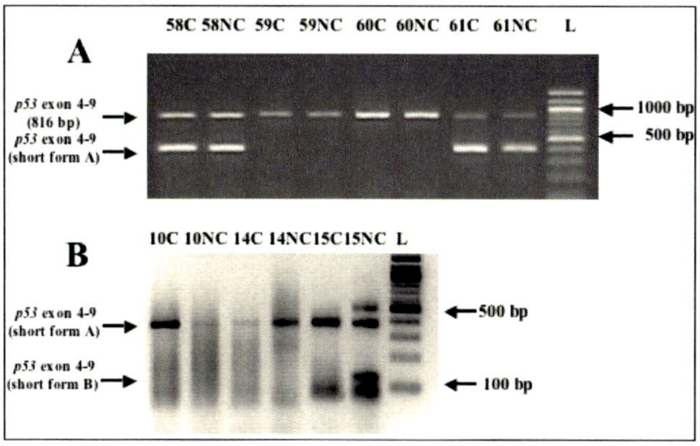

Figure 5. RT-PCR analysis of *p53* exons 4-9. Agarose gel electrophoresis of RT-PCR amplification products of exons 4-9 (A) and (B) reveals the presence of short *p53* transcripts. Total RNA was isolted from cancerous (C) and non-cancerous (NC) lung specimens. L, 100bp DNA ladder. RT-PCR reactions were performed as described in Fig. 4. The primers used to amplify *p53* exons 4-9 were: p53ex4s: 5′ AAGCAA-TGGATGATTTGATGC 3′ (sense); p53ex9a: 5′ GGAGAGGAGCTGGTGTTGTTGG 3′ (antisense). Cycling conditions were: 94 °C x 2 min; 94 °C x 45s; 57 °C x 45s; 72 °C x 1.5 min, 40 cycles, followed by a step at 72 °C for 10 min. cancerous (C) and non-cancerous (NC) lung specimens. L, 100bp DNA ladder.

75 and part of the sequence of exon 9 codons 309-315 (Figure 6) i.e. exons 5-8 and part of exon 4 were not detected. Thus, in all analyzed short *p53* products we detected serious splicing anomalies.

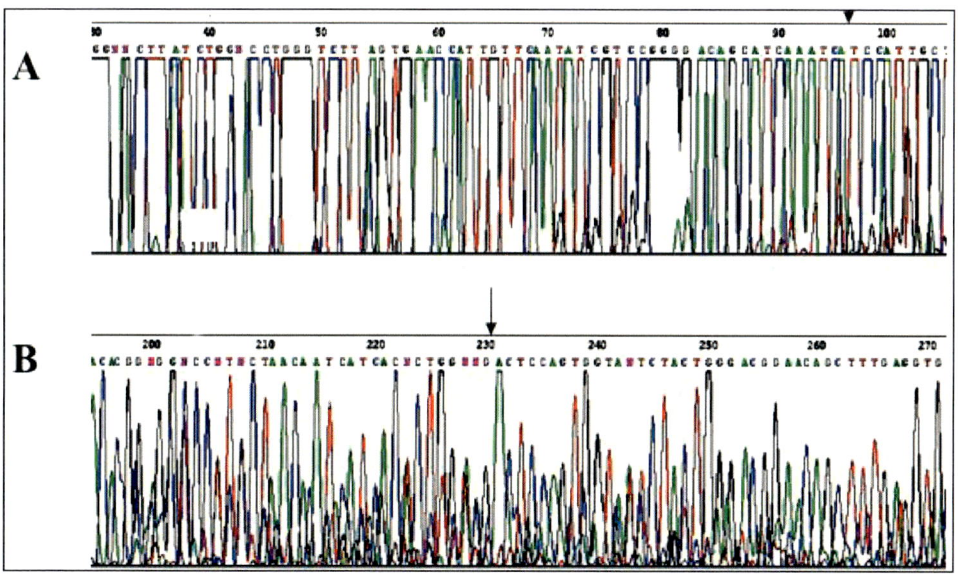

Figure 6. Nucleotide sequence analysis of A and B *p53* splice variants. The RT-PCR fragments were gel extracted, purified and sequenced on an ABI373A DNA sequencer (Applied Biosystems) using the Taq-cycle sequencing method. Only the sequence around the splicing abnormalities is shown. The arrow indicates the novel splicing event.

Implications of Asymmetric Methylation in Genetic Instability, Gene Expression and Aging

Asymmetric Methylation and Genetic Instability in *p53*

Transient asymmetric methylation at CC sequences could lead to gene inactivation not only in plants, but also in mammals i.e. in genetically modified genomes [28]. In lung cancer, non-random asymmetric methylation is frequent in *p53* exon 5, mostly at specific codons which are also sources of an unexpectedly high rate of silent G:C>A:T transitions [31]. Moreover, we presently provide experimental evidence that the same type of modification is also present in oral carcinoma specimens. However, the site-specific instability of *p53,* observed as frequent silent and nonsense non-CpG G:C>A:T transitions, is also frequent in breast, bladder, oral, and other cancers of category A presently discussed, while the same mutation analysis in other cancers does not reveal any instability at these sites (i.e. colon/colorectal). Since the high frequency of silent mutations at specific codons probably reveals unstable DNA sequences, independent of protein selection considerations, methylation could be the underlying cause for the observed genetic instability increase at CC sequences in all cancers presently classified as category A and possibly B. Due to the degeneracy of the code, the genetic instability at CC sequences is magnified. Nevertheless, the selective instability of cytosines at CC sequences has been previously verified for all cancers using different analytical procedures, which were also taking into account the frequency of the individual codons in exons 5-8 and the degeneracy of the code [31]. This type of analysis is also in support of the high instability of *p53* at GC-rich sites only in certain cancers.

It has been previously showed that in breast cancer, certain codons which are seldom mutated in most other cancers, give frequent low impact missense non-CpG G:C>A:T transitions, in addition to silent, or nonsense mutations. The later are missing most of the DNA binding- and tetramerization domains of TP53 and are probably non-functional, unless over-expressed [32]. All these mutations, and particularly those corresponding to the non-expressed codons 126-132 in the breast, are probably functionally neutral and reveal the presence of extensive asymmetric epigenetic modification in breast cancer.

The detailed analysis of the distribution of the G:C>A:T silent and nonsense transitions in exons 5-8 of *p53* presented above verifies the stochastic character of these mutations, and their relatively high frequency only in specific cancers. In addition, analysis of the relative frequencies of CpG and non-CpG cytosines shows that exogenous genotoxic factors which might influence the rate of mutagenesis at methylated G:C pairs, is not sufficient to explain the observed mutation differences particularly in exon 5 of *p53* (e.g. in breast cancer, sarcomas). It is therefore evident that this stochastic process does not reflect cytosine destabilization associated to genotoxic modification of complementary guanosines. The total transition rates of non-CpG cytosines become approximately equal to that of methylated (CpG) cytosines mostly in certain cancers of the first category, and this process is independent of the simultaneous increase of instability at CpG cytosines. All the above considerations concerning the characteristics of genetic instability in specific cancers are in compliance with a model of tissue-selective specific asymmetric methylation of cytosines, as opposed to non-specific mutagenesis of quiescent stem cells [17]. The presence of asymmetric methylation at specific codons in cancers of category B in particular, is supportive of a focused mutation profile. For example in codon 177, 4 C>T transitions have been reported in brain and 3 G>A mutations in prostate cancer. In addition, contrary to category A cancers, C>T mutagenesis in these tissues is not characterized by widespread of non-CpG G:C>A:T transitions, probably indicating the absence of spreading of asymmetric methylation. The process of methylation spreading in transcribed regions has been previously suggested by other authors [34,35] but little experimental evidence exists until presently.

The absence of a genetic instability profile (frequent non-CpG G:C>A:T transitions in colon/colorectal and hemopoietic cancers) might either reflect the absence of asymmetric methylation in these cancers or an extremely efficient system of DNA repair. However, it is well known that the inefficiency of DNA repair mechanisms is one of the primary causes of carcinogenesis in colon/colorectal cancers [36]. It would thus be expected that asymmetric methylation, if present, would probably be reflected in the mutation spectrum of *p53* in these cancers among carriers of inefficient mismatch repair systems.

Asymmetric Methylation and *p53* Transcription

Tissue-specific TP53 activity might be a complex function of the expression of specific *p53* isoforms, some of which are selectively expressed in different tissues [7]. Particularly in the breast, specific *p53* isoforms are not expressed. These data will probably contribute to the resolution of the controversial findings on the role of this gene in cancer and the complex results obtained by immunoprecipitation experiments [9]. However, they also reveal that the regulatory mechanisms governing the critical, tissue-selective expression of *p53* still remain largely unknown. It is possible that epigenetic regulatory mechanisms might be involved in this process.

In a recent publication, we reported that the loss of Δ133p53 isoforms is the principle cause of breast cancer in the Li-Fraumeni syndrome [32]. It should also then be expected, that the suggested patterns of asymmetric methylation clusters in breast as well as in other cancers, which are distributed along other codons which have a considerable impact in Li-Fraumeni-related breast cancer, are probably also related to the expression TP53 isoforms and splicing.

Although the role of asymmetric cytosine methylation on retrotransposon inactivation [37] and embryonic stem cells [29] has been previously addressed, very little information is presently available on the effect of asymmetric methylation in gene silencing in mammals. In a recent report, Inoue and Oishi [38] determined that the presence of asymmetric methylation in the promoter region of human synaptotagmin XI (syt11) gene significantly reduced the binding of an Sp family protein, to the corresponding region and reduced the expression of this gene. Although asymmetric methylation in *p53* is not observed in the promoter region, the presence of a Sp1/3 consensus sequence in codons 151-157 might be related to the regulation of the different TP53 isoform expression.

Transcriptional irregularities, such as under representation of exons 5 and 6 and the expression of specific truncated transcripts missing exon 5, are presently shown to be frequent in lung carcinomas which exhibit asymmetric cytosine methylation in *p53*. This event could be related to the spreading of methylation which appears to be frequent in the corresponding specimens. Although there are several reports on the production of irregular *p53* RNA transcripts in several cancers [39-41], this is the first report in which the corresponding coding sequence of this gene is shown to be asymmetrically methylated.

Taken together, the experimental and mutation analysis results presented above provide conclusive evidence that asymmetric epigenetic modification of the gene in specific regions is probably associated with the tissue-specific expression of *p53*. However, it still remains unknown whether spreading is also a regulatory or a cancer-related mechanism.

Genetic Instability in *p53* and Aging

The frequency of transitions at non-CpG G:C base pairs is very low among young patients and scarce among *p53* germline mutations carriers (Li-Fraumeni and Li-Fraumeni like syndromes) as well as among carriers of other germline cancer-predisposing mutations (*brca1/2*) [42]. We presently showed that although these mutations are not absent among young individuals in sporadic cancers, they show a considerable increase over the ages of 50-70 years. Thus, these transitions appear to characterize carcinogenesis during aging. The frequency of all other mutations appears to increase earlier in life, at the ages of 40-50 years.

Different parameters and repair mechanisms are probably involved in the creation of mutations other than G:C>A:T transitions [43]. Specific sequence characteristics such as repeats, inverted repeats or enhancement of these elements have been shown to contribute to silent transversions and A:T>G:C transitions probably associated with inefficient repair mechanisms [44]. Based on our data, this type of genetic instability might be associated with inefficient repair mechanism and would probably be expected to exert its deleterious effect early in life (also previously observed by Ozgenc et al [45]).

The absence of these transitions in some solid tumors such as colon/colorectal carcinomas, reveals that the increase of G:C>A:T transitions is not an absolute requirement, or a result of the process of carcinogenesis. An age-related increase of deletions at methylated sequences has also been observed in the aging heart of mice [46]. Thus, genetic instability, probably also associated with inefficient methylation mechanisms [47] and production of truncated transcripts [48,49] and proteins [50], is most likely a universal process related to aging as well as cancer.

Conclusion

Analysis of *p53* silent and nonsense mutations, which are neutral with respect to the resulting TP53 products in different cancers, strongly supports the presence of site- and tissue-specific asymmetric methylation of cytosine in exons 5-8 of this gene. During aging, the delayed C>T silent and nonsense transition increase at both CpG and non-CpG sequences, compared to other types of mutations, might be due to independent mechanism by which these mutations arise. The asymmetric methylation spreading which is suggested by these mutations is in close proximity with codon 133, the initiation codon of several alternative *p53* transcripts and a Sp1/3 consensus sequence which might influence alternative splicing of this gene. Experimental evidence from lung cancer, presently reported, reveals the frequent expression of variable transcripts, some missing the 5′ end of the gene, and others including only small sequences of the DNA binding region. Several previous publications have also reported on the frequent presence of irregular *p53* transcripts in several tissues. It is concluded that besides the pronounced effect of specific mutations on TP53 activity, mostly at hotspots, the probable presence of asymmetric cytosine methylation might be related to considerable mutagenesis and transcriptional instability in this gene.

REFERENCES

[1] Ballestar E, Wolffe AP. Methyl-CpG-binding proteins. Targeting specific gene repression. *Eur J Biochem* 2001 Jan;268(1):1-6.

[2] Herman JG, Merlo A, Mao L, Lapidus RG, Issa JP, Davidson NE, et al. Inactivation of the CDKN2/p16/MTS1 gene is frequently associated with aberrant DNA methylation in all common human cancers. *Cancer Res* 1995 Oct 15;55(20):4525-30.

[3] Nguyen CT, Gonzales FA, Jones PA. Altered chromatin structure associated with methylation-induced gene silencing in cancer cells: correlation of accessibility, methylation, MeCP2 binding and acetylation. *Nucleic Acids Res* 2001 Nov 15;29(22):4598-606.

[4] Esteller M. Aberrant DNA methylation as a cancer-inducing mechanism. *Annu Rev Pharmacol Toxicol* 2005;45:629-56.

[5] Hirotsune S, Yoshida N, Chen A, Garrett L, Sugiyama F, Takahashi S, et al. An expressed pseudogene regulates the messenger-RNA stability of its homologous coding gene. *Nature* 2003 May 1;423(6935):91-6.

[6] Cuccurese M, Russo G, Russo A, Pietropaolo C. Alternative splicing and nonsense-mediated mRNA decay regulate mammalian ribosomal gene expression. *Nucleic Acids Res* 2005;33(18):5965-77.

[7] Bourdon JC, Fernandes K, Murray-Zmijewski F, Liu G, Diot A, Xirodimas DP, et al. p53 isoforms can regulate p53 transcriptional activity. *Genes Dev* 2005 Sep 15;19(18):2122-37.

[8] Appella E, Anderson CW. Post-translational modifications and activation of p53 by genotoxic stresses. *Eur J Biochem* 2001 May;268(10):2764-72.

[9] Hashimoto T, Tokuchi Y, Hayashi M, Kobayashi Y, Nishida K, Hayashi S, et al. p53 null mutations undetected by immunohistochemical staining predict a poor outcome with early-stage non-small cell lung carcinomas. *Cancer Res* 1999 Nov 1;59(21):5572-7.

[10] Mechanic LE, Marrogi AJ, Welsh JA, Bowman ED, Khan MA, Enewold L, et al. Polymorphisms in XPD and TP53 and mutation in human lung cancer. *Carcinogenesis* 2005 Mar;26(3):597-604.

[11] Strauss BS. Role in tumorigenesis of silent mutations in the TP53 gene. *Mutat Res* 2000 Dec 20;457(1-2):93-104.

[12] Strauss BS. Silent and multiple mutations in p53 and the question of the hypermutability of tumors. *Carcinogenesis* 1997 Aug;18(8):1445-52.

[13] Rodin SN, Rodin AS, Juhasz A, Holmquist GP. Cancerous hyper-mutagenesis in p53 genes is possibly associated with transcriptional bypass of DNA lesions. *Mutat Res* 2002 Dec 29;510(1-2):153-68.

[14] Soussi T, Kato S, Levy PP, Ishioka C. Reassessment of the TP53 mutation database in human disease by data mining with a library of TP53 missense mutations. *Hum Mutat* 2005 Jan;25(1):6-17.

[15] Toyooka S, Tsuda T, Gazdar AF. The TP53 gene, tobacco exposure, and lung cancer. *Hum Mutat* 2003 Mar;21(3):229-39.

[16] Rodin SN, Rodin AS. Human lung cancer and p53: the interplay between mutagenesis and selection. *Proc Natl Acad Sci U S A* 2000 Oct 24;97(22):12244-9.

[17] Rodin SN, Rodin AS. Strand asymmetry of CpG transitions as indicator of G1 phase-dependent origin of multiple tumorigenic p53 mutations in stem cells. *Proc Natl Acad Sci U S A* 1998 Sep 29;95(20):11927-32.

[18] Glazko GV, Koonin EV, Rogozin IB. Mutation hotspots in the p53 gene in tumors of different origin: correlation with evolutionary conservation and signs of positive selection. *Biochim Biophys Acta* 2004 Aug 12;1679(2):95-106.

[19] Zhu WG, Srinivasan K, Dai Z, Duan W, Druhan LJ, Ding H, et al. Methylation of adjacent CpG sites affects Sp1/Sp3 binding and activity in the p21(Cip1) promoter. *Mol Cell Biol* 2003 Jun;23(12):4056-65.

[20] Pfeifer GP. p53 mutational spectra and the role of methylated CpG sequences. *Mutat Res* 2000 May 30;450(1-2):155-66.

[21] Smith NG, Hurst LD. The causes of synonymous rate variation in the rodent genome. Can substitution rates be used to estimate the sex bias in mutation rate? *Genetics* 1999 Jun;152(2):661-73.

[22] Agirre X, Vizmanos JL, Calasanz MJ, Garcia-Delgado M, Larrayoz MJ, Novo FJ. Methylation of CpG dinucleotides and/or CCWGG motifs at the promoter of TP53 correlates with decreased gene expression in a subset of acute lymphoblastic leukemia patients. *Oncogene* 2003 Feb 20;22(7):1070-2.

[23] Kouidou S, Malousi A, Maglaveras N. Methylation and repeats in silent and nonsense mutations of p53. *Mutat Res* 2006 Jul 25;599(1-2):167-77.

[24] Okano M, Bell DW, Haber DA, Li E. DNA methyltransferases Dnmt3a and Dnmt3b are essential for de novo methylation and mammalian development. *Cell* 1999 Oct 29;99(3):247-57.

[25] Bestor TH, Ingram VM. Two DNA methyltransferases from murine erythroleukemia cells: purification, sequence specificity, and mode of interaction with DNA. *Proc Natl Acad Sci U S A* 1983 Sep;80(18):5559-63.

[26] Mund C, Musch T, Strodicke M, Assmann B, Li E, Lyko F. Comparative analysis of DNA methylation patterns in transgenic Drosophila overexpressing mouse DNA methyltransferases. *Biochem J* 2004 Mar 15;378(Pt 3):763-8.

[27] Cao X, Jacobsen SE. Locus-specific control of asymmetric and CpNpG methylation by the DRM and CMT3 methyltransferase genes. *Proc Natl Acad Sci U S A* 2002 Dec 10;99 Suppl 4:16491-8.

[28] Brooks AR, Harkins RN, Wang P, Qian HS, Liu P, Rubanyi GM. Transcriptional silencing is associated with extensive methylation of the CMV promoter following adenoviral gene delivery to muscle. *J Gene Med* 2004 Apr;6(4):395-404.

[29] Ramsahoye BH, Biniszkiewicz D, Lyko F, Clark V, Bird AP, Jaenisch R. Non-CpG methylation is prevalent in embryonic stem cells and may be mediated by DNA methyltransferase 3a. *Proc Natl Acad Sci U S A* 2000 May 9;97(10):5237-42.

[30] Buslei R, Kreutzer J, Hofmann B, Schmidt V, Siebzehnrubl F, Hahnen E, et al. Abundant hypermethylation of SOCS-1 in clinically silent pituitary adenomas. *Acta Neuropathol (Berl)* 2006 Mar;111(3):264-71.

[31] Kouidou S, Agidou T, Kyrkou A, Andreou A, Katopodi T, Georgiou E, et al. Non-CpG cytosine methylation of p53 exon 5 in non-small cell lung carcinoma. *Lung Cancer* 2005 Dec;50(3):299-307.

[32] Kouidou S, Malousi A, Kyventidis A, Fragou A, Maglaveras N. G:C>A:T mutations and potential epigenetic regulation of *p53* in breast cancer. *Breast Cancer Res Treat* 2007 May 16; [Epub ahead of print].

[33] Pfeifer GP, You YH, Besaratinia A. Mutations induced by ultraviolet light. *Mutat Res* 2005 Apr 1;571(1-2):19-31.

[34] Nguyen C, Liang G, Nguyen TT, Tsao-Wei D, Groshen S, Lubbert M, et al. Susceptibility of nonpromoter CpG islands to de novo methylation in normal and neoplastic cells. *J Natl Cancer Inst* 2001 Oct 3;93(19):1465-72.

[35] Tollefsbol TO, Hutchison CA, III. Control of methylation spreading in synthetic DNA sequences by the murine DNA methyltransferase. *J Mol Biol* 1997 Jun 20;269(4):494-504.

[36] Grady WM. Genomic instability and colon cancer. *Cancer Metastasis Rev* 2004 Jan;23(1-2):11-27.

[37] Kuhlmann M, Borisova BE, Kaller M, Larsson P, Stach D, Na J, et al. Silencing of retrotransposons in Dictyostelium by DNA methylation and RNAi. *Nucleic Acids Res* 2005;33(19):6405-17.

[38] Inoue S, Oishi M. Effects of methylation of non-CpG sequence in the promoter region on the expression of human synaptotagmin XI (syt11). *Gene* 2005 Mar 28;348:123-34.

[39] DiCioccio RA, Werness BA, Peng R, Allen HJ, Piver MS. Correlation of TP53 mutations and p53 expression in ovarian tumors. *Cancer Genet Cytogenet* 1998 Sep;105(2):93-102.

[40] Eicheler W, Zips D, Dorfler A, Grenman R, Baumann M. Splicing mutations in TP53 in human squamous cell carcinoma lines influence immunohistochemical detection. *J Histochem Cytochem* 2002 Feb;50(2):197-204.

[41] McIver B, Grebe SK, Wang L, Hay ID, Yokomizo A, Liu W, et al. FHIT and TSG101 in thyroid tumours: aberrant transcripts reflect rare abnormal RNA processing events of uncertain pathogenetic or clinical significance. *Clin Endocrinol (Oxf)* 2000 Jun;52(6):749-57.

[42] Greenblatt MS, Chappuis PO, Bond JP, Hamel N, Foulkes WD. TP53 mutations in breast cancer associated with BRCA1 or BRCA2 germ-line mutations: distinctive spectrum and structural distribution. *Cancer Res* 2001 May 15;61(10):4092-7.

[43] Mills KD, Ferguson DO, Alt FW. The role of DNA breaks in genomic instability and tumorigenesis. *Immunol Rev* 2003 Aug;194:77-95.

[44] Hebert ML, Wells RD. Roles of double-strand breaks, nicks, and gaps in stimulating deletions of CTG.CAG repeats by intramolecular DNA repair. *J Mol Biol* 2005 Nov 11;353(5):961-79.

[45] Ozgenc A, Loeb LA. Current advances in unraveling the function of the Werner syndrome protein. *Mutat Res* 2005 Sep 4;577(1-2):237-51.

[46] Dolle ME, Snyder WK, Gossen JA, Lohman PH, Vijg J. Distinct spectra of somatic mutations accumulated with age in mouse heart and small intestine. *Proc Natl Acad Sci U S A* 2000 Jul 18;97(15):8403-8.

[47] Agrelo R. A new molecular model of cellular aging based on Werner syndrome. *Med Hypotheses* 2007 68 (4):770-80.

[48] Delacourte A, Sergeant N, Buee L. In vitro models of age-related neurodegenerative disorders. *Exp Gerontol* 2003 Nov;38(11-12):1309-17.

[49] Gentry A, Venkatachalam S. Complicating the role of p53 in aging. *Aging Cell* 2005 Jun;4(3):157-60.

[50] Liu B, Wang J, Chan KM, Tjia WM, Deng W, Guan X, et al. Genomic instability in laminopathy-based premature aging. *Nat Med* 2005 Jul;11(7):780-5.

INDEX

A

B

C

D

E

H

I

J

K

L

M

N

O

P

Q

R

S

T

U

V

W

X

Y

Z